Ice Sheets and Late Qu[illegible]
Environmental Cha[illegible]

Ice Sheets and Late Quaternary Environmental Change

Martin J. Siegert
Bristol Glaciology Centre,
School of Geographical Sciences
University of Bristol

JOHN WILEY & SONS, LTD
Chichester • New York • Weinheim • Brisbane • Singapore • Toronto

National 01243 779777
International (+44) 1243 779777
e-mail (for orders and customer service enquiries): cs-books@wiley.co.uk.
Visit our Home Page on http://www.wiley.co.uk
or http://www.wiley.com

Other Wiley Editorial Offices

John Wiley & Sons, Inc., 605 Third Avenue,
New York, NY 10158-0012, USA

WILEY-VCH Verlag GmbH, Pappelallee 3,
D-69469 Weinheim, Germany

John Wiley & Sons Australia, Ltd, 33 Park Road, Milton,
Queensland 4064, Australia

John Wiley & Sons (Asia) Pte Ltd, 2 Clementi Loop #02-01,
Jin Xing Distripark, Singapore 129809

John Wiley & Sons (Canada) Ltd, 22 Worcester Road,
Rexdale, Ontario M9W 1L1, Canada

Library of Congress Cataloging-in-Publication Data

Siegert, Martin J.
Ice sheets and Late Quaternary environmental change / by Martin J. Siegert.
p. cm.
Includes bibliographical references and index.
ISBN 0-471-98569-4 (alk. paper) — ISBN 0-471-98570-8 (alk. paper)
1. Glacial epoch. 2. Global environmental change. 3. Ice sheets. 4. Geology, Stratigraphic—Holocene. I. Title.

QE697 .S524 2001
551.7'92—dc21 00-047749

British Library Cataloguing in Publication Data

A catalogue record for this book is available from the British Library

ISBN 0-471-98569-4 (cloth)
ISBN 0-471-98570-8 (paperback)

Typeset in 10/12pt Times by Mayhew Typesetting, Rhayader, Powys
Printed and bound in Great Britain by Bookcraft (Bath) Ltd, Midsomer Norton
This book is printed on acid-free paper responsibly manufactured from sustainable forestry, in which at least two trees are planted for each one used for paper production.

Contents

Preface

In 1981 Denton and Hughes published *The last great ice sheets*; a seminal work detailing the configuration of Late Quaternary ice sheets based mainly on interpretation of geological evidence. Since then, there have been a number of advances in subjects involved in the reconstruction of former ice sheets. A first is an abundance of recent geological evidence (from land, sea and ocean) that supplements and adds to that available in 1981. A second is the wide use of numerical ice sheet modelling as a tool for quantifying the growth, decay and flow of former ice sheets. A third is an appreciation for how ice sheets, climate and oceans interact. Because of these developments (and probably some others), several of the ice sheet reconstructions detailed in *The last great ice sheets* have been revised. Despite this, however, many scientists today refer to Denton and Hughes' reconstructions because, I believe, the most recent ideas on Late Quaternary ice sheets have not, until now, been collated.

This book has three main aims. First to detail how Late Quaternary ice sheets can be reconstructed. Second to present the dimensions and dynamics of these former ice sheets and show how these changed over the last ice age. Third, to indicate how Late Quaternary ice sheets were an interactive element of the global environment. Although the book is intended as an undergraduate text I hope the work presented in it will also benefit researchers requiring information about recent advances in our comprehension of Late Quaternary ice sheets.

We currently know enough about Late Quaternary ice sheets to realise that our knowledge is incomplete. So, although few things are certain in science, one thing I can predict with confidence is that the information in this book will require revision as new evidence presents itself. Until then, I hope that this book will provide a useful guide to the ice sheets of the Late Quaternary.

Martin J. Siegert
Bristol Glaciology Centre, August 2000

Acknowledgements

This book was conceived while I was at the Centre for Glaciology, University of Wales, Aberystwyth, between 1994–1998. The book was written in the subsequent two years at the Bristol Glaciology Centre, University of Bristol. I extend my sincere thanks to the directors and staff of these centres for sharing with me their knowledge and enthusiasm for glaciology. Without their encouragement and assistance I very much doubt this book would have been possible. In particular, a special note of thanks is owed to Julian Dowdeswell for not only introducing me to glaciology, but for guidance and encouragement throughout my academic career.

I thank Richard Hodgkins, Quaternary Research Group, Royal Holloway, University of London, for providing a constructive review, and to two anonymous referees for commenting on the original outline of the book. I also thank the editors and staff at John Wiley for making the publication process as painless as possible.

A special word of acknowledgement is given to Jonathan Tooby, who has drawn many of the book's figures, always to his extremely high standard.

I finally thank Maggie, and my family, for their unconditional support.

ILLUSTRATIONS

Most of the illustrations in this book have either been adapted or taken from published sources. Every effort has been taken to seek permission from copyright holders to use previously published illustrations. In cases where permission has been received, the name of the copyright holder is included in the figure caption. However, when no reply to the request for permission was received it is assumed that there is no objection by the copyright holder to the use of material. In these cases, and where previously published diagrams have been redrawn and significantly altered, full reference to the original source has been made in the figure caption.

I am grateful to the following copyright holders for permission to use figures: Academic Press (Figs. 9.4, 11.1, 11.13, 11.14, 12.9, 12.10, 12.12, 13.2, 13.3); Alfred-Wegener-Institute (Figs. 9.10a-b); American Association for the Advancement of Science (Figs. 9.7, 10.10); American Geophysical Union (Figs. 8.1, 8.5, 8.6, 8.7, 8.9, 12.13); Annual Reviews (Fig. 7.1); Blackwell Science (Figs. 5.3); Cambridge University Press (Fig. 9.8); Canadian Quaternary Association (Figs. 12.3, 12.4, 12.5); Elsevier Science (Figs. 5.2, 5.6, 7.2, 9.3, 10.1a, 10.2, 10.8, 10.9, 11.4, 11.7, 11.9, 12.14, 13.1, 13.5); Geological Society of America (Figs. 8.2b, 8.8, 8.10, 9.5, 9.6); Geological Society of Denmark (Figs. 10.1b, 10.6, 10.7); International Glaciological Society (Figs. 9.10, 10.4, 11.3); Macmillan Magazines (Figs. 1.6, 2.4, 2.5, 8.3, 8.4, 10.3, 12.6, 12.8); NRC Research Press (Figs. 12.1 and 12.11); Springer-Verlag GmbH & Co. KG (Fig. 9.9); Taylor and Francis (Fig. 5.5); Taylor and Francis AS (Scandinavia) (Figs. 10.5 and 11.6a); UCL Press (Fig. 4.5).

The following authors are thanked for giving their personal permission to use illustrations: Richard Alley, Peter Clark, Phillipe Huybrechts, Anne Letréguilly and Johan Kleman. I am particularly grateful to Shawn Marshall for providing electronic versions of Figures 12.1, 12.9, 12.10, 12.11, 12.12 and 12.13; to Justin Taylor for Figures 6.1 and 6.5 and to Jonathan Bamber for Figures 9.1b and 10.1c.

CHAPTER 1

Causes of Ice Ages

INTRODUCTION

The last ice age occurred between about 120 000 and 10 000 years ago in a period known generally as the Late Quaternary (Figure 1.1). Although ice sheets grew and shrank several times during the Late Quaternary, they reached their largest size at around 21 000 years ago; a time referred to as the Last Glacial Maximum (LGM). Ice sheets at the LGM captured so much water from the oceans that global sea level was around 120–135 metres lower than its present day level. Glacier expansion occurred predominantly in the high- to mid-latitude Northern Hemisphere, where continental-scale ice sheets grew across North America and Europe. In some regions, such as Greenland and Antarctica, LGM ice sheets expanded from ice masses that had survived the preceding interglacial. The growth of ice was also witnessed in the Southern Hemisphere mid-latitudes (e.g. Patagonia and New Zealand). However, outside of Antarctica, the lack of land surface (and continental shelf) limited the growth of ice in this half of the globe. The formation of Late Quaternary ice sheets caused alteration to the chemistry and circulation of the oceans, the flow of air within the atmosphere, the reflection of sunlight from the Earth's surface as well as the sea level. The environment of the planet was therefore very different at the LGM compared with today.

This book is primarily concerned with how Late Quaternary ice sheets can be reconstructed. To do this, changes to the atmosphere and oceans are reviewed to ascertain the past environment of formerly glaciated regions. Glacial geological evidence is assessed as a first-order guide to the extent and dynamics of the ice sheets. This information is supplemented by numerical modelling information so that a quantitative representation of Late Quaternary ice sheets can be provided.

It should be acknowledged from the outset that climate change during the Late Quaternary affected far more than glaciers. Many important aspects of Quaternary change, such as aeolian activity (the development of 'loess'), changes to the global biosphere, variation in river systems and alteration to periglacial systems, are dealt with well in other text books (e.g. Dawson, 1992; Lowe and Walker, 1997) and are not investigated here.

In terms of geological time, 21 000 years is a very short time for the climate to change from that supporting huge mid-latitude ice sheets to the present situation in which there are none (Figure 1.1). In order to determine the processes that cause climate change in the future, it is essential that the causes of recent global-scale environmental change be evaluated. The first step in this evaluation is to determine how large ice sheets are able to form.

GROWTH AND DECAY OF TERRESTRIAL GLACIERS

Surveys of very recent glacier behaviour (over the past 50 years) have resulted in an understanding of how glaciers react if their environment is changed. However, it is important to note from the beginning of this book that not only will ice sheets (and glaciers) *respond* to climate change, their existence will interact with and *affect* environmental change. This interaction between ice and climate will be developed in later chapters.

Glaciers are fed with snow which, after a number of years of compression and firnification, turns to dense ice. The region in which ice accumulates is known as the 'accumulation zone', and is located in the highest, coldest regions of the ice mass. At the other end of the glacier, at lower elevations where it is warmer, ice is lost due to surface melting and, if the glacier terminates in water, iceberg calving. This region of net ice loss is known as the ablation zone. The line between these zones is where there is neither ablation nor accumulation of ice, and is called the 'equilibrium line'. The altitude of this line is referred to as the 'equilibrium line altitude' or, more commonly, the ELA. It is the position of the ELA that governs the surface areas on the glacier where accumulation and ablation occur. The flow of ice within glaciers acts to transport mass from the accumulation zone to the ablation zone. Thus, when the volume of ice accumulation equals the volume of ice ablated and the volume of ice transported from the accumulation zone to the ablation zone, the glacier is said to be 'in balance'. When a glacier is like this, it remains in a stable, steady position, neither growing nor shrinking. In this case, the ELA is in a position that permits this 'steady-state' to exist.

Glacier growth occurs when the annual mass balance is positive for a number of years. As a rule, ice accumulates in the winter and ablates in the summer (Figure 1.2). Generally, for glacier mass balance to become positive, the summer melt season needs to be cooler resulting in less melting of ice. The net result of summer cooling is a lowering of the ELA. In this situation, the accumulation area is increased and the ablation region is decreased. This causes

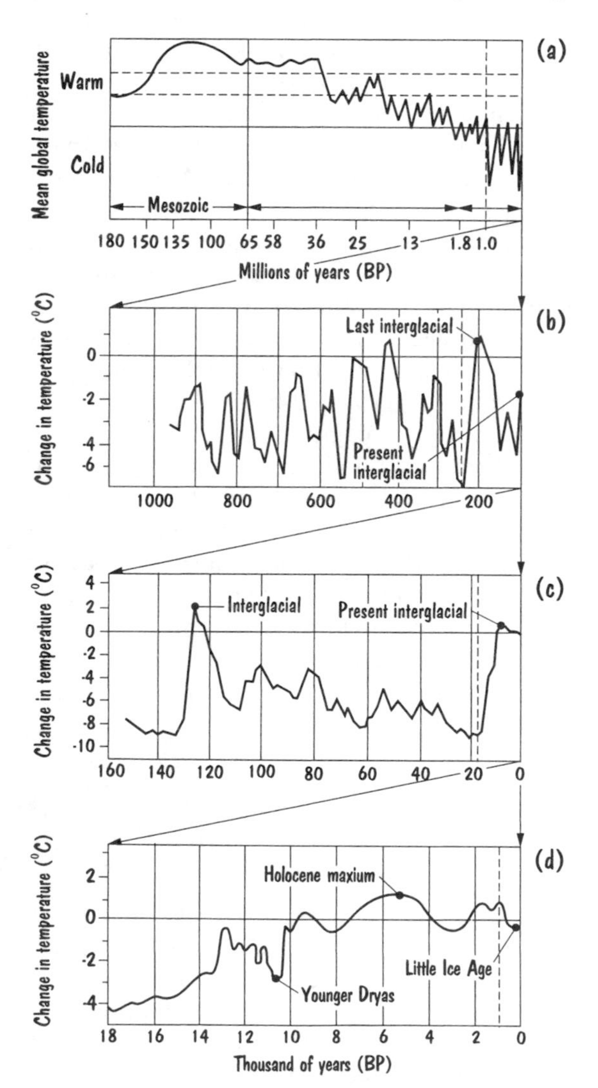

Figure 1.1 Global temperature variation since (a) 180 000 000 years ago, (b) 1 000 000 years ago, (c) 160 000 years ago and (d) 18 000 years ago. The Quaternary period; from 1.8 million years ago to present day, can be split into two subdivisions, the Holocene (representing the last 10 000 years), and the Pleistocene (between 1.8 Ma and 10 ka BP). Last glacial period; between 130 and 10 ka BP can be termed the *Late Quaternary*. However, it is also referred to by formal chrono-stratigraphic names such as the *Weichselian* in northern Europe, *Devensian* in the UK, *Würm* in central Europe, and the *Wisconsin* in North America. The last glacial maximum (LGM) was recorded at about 21 000 years ago. This glacial(ism) is called (in Northern Europe) the Late Weichselian between about 40–10 000 years ago. The last ice sheets had disintegrated by 10–7000 years ago. Adapted from Boulton (1993).

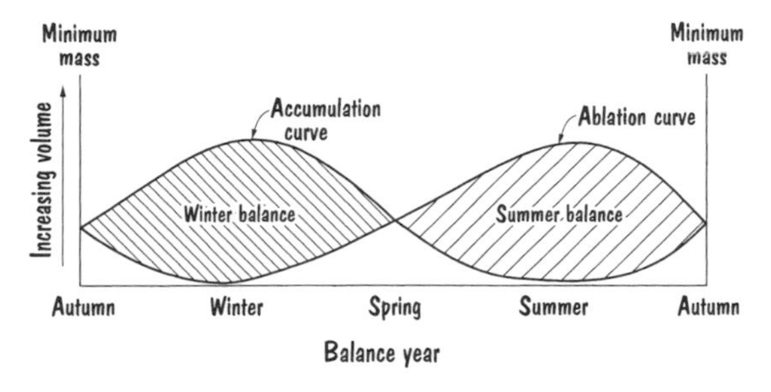

Figure 1.2 Glacier mass balance. Glaciers are fed mainly by the surface accumulation of snow. There are several mechanisms by which glaciers receive mass and these include direct snowfall, wind-blown snow, avalanches and add-freezing of water. The mass added to a glacier in one year is known as accumulation. Glaciers lose mass by melting and, if the glacier ends in marine or lake waters, by iceberg production, known collectively as ablation mechanisms. A parameter known as glacier 'mass balance' is defined as the 'algebraic sum of accumulation and ablation' (Drewry, 1986). If accumulation exceeds ablation over a balance year, from the end of one summer to the next, then the glacier has a positive net mass balance. If ablation is the greater term, then net mass balance is negative. Adapted from Sugden and John (1976).

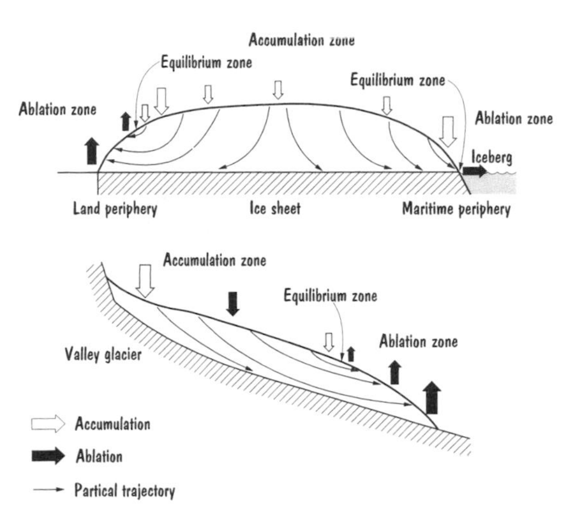

Figure 1.3 Glacier flow and surface mass balance for ice sheets (upper figure) and glaciers (lower figure). See Chapter 3 for details on the flow of ice-sheets. Adapted from Sugden and John (1976).

more flow of ice into the ablation region than gets ablated and, so, the glacier advances. Conversely, warmer conditions in the summer season results in a raising of the ELA, more melting across a wider area of the glacier surface and, hence, if this happens for several years, the glacier will retreat.

This simple relationship between summer air temperatures, the ELA and mass balance is true for large terrestrial ice sheets as much as it is for small glaciers. A simple analysis of the hypsometry of land masses shows that as the ELA is lowered, so more and more land is available to be glacierised. It is therefore easy to comprehend how a reduction in summer air temperatures across a wide mountainous area would result in the large-scale expansion of glaciers. Further reduction in the value of the ELA may cause these glaciers to coalesce and build up as single, large bodies of ice. And so it is that terrestrial ice sheets are initiated across mountain regions, when the ELA is forced down by a reduction in air temperature. Increase in summer air temperatures causes an increase in the ELA and, therefore an increase in net ablation. Thus, the growth and decay of ice sheets on land is related to the position of the ELA and, because of this, to surface air temperatures.

However, it should be noted here that the mass balance of a glacier is affected by changes to the flow of ice (Figure 1.3 and Figure 3.1). If circumstances occur in which increased rate of ice flow causes more ice to be transported from the accumulation zone to the ablation zone, the advance of the glacier margin will eventually result in net ablation and subsequent retreat. This leaves the glacier with less ice than it had prior to the increase in ice flow. This unsteady flow of glaciers is called glacier surging (see Chapter 3) and has been observed to occur in many glaciers and, recently, sections of relatively large ice caps (e.g. Sharp, 1988; Joughin et al., 1996; Dowdeswell et al., 1999). Glacier surging is an extremely important process when one considers that a sudden surge of a large ice sheet could result in the net loss of huge quantities of ice to the ocean,

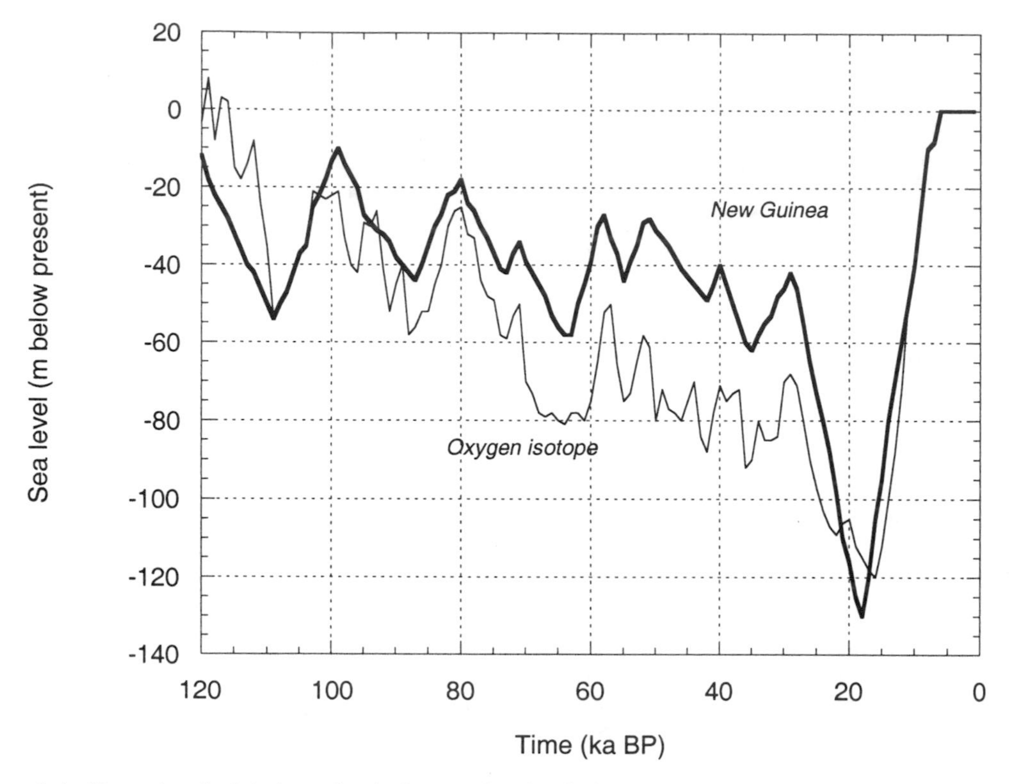

Figure 1.4 Records of global sea-level change. Sea-level change over the past 120 000 years from New Guinea terraces and sea-floor sediment $\delta^{18}O$ records. Adapted from Shackleton (1987).

resulting in a potential sudden rise in global sea level.

Although changes in the ELA can cause terrestrial regions to become glaciated, the formation of marine-based ice sheets requires additional processes. These are (1) a reduction in the sea level and (2) a reduction in the rate of ice lost to the oceans (by iceberg calving and melting of floating ice). Lowering sea level may have the effect of allowing some of the sea floor to be exposed sub-aerially, whereupon terrestrial ice-formation can occur. Glaciers that terminate grounded in water will calve icebergs at a rate that is linearly related to the depth of water (Brown et al., 1982, Pelto and Warren, 1990; van der Veen, 1996). Thus, a reduction in sea level may also result in the lowering of the rate of iceberg calving and, thus, allow ice growth further into the sea. Iceberg calving can also be reduced if a permanent cover of sea-ice is allowed to thicken (due to surface accumulation of ice) into an ice shelf. In this case, the buttressing effect of the ice shelf will curtail iceberg calving, and so cancel the ablation mechanism of the grounded ice body, thus encouraging the growth of grounded ice into deeper water. If we understand what causes the conditions that permit terrestrial and marine ice sheets to build up, we may subsequently understand the causes of ice ages.

Sea-level reduction (Figure 1.4) is caused by an increase in global ice volume, transferring water from oceans to the continent. A lowering of sea level will encourage marine ice sheets to grow more, which will cause more sea-level fall. A simple positive feedback process is thus defined. Since marine ice sheets cannot initiate sea-level fall, its cause must therefore be due to the growth of terrestrial ice masses. Since these are formed when the surface air temperature is

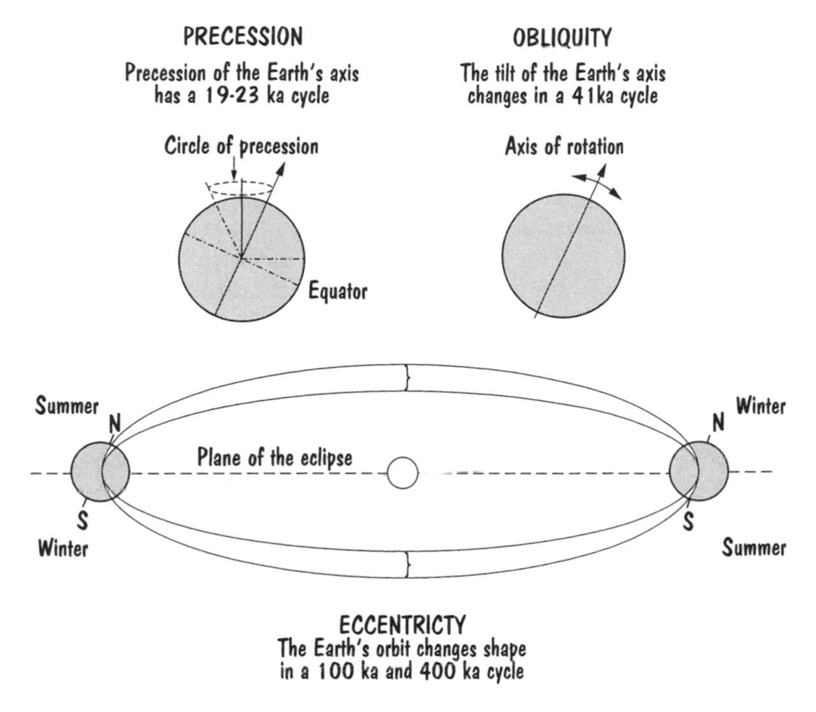

Figure 1.5 Milankovitch theory of orbital variations as a force of ice ages. Adapted from Boulton (1993).

reduced, the primary causes of ice ages must be related to climate change.

MILANKOVITCH THEORY

If we determine what causes the mean annual air temperature to fall causing the ELA to lower, we can identify why ice ages happen. By far the most cited theory relating climate change with ice sheet growth was discussed initially by Croll (in the late nineteenth century) and later elaborated by Milankovitch (in 1930). The theory is often referred to as the Milankovitch Theory (Figure 1.5). In this theory cyclical changes in three of the Earth's orbital variants, namely (i) orbital eccentricity, (ii) tilt of the axis (obliquity), and (iii) precession of the equinoxes, cause an alteration in solar radiation received at the Earth's surface and, hence, are thought to force a change in air temperature.

(i) Orbital eccentricity. Changes in orbital eccentricity of the Earth vary in time between a highly elliptical orbit, and a circular orbit. The periodicity of the cycle is 95 800 years (~100 ka). It should be noted that, in any one year, change in the orbital eccentricity does *not* result in a change in incoming solar radiation. Instead, the effect is to cause contrast between the seasonality of one hemisphere compared with the other. This is the most contentious of the orbital variations. Proponents of the theory claim it has the largest effect on ice sheet growth, but actually is responsible for only very small changes in solar insolation. As will be discussed later, there are several new theories on the viability of this 100 000-year signal as a major forcing process for ice ages (Ridgewell et al., 1999; Roe and Allen, 1999).

(ii) Changes in axis inclination (obliquity). The tilt of the Earth's spin axis varies between 21.39° and 24.36°. The present value is 23.44°. The periodicity of this cycle is 41 000 years. Decreases in the axial tilt result in cooler summers in the more polar regions (high and mid latitudes), but does not change significantly the amount of solar radiation at low latitudes.

(iii) Precession of the equinoxes. This orbital variation affects the timing of perihelion and aphelion (the position where the Earth is closest and furthest from the Sun). This is caused by the rotation (in a conical manner) in the Earth's spin axis as the planet orbits around the Sun. The period of this movement is 21 700 years, over which time the Northern Hemisphere is tilted towards the Sun at successively different points in the Earth's elliptical orbit. At present, this takes place during the Northern Hemisphere summer.

Milankovitch analysed, through the mathematics that describes the Earth's orbit around the Sun, the change in solar insolation caused by each of the orbital variations described above. These individual changes combine to yield the total modification in solar insolation due to time-dependent variations in Earth's orbit (Figure 1.5). The main problem with Milankovitch Theory is that gross values of incoming solar radiation are not really affected (i.e., if less radiation is received in the Northern Hemisphere, then more may be received in the Southern Hemisphere). Subsequently, there are two possible important circumstances for widespread glaciation, which concern the contrast in insolation between seasons.

First, ice growth (glaciation) is when a reduction in solar insolation in the summer causes cooler temperatures and so less melting of snow and ice. This is followed by a warm winter, when increased precipitation from the oceans falls onto high latitudes as snow.

Second, ice recession (deglaciation) is when an increase in solar insolation in the summer causes relatively warm conditions which results in an increase in the rate of ablation. In addition, lower solar insolation values in the winter may cause colder, drier conditions and, therefore, a decrease in the amount of snow fall over the ice mass.

The most important season is the summer, when melting can take place. Milankovitch calculations provide information on the variation of incident solar radiation as a function of season and latitude. From this a complex relation between latitude and solar insolation through the last 130 000 years has been developed. However, on their own, the solar insolation changes predicted by Milankovitch are small, and not generally considered significant enough to initiate ice ages. Some sort of amplification of the climate response to orbital variation is therefore required.

Dalgleish et al. (2000) analysed the variance in solar insolation caused by each of the orbital parameters with respect to the global ice volume record (Chapter 2). They concluded that the last 600 000 years is characterised mostly by the build-up of ice. The general steady increase in ice volume is punctuated by relatively brief periods of ice decay. Dalgleish et al. (2000) calculated that these deglaciation events are associated with obliquity and precession of equinoxes rather than any noticeable change in the eccentricity. This work is important because it highlights that the eccentricity orbital parameter may not be as important a control on glacial cycles as thought previously.

AMPLIFYING FACTORS AND FEEDBACK MECHANISMS

Sea level

One feedback mechanism, mentioned earlier, is the link between sea level and ice sheet volume. To recap, ice growth in terrestrial regions causes sea-level reduction and, so, enables marine regions to become glaciated. This in turn lowers the sea level further and, hence, allows an additional expansion of marine-based ice. In addition, as sea level is lowered the position of the ELA with respect to sea level will remain unchanged. However, the relative elevation of the land surface will increase. This results in more land above the ELA, and so an increase in the accumulation area. The subsequent growth of ice leads to further sea-level fall and so a simple feedback system is initiated.

There are several other feedback mechanisms between ice sheets, oceans and the atmosphere. A very good summary of the nature of this interaction is provided in Clark et al. (1999). They contend that ice sheets modulate ocean surface temperatures, the ocean circulation, the biosphere, surface albedo and the continental water balance. In turn, these changes 'feed back' and cause changes to the ice sheet configurations. Several of these feedback mechanisms are described below.

Surface albedo

A second feedback system involves the reflection properties of the ice surface and the radiation budget of the Earth. The reflection of the Sun's radiation from the Earth's surface is related to the surface albedo. If the albedo is high, more radiation will be reflected back into space. If the albedo is low, more radiation is absorbed by the planet's surface. Greater surface reflection of radiation results in a reduction in the amount of radiation absorbed. Snow and ice has an extremely high albedo. Therefore, as snow fields and ice sheets expand as a result of adjustments in the solar-insolation-induced air temperature, so too will the increased surface albedo cause an increase in the reflection of solar radiation and, so, a further reduction in air temperature. Since this process will reduce the amount of summer warmth over snow-covered continents, it acts to reduce the rate of melting of ice. At the time of the LGM, when large sections of the Northern Hemisphere were permanently covered with snow and ice, the surface albedo of the planet was drastically different from that of today.

Carbon dioxide

Atmospheric carbon dioxide (CO_2) has an impact on climate because it absorbs/reflects longwave radiation emitted from the Earth's surface (i.e. the Greenhouse effect). For reasons not yet fully understood, when it gets colder, atmospheric CO_2 is depleted (Archer et al., 2000). As the atmosphere cools in summers at the beginning of a glacial cycle (due to solar insolation changes), the CO_2 level drops and, because of this, so too does the absorption of longwave radiation. Hence heat is lost from the Earth yielding further cooling. The role of CO_2 on Late Quaternary ice sheets is currently not understood well (see Chapter 2). However, there are several potential climatic feedback mechanisms that involve CO_2 that may be relevant to the growth of ice sheets.

Levis et al. (1999) showed how CO_2 and surface albedo feedbacks may be linked, by running a model of the biosphere and climate at the LGM. The model results showed that CO_2 changes over a glacial cycle will affect the biospheric surface albedo. Their work indicated that a positive feedback, caused by albedo changes induced by vegetation cover, contributes to cooling at the mid and high latitudes at the LGM.

In addition, the growth of Antarctic sea-ice extent during glacial times would have covered a much larger portion of the southern ocean than today (Armand, 2000), inhibiting further the exchange of CO_2 between the ocean and atmosphere. Stephens and Keeling (2000) suggest that this process may be responsible for about 80 per cent of the total CO_2 drop recorded in atmospheric proxy records during the last glacial. As Holland et al. (2000) point out from modelling the ice-ocean-atmospheric system, changes in the sea-ice extent (especially in the Northern Hemisphere) are likely to have an effect on the thermohaline circulation (Chapter 8) and, therefore, the climate. However, whilst Stephens and Keeling (2000) indicate that CO_2 depletion is caused by sea-ice growth, Holland et al. (2000) suggest that

CO_2 actually forces climate, sea-ice and oceanic change.

Ice sheet elevation

An additional feedback system involves the growth of large ice sheets. As ice sheets grow, the surface elevation on which snow accumulates increases. For example, the mean elevation over central East Antarctica is over three kilometres above sea level. Because of this, ice growth will result in an increase in the area of the ice sheet above the ELA. The increase in the accumulation zone will result in further glacier growth.

Ocean circulation

There may also be a relationship between ice sheets and the circulation of the world's oceans. Currently, the North Atlantic provides an important source of heat to Western Europe because warm water flows from the mid-Atlantic northwards. If this flow of water is halted, the North Atlantic will cool and the heat supplied to Western Europe will be taken away. This would result in colder conditions over Europe and, hence, a reduction in the ELA. The interaction between ice sheets and oceans is discussed in Chapters 8 and 12.

Concluding remarks on feedback and ice-age forcing

It should be noted that there are likely to be other feedback processes, and some of the above feedbacks may be interactive. Through the interpretation of proxy records of ice volume and past climate change (Chapter 2), these feedback systems and the timescales over which they operate can be understood.

In summary, once an ice age has begun, a number of feedback processes can act to deepen the glacial activity, regardless of the solar radiation inputs. For example, at 21 000 years ago, the solar insolation values for the last glacial maximum were similar to that at present (i.e., a deglacial period). Thus, the ice age was maintained by processes external to orbital variations. Thus, the climate at 21 000 years ago was greatly influenced by both preceding global climate change and the ice sheet coverage present at that time.

Most people believe, to some extent, Milankovitch cyclicity because of the link with dated indicators of global ice volume (e.g., oxygen isotope curves and carbon dioxide curves – Chapter 2). These data suggest that global ice volume varies in a cyclical manner with 100, 40 and 20 ka periods visible within the signal. However, opinions are divided as to actual level of influence that solar insolation has on ice ages (e.g. Roe and Allen, 1999).

The problems with Milankovitch can be summarised as follows: (i) the 100 ka cycle should, according to the mathematics, produce the lowest variations in solar insolation values. Yet it is the 100 ka cycle which dominates in the Quaternary glacial-geologic record; (ii) prior to about 800 000 years ago, although ice ages occurred, this 100 ka cycle was absent. If orbital variations occur all the time, then we would expect regularity within the timing of ice ages prior to 800 000 years ago.

Recently the validity of the 100 000-year Milankovitch cycle has been questioned. It has been claimed that a hitherto neglected orbital parameter, the inclination of the Earth's orbital plane, may be of importance (Muller and MacDonald, 1997).

RECENT THEORIES FOR THE 100 000 YEAR GLACIAL CYCLES

Inclination of Earth's orbital plane

The inclination in the Earth's orbital plane should vary with a 100 000-year cyclicity. However, such changes will not affect the incident solar radiation directly. Instead, the climate cycle may be affected through variations in the accretion rate of interplanetary dust related to Earth's orbit inclination (Muller and MacDonald, 1997). Some of this dust falls into the atmosphere, and subsequently to the sea, where it may accumulate on the sea floor. Such extra-terrestrial material (^{3}He) has been observed in sea-floor sediment cores, and occurs with a 100 000-year periodicity. Interestingly, the flux of dust to the Earth, calculated from the sea-floor data, is only significant over 100 000-year cycles for the Quaternary, and is much less for the previous 70 million years.

So, the question must be posed: how can extra-terrestrial ^{3}He dust particles affect the Earth's climate? In answer, fine grained ^{3}He probably will not have much effect. However, we can theorise as to how it may signify a further extra-terrestrial forcing of glacier growth.

^{3}He dust particles are very small, and will survive entry into the Earth's atmosphere intact. However, larger particles that may be associated with extra-terrestrial ^{3}He, could be vaporised on entry within the Mesosphere. Such vapours, and condensates, within the upper atmosphere, may play an effective role in changing the Earth's climate. It is therefore the reaction of extra-terrestrial material as it enters our atmosphere that may spark the feedback processes at the start of an ice age. Obviously more research needs to be performed on this idea, and as yet there are no signs that it is less or more likely a true explanation for the 100 000-year glacial cycle (Roe and Allen, 1999).

Potential cyclicity within the thermohaline ocean conveyor

Denton (2000) postulated that the ocean conveyor system oscillates with a periodicity of 100 000 years and, consequently, may act as a control on ice sheet extent over these periods. Appreciation of this theory requires an understanding of the ocean thermohaline system (see Chapter 8). His theory is that the thermohaline circulation system, if halted during the first phase of a glacial cycle, is unable to reinstate itself because the large amount of freshwater icebergs in the North Atlantic do not permit dense salty 'deep water' to form. This maintains a halt in the ocean conveyor for a long period, during which time ice sheets build up. After about 100 000 years the ice sheets are so large that the marine-based portions collapse into the sea due, presumably, to their unstable nature. This releases huge amounts of icebergs into the Nordic Seas. However, when it is done, the Nordic Seas are left without any icebergs for the first time in 100 000 years, and so 'deep water' is allowed to initiate, so starting the ocean conveyor and ending the glacial cycle. As sea-level rises, so the West Antarctic Ice Sheet collapses and a new cycle begins. It should be mentioned that this theory has no real field evidence to support it. However, because the oceans are undoubtedly a major influence on global climate, its place here as a potential control on glacial cycles is appropriate.

RECENT EVIDENCE FROM THE LAST INTERGLACIAL

Imbrie et al. (1992) site several datasets from the Northern Hemisphere as evidence showing it to respond quickly to changes in solar insolation. This, they conclude, provides evidence that the Northern Hemisphere initiates solar-induced climate change. However, the whole debate about causes of ice ages has taken a recent twist after a highly accurate measurement of the date of the penultimate deglaciation. Henderson and Slowey (2000) calculate that the last interglacial can be dated at 135 000 years ago, some 8000 years earlier than thought previously (Imbrie et al., 1992). What makes this date significant are the orbital insolation values calculated at this time. The traditional notion that it is solar changes in the Northern Hemisphere that controls the growth and decay of ice is put into question because, at 135 000 years ago, solar insolation was maximum in the

Southern Hemisphere and minimum in the Northern (Figure 1.6). The implications of this finding are not yet fully understood. However, we may speculate that the Southern Hemisphere may play an important role in the regulation of ice-age cycles, perhaps associated with feedbacks initiated by the Pacific Ocean's response to solar changes.

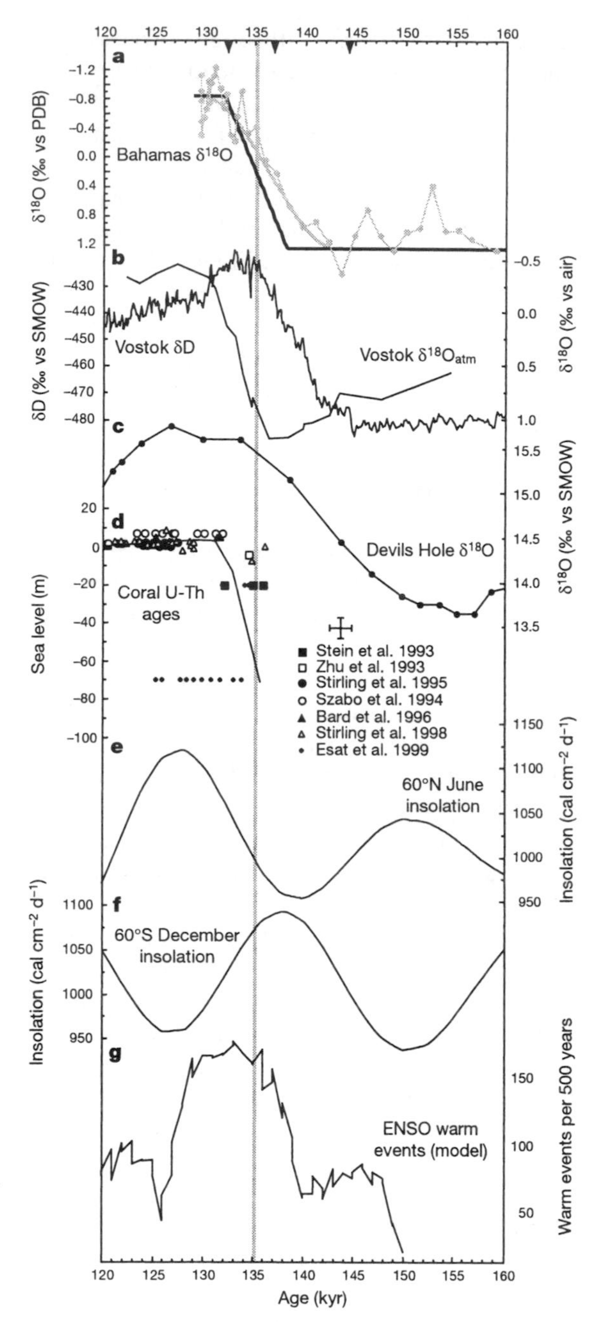

Figure 1.6 The timing of the last interglacial. (a) δ^{18}O from the Bahamas. (b) δ^{18}O and δD from the Vostok ice core. (c) δ^{18}O from the Devil's Hole calcite vein. (d) Sea level. (e) Summer insolation at 60°N. (f) Summer insolation at 60°S. (g) Modelled number of warm ENSO (El Ninõ) events per 500 years. Reprinted with permission from *Nature* vol 404, pp. 61–66, copyright (2000) Macmillan Magazines Limited.

CHAPTER 2

Indicators of Ice Volume and Climate Change

Our current understanding of past glacial activity is only possible because of records of ice volume and climate change. These records allow us to comprehend the timescales over which ice ages operate, and the rates of ice sheet growth and decay during these periods.

A. RECORDS OF GLOBAL CLIMATE CHANGE

Introduction

For specific climate conditions, the balance of a number of isotopic constituents (e.g., those of oxygen and carbon dioxide) within the atmosphere, cryosphere and oceans will also be specific (Figure 2.1). Climate change will itself induce change in the balance of these components. To determine information about past climate conditions, we look for continuous sequences of material dating back to the last ice age (and before), which store within them the legacy of a previous isotopic constituent balance that existed within the atmosphere/ocean/cryosphere. There are two such places where good records can be found; at the centre of ice sheets and on the floor of oceans.

Ocean floor sediments at abyssal depths

Sediment deposits on the ocean floor provide a continuous record of material dating back, in some places, several millions of years, because although sedimentation rates here are low, they are regular and erosion is very low or negligible. Skeletal remains (calcareous organisms), which comprise most of the sediment, can be used to determine information about the past climate. This is because the shells of these creatures will contain the same isotopic balance as the water in which they lived. It should be noted that when water depths increase beyond four kilometres calcium carbonate dissolves in seawater. Consequently, most cores taken are in ocean water which is shallower than four kilometres. Theoretically, the only real limit to the age of the ocean floor sediments, is the age of the oceanic plate itself (which is of the order of millions of years). The use of sea-floor sediment cores as proxy records of past climate change is largely due to the ocean drilling programme (ODP), which has recovered hundreds of deep sediment cores from the floors of the world's oceans.

Deep ice cores

Near the centre of a large ice sheet, the horizontal velocity of ice will be zero (Figure 2.2). In such places, each accumulation of snow will lie on top of the previous layer. The deformation of this ice is by vertical compression rather than shearing. Thus, the deeper down the ice column, the thinner the annual layers become. Therefore at the centre of ice sheets the age of ice is a function of depth and the accumulation of ice only. Ice, and air bubbles held as hydrates

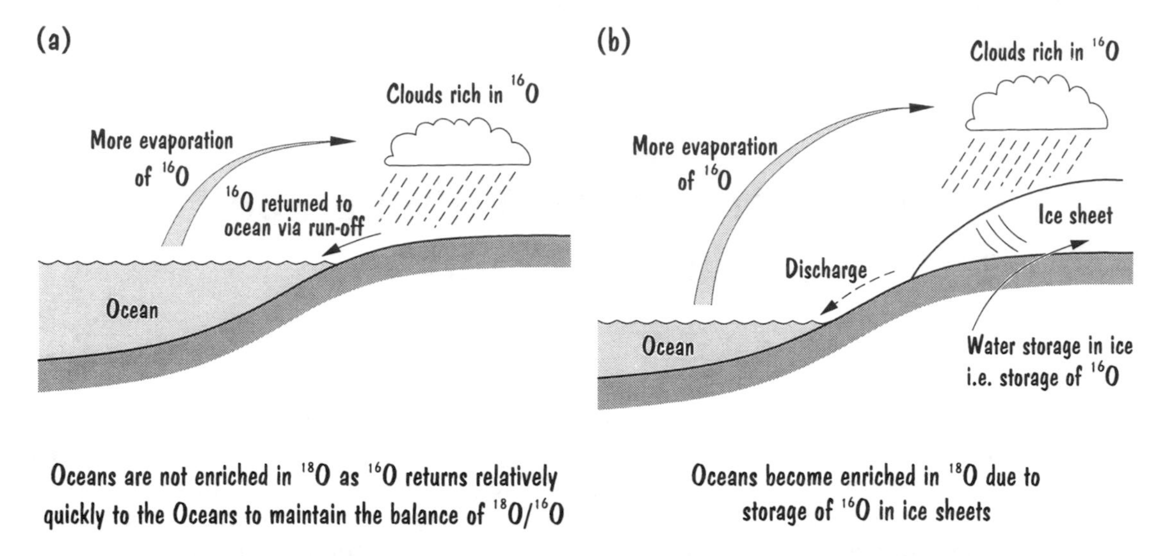

Figure 2.1 Sea level, ice ages and oxygen isotopes. The left hand panel represents periods of interglacials where sea-levels are at their present positions. The right hand panel shows the situation in times of glaciation. During glaciation water is transferred from the oceans to the ice sheets resulting in sea level lowering and ice sheet growth. Since ^{16}O evaporates more readily that ^{18}O, ice sheets become enriched with ^{16}O and oceans become enriched in ^{18}O. Adapted from Bennett and Glasser (1996).

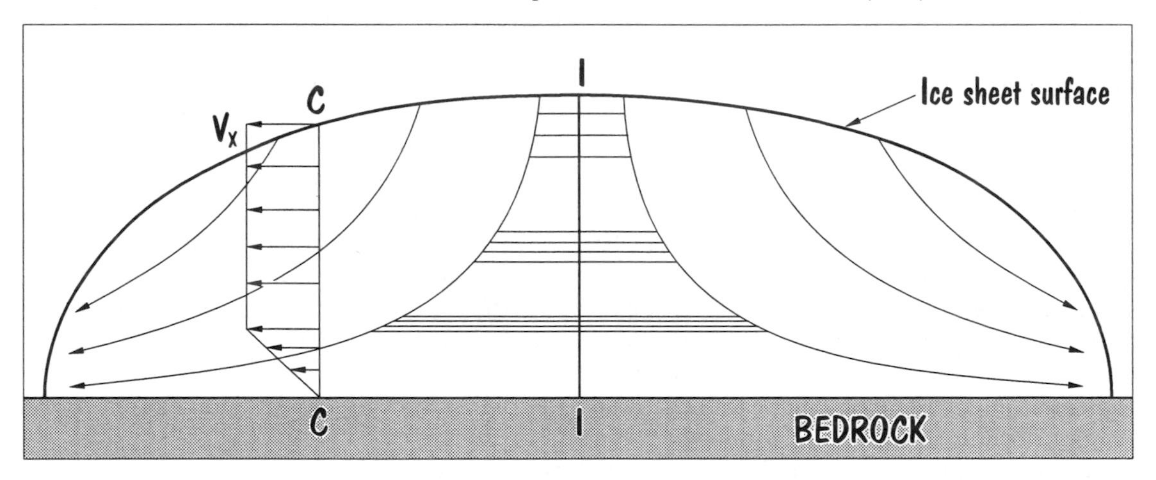

Figure 2.2 Location of ice cores at the centre of large ice sheets. There are two main methods of dating ice, which combine to yield the depth-age relationship of ice cores. First there are stratigraphic markers (such as dust or ash layers – see later in Chapter 2) which pin-point the age of ice at a certain depth. Between these points, numerical modelling has to be used to determine a continuous depth-age relationship. The ice at the centre of an ice sheet (at the ice/flow divide – site I – where the horizontal component of velocity is very small (negligible)), accumulates as if in a column, with very old ice at the base and younger thicker units above. Ice cores taken close to these can yield an uninterrupted sequence of ice, dating back tens of thousands of years. The theory for calculating the age of this ice is simple: Each yearly accumulated ice layer is buried by the subsequent year's ice. The ice is then continually thinned as more ice is piled on top. Since ice is 'incompressible', to maintain the volume of ice, lateral spreading has to take place at depth. At site C, the flow of ice is more complicated and so the age–depth relationship must account for this. We use knowledge about the rheology of ice to determine how it will deform with depth in order to establish the depth-dependent age of the ice column across central and non central regions. Adapted from Dansgaard (1969).

within ice, trap the constituent isotopic components of the atmosphere at the time at which the original snow fell on the ice surface. Thus, taking a core of ice at the centre of ice sheets provides us with samples of air of age increasing with depth in the ice column. The limit to the age of ice core information is the age of the base of the ice sheet. The deepest ice core, holding the longest record of past climate change is the Vostok ice core in central East Antarctica (Petit et al., 1999). Analysis of this core has provided knowledge of not only the last ice age climate, but the previous four ice ages, spanning back over 420 000 years. Further ice cores are planned at Dome C (East Antarctica), and Dronning Maud Land, by the European Ice Core Project in Antarctica (EPICA). Recent ice cores from the Greenland Ice Sheet include the GRIP and GISP2 cores. These cores provide a high resolution record of the glacial-interglacial cycle, and show major environmental instabilities during the last deglaciation (e.g. GRIP project members, 1993; Grootes et al., 1993; Stuiver and Grootes, 2000) (see Chapter 10).

Oxygen isotope records

Isotopic concentrations and ice cores

Evaporation of seawater releases three isotopes of oxygen into the atmosphere (^{16}O, ^{17}O and ^{18}O). In this process, the rate of evaporation is related to the density (i.e., the atomic weight of the isotopes). Thus, there is a preference for the evaporation of lighter isotopes. This rate of evaporation is also related to the temperature of the water (e.g., cold water temperatures will allow relatively more ^{16}O to evaporate than ^{18}O). During periods of glaciation, water is transferred, via evaporation, from the cold sea to the ice sheets. Thus, the snow falling will be enriched with ^{16}O, and the oceans enriched in ^{18}O. Conversely, during interglacials; (i) the ice sheets (enriched in ^{16}O) will melt, reducing the relative concentration of ^{18}O in the ocean and (ii) the warm sea will allow more ^{18}O to be evaporated, further reducing ocean ^{18}O and increasing the ice sheet's concentration of ^{18}O. Thus, ice cores store information, within the ice chemistry, about global oxygen isotope variations.

Isotopic variations in ^{18}O within ice sheets bear an approximate inverse relationship with fluctuations from that derived from deep-sea sediment cores. The biggest problem with ice cores is concerned with the flow of ice with depth in the ice sheet (Figure 2.2). However, we can use knowledge from measurement and modelling of ice masses to tell us information about:

- where on the ice sheet surface to take ice cores
- how deep the ice is at the coring site
- what the basal topography is like
- how the age of ice varies with depth.

Figure 2.3 shows the $\delta^{18}O$ signal from the GISP2 ice core in central Greenland. Significant changes in the $\delta^{18}O$ signal correspond to different 'stages'. These are known as oxygen isotope stages, and allow the Late Quaternary to be subdivided into periods that comprise glacial-interglacial cycles (Figure 2.3). For comparison, Figure 2.4 shows the chemical signature of the 3.5 km deep Vostok ice core from the Antarctic Ice Sheet.

Isotopic concentrations and sea floor sediments

The stable oxygen isotope concentration within calcareous organisms (foraminifera) is a function of (1) the isotopic composition of the ocean water, and (2) the temperature of the water, during the life-time of the foraminifera. For every 1°C fall in sea temperature (which may occur with depth), there is a relative enrichment within faunal shell debris of 0.02 parts per ml (0.02‰) in the proportion of ^{18}O to ^{16}O. This has to be corrected when analysing oxygen

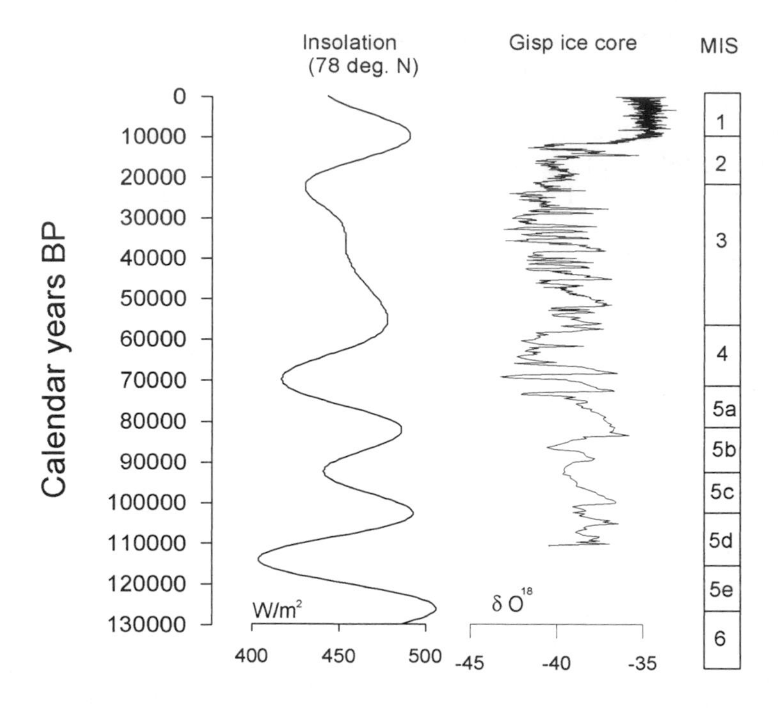

Figure 2.3 GISP2 ice core records and solar forcing. (a) Summer solar insolation values at 78°N. (b) GISP2 δ^{18}O ice core record. (c) Marine oxygen isotope (MIS) stages.

isotope signals from ocean sediments, to establish information about global ice volumes. The ratios of oxygen isotopes are calculated as the relative deviations from mean ratios of a standard value. Thus, the deviations are calculated in positive or negative values relative to a standard zero value (defined as the Vienna Standard Ocean Water, or SMOW). For example, a ^{18}O value of –4‰ corresponds to a sample that is 4 parts per ml deficient in ^{18}O relative to the standard value.

A number of problems exist which complicate the interpretation of oxygen isotope ratios within sea floor sediments as follows:

- Bioturbation with regions of high sedimentation. This problem results in a reduction in the resolution of the information within the core to a maximum of about 1000 years. Therefore, high-resolution information is unlikely to be acquired from deep-sea sediment cores.
- Isotopic equilibrium. For various reasons, the shells of some forams do not exist in isotopic equilibrium with that of the ocean. Thus, studies using mixed samples of forams may yield peculiar results. This problem can be overcome to an extent by picking out specific species of forams that are known to provide reliable results.
- Physical transportation. Submarine slides may transport shells from forams living in shallow water (which have an isotopic ratio in balance with shallow water) to greater depths, where the isotopic balance is different. This problem is overcome by analysing the provenance of forams.

Forams are either benthonic (living at depth within the water) or planktonic (living at, or close to, the surface of water). Thus, by separating these forms of forams within the sea floor sediment, information can be gained about global ice volume changes from benthonic

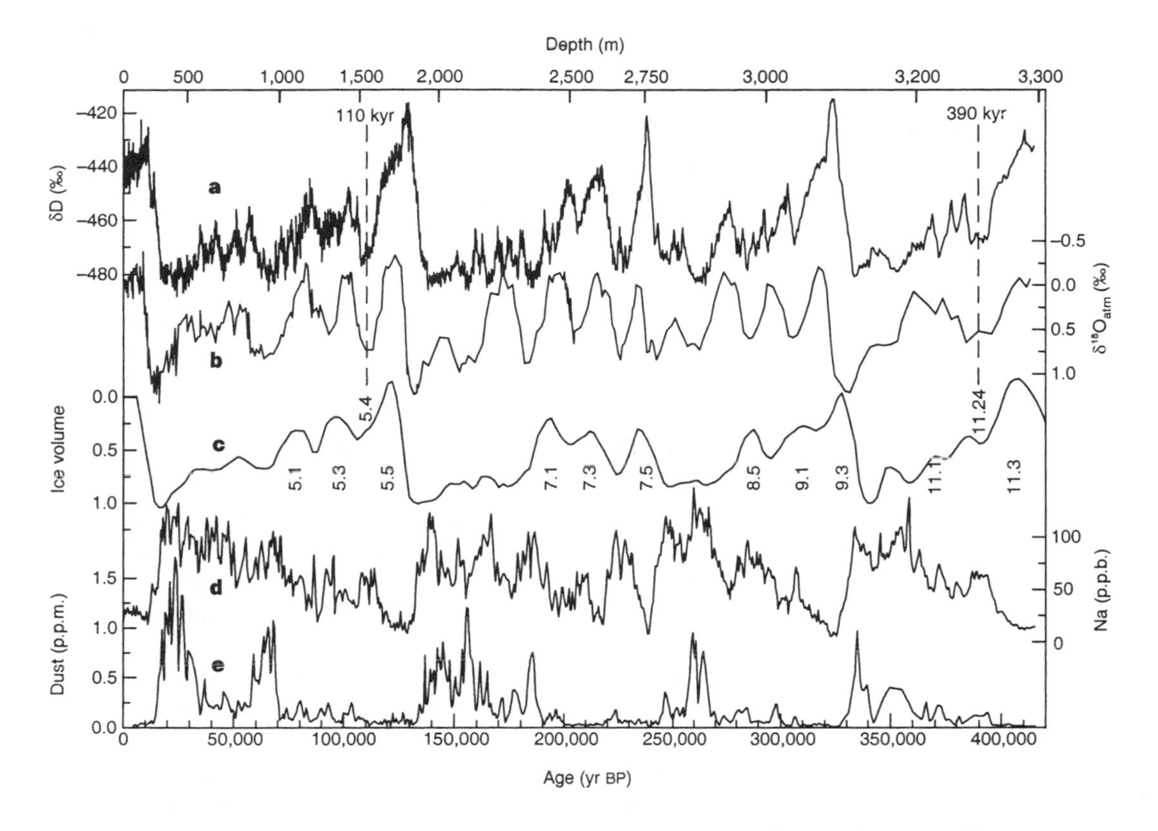

Figure 2.4 Vostok ice core chemical records for the past 420 000 years. (a) δD levels. (b) δ^{18}O variations. (c) Ice-volume proxy records and oxygen isotope stages. (d) Na concentrations. (e) Dust concentrations. Reprinted with permission from *Nature* 399, pp. 429–436, copyright (1999) Macmillan Magazines Limited.

forams, and ocean surface conditions from planktonic forams.

Deep-sea oxygen isotope data and Milankovitch forcing

Shackleton (1987) reviewed how the periodicity of oxygen isotope data from deep-sea cores was in accordance with those calculated from orbital variations. It is this similarity between the oxygen isotope and orbital predictions that lead many scientists to recognise Milankovitch Theory as a plausible idea. The marine oxygen isotope record shows that ice sheet build-up takes a considerable period relative to the short, rapid decay of ice during deglaciation.

Mg/Ca palaeothermometry from sea floor sediments

The partition coefficient of Mg^{2+} into inorganic calcite in the open sea is related to water temperature. Moreover, the uptake of Mg in marine micro-organisms is also related to the water temperature. Thus, measuring the Mg/Ca concentrations of the calcite shells deposited by planktonic forams over geological time establishes a proxy for the ancient temperature of the ocean. Recent investigations using this technique have detailed Quaternary sea surface temperatures (Hastings et al., 1998), as well as deep sea temperatures further back across the Cenozoic (Lear et al., 2000). Although this technique is extremely new, results so far have shown a very good correlation with the $\delta^{18}O$ proxy records of global ice volume. This method may therefore, in combination with $\delta^{18}O$ records, provide an additional measurement to gain information on sea temperatures over glacial cycles (Elderfield and Ganssen, 2000).

Further information from ice cores

Snowfall analysis from ice core studies

Ice cores can provide information about the past rate of snow accumulation from the following features:

(a) The thickness of ice layers. However, layers of ice are difficult to distinguish when samples are taken from several hundred metres below the surface due to the vertical compression that occurs there.
(b) Beryllium-10 (^{10}Be) concentrations. In the atmosphere ^{10}Be is created by cosmic radiation (solar activity). It becomes attached to aerosols quickly after formation whereupon it falls to the Earth's surface. The concentration of ^{10}Be received at the Earth's surface is controlled by its rate of production in the atmosphere (i.e. the cosmic-ray flux). In an ice core, the concentration of ^{10}Be (measurable by accelerator mass spectrometry) is also controlled by the accumulation rate of ice. Therefore ^{10}Be within an ice core is useful in two ways. First it can be used as a stratigraphic marker to compare ice cores. Second, if the past production of ^{10}Be can be determined, variations in the concentration of ^{10}Be measurable within the ice can be used to calculate long-term changes in the amount of snow fall. High concentrations of ^{10}Be suggest low accumulation rates, whilst low concentrations of ^{10}Be have been 'diluted' by high rates of past snow fall (Paterson, 1994).

Carbon dioxide analysis from ice core studies

Atmospheric CO_2 concentrations can be measured from air bubbles trapped within the ice (Figure 2.5). The form of the CO_2 curve is remarkably similar to the oxygen isotope curve derived from both ice cores and sea floor sediments. The level of CO_2 will vary with temperature. During glaciations, the CO_2 value will be low whilst, within interstadials, the value is high. However, it is still not resolved whether the CO_2 record (1) lags behind the $\delta^{18}O$ signal, (2) is synchronous with the $\delta^{18}O$ signal, or (3) occurs before the $\delta^{18}O$ variations. This represents a major scientific question at present. If this problem can be resolved we will arrive at one of two conclusions: (a) that CO_2 is an effect of glaciations rather than a cause or (b) that CO_2 variations force glacial activity. Once the role of CO_2 on ice sheet behaviour is established, the relative importance of the numerous potential feedback processes (Chapter 1) that involve this atmospheric gas on Quaternary climate change can be evaluated.

Recent work concerning the glacial/interglacial cycles of CO_2, as recorded in ice cores, is summarised in Archer et al. (2000). They indicate that the importance of CO_2 as a potential forcer of past climate change comes from the last two deglaciations when the CO_2 level is known to have increased prior to the decay of ice as shown in the $\delta^{18}O$ signal. It has long been recognised that the most likely driver of CO_2 change over a glacial cycle is the ocean or, more specifically, the uptake of CO_2 in oceanic $CaCO_3$. Although CO_2 is more readily soluble in cold water, the relatively small contrast in glacial-interglacial sea-surface temperatures precludes this process as a potential explanation for the changes. Ocean processes involving biological productivity and the pH of the ocean are more likely to be responsible for long-term variations in the exchange of CO_2 between the air and ocean (Archer et al., 2000). However, each of these processes may be forced by external conditions. So, although it is known that CO_2 varies across a glacial cycle, and that this

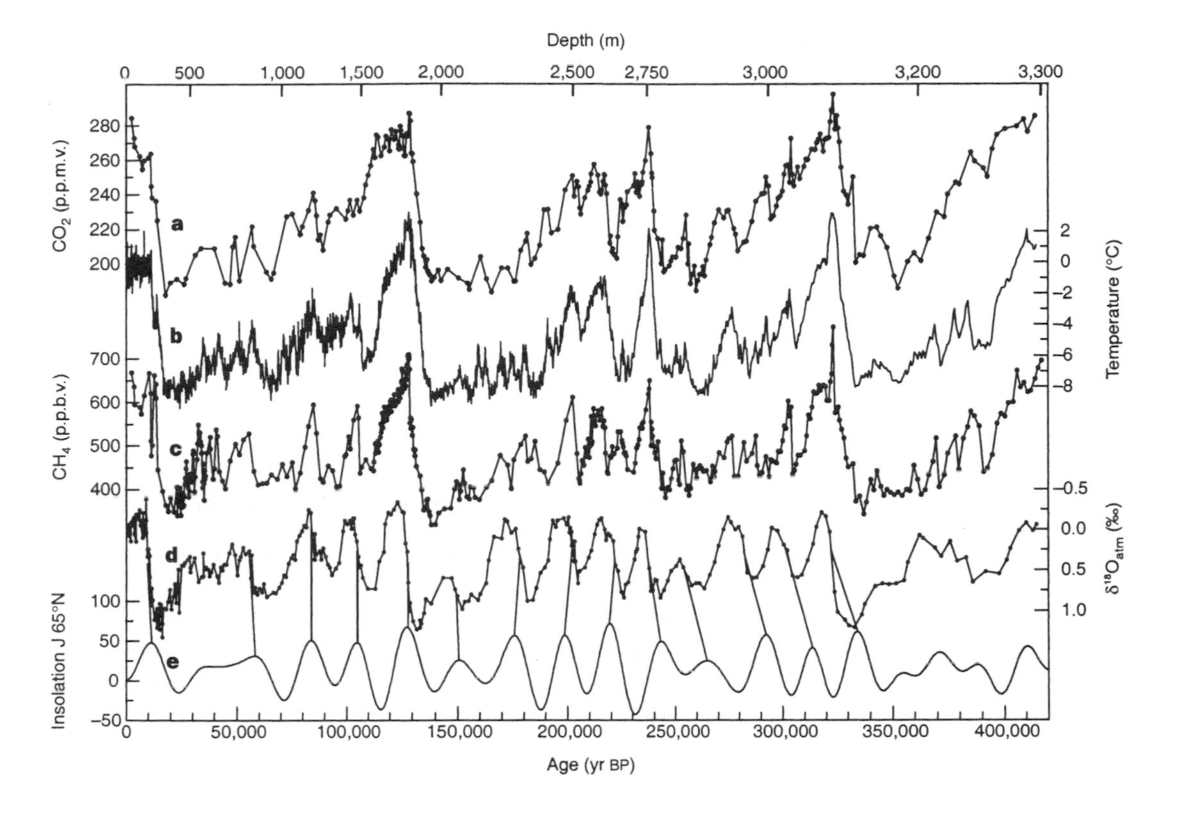

Figure 2.5 Vostok ice core gas records for the past 420 000 years. (a) CO_2 levels. (b) Reconstructed air temperature. (c) CH_4 values. (d) $\delta^{18}O$ levels (as in Figure 2.4). (e) Summer solar insolation at 65°N. Reprinted with permission from *Nature* 399, pp. 429–436, copyright (1999) Macmillan Magazines Limited.

variation is likely to be involved in the amplification of climate variation, it is still not clear what ocean processes are responsible for the uptake and delivery of CO_2 across glacial cycles.

The role of CO_2 in ice-age forcing is further investigated using a numerical model of the atmosphere over the last glacial-interglacial cycle (Loutre and Berger, 2000). These experiments used the known CO_2 variation from the Vostok ice core to force the Earth's climate, but with no solar insolation variations accounted for. The results indicated that although mean air temperatures are affected by variation in CO_2, the model was not capable of reproducing the ice-age climate cycle when forced by CO_2 variations alone. This model appears to support the view that CO_2 variations may not themselves be external forcers of climate change, but may be influential as a feedback mechanism (outlined in Chapter 1).

Chemical signatures in ice cores

Deep ice cores are now used in a variety of ways to obtain information about ice sheets and their former environment. Since the industrial revolution, the chemical composition of the atmosphere has been altered dramatically. Ice cores provide a means by which the relationship between these chemicals and past global change can be examined. Such investigations have found that there is a remarkable association between the chemistry of the atmosphere and global ice volume. For example, Legrand and Mayewski (1997) show that concentrations of Na^+, Cl^-/Na^+, Ca^{2+} and SO_4^{2-} are related strongly to the $\delta^{18}O$ signal and, therefore, to global ice volume. Thus, there is a link between atmospheric chemistry and global climate change (for example the ocean-atmosphere sulphur cycle – Legrand et al., 1991).

Dust in ice cores

During periods of glaciation, atmospheric dust concentrations increase due to glacially-derived rock flour in the outwash plains of large ice sheets. This fine grained silt, clay and sand is transported by aeolian processes. This material is observed in (a) the formation of Loess found commonly around the world and (b) dust levels found within ice cores. Petit et al. (1999) analysed dust levels in the Vostok ice core, and showed conclusively that dust concentrations have strong periodicities of 100 and 41 ka (Figure 2.4). This would mean that dust concentrations are linked to air temperatures in some way. However, analysis of the GRIP ice core shows this is not the case in Greenland. Fuhrer et al. (1999) showed that, if the source area for the dust is Asia, as seems most likely (e.g. Svensson et al., 2000), then significant changes in the wind speed are required at the source, some time prior to deposition, to account for the variation in dust received in Greenland. They also demonstrate that these wind-speed changes are simultaneous with air temperature changes in Greenland, which suggests that the atmospheric circulation system was affected by large-scale rapid alterations during the last glacial cycle. The stable climate around Antarctica may, therefore, be responsible for the strong correlation between dust and temperature in the Vostok ice core.

Ash and volcanic activity recorded in ice cores

Large volcanic events throw ash and other material into the atmosphere, which can be recorded in ice cores through the identification of ash layers and the acidic aerosol product from volcanic activity. For example, Petit et al. (1999) found three ash layers within the Vostok ice core, 3.3 km below the ice sheet surface. This is an extreme example, and highlights the usefulness of identifying stratigraphic markers from ash layering in ice cores. In addition, acidic layers of ice can be recorded in the ice-core chemical signature, and from airborne radar sounding across the centre of ice sheets because internal radar reflections at depths greater than one kilometre are caused by such acidic ice (e.g. Millar, 1981). Analysis of these radar layers has shown that there appears to be no obvious relationship between volcanic activity and ice age cycles (Millar, 1981).

Cosmogenic radionuclides in ice cores

The production rate of ^{36}Cl in the atmosphere increases non-linearly with a decrease in the geomagnetic field. Since the intensity history of the Earth's magnetic field has been reconstructed between 10 000 and 75 000 years ago, the dates of global magnetic field 'events' such as that recorded in lake floor sediments in Mono Lake (Denham and Cox, 1971), can be established. Thus, by analysing ice cores for ^{36}Cl concentrations, these geomagnetic events can be utilised as starigraphic markers in a similar way to dust and volcanic ash layers (Wagner et al., 2000).

Carbonate records

Records of $\delta^{18}O$ have also been obtained from calcite veins and carbonate deposits. One good example is the Devil's Hole subaqueous cave, located within an aquifer in Nevada (USA) (Winograd et al., 1992). The walls of the cave are covered with vein-type calcite that preserves a record of oxygen isotope variation in palaeo-precipitation and, hence, past environmental change. Because of the slow rates of calcite precipitation, the 500 000 year record from Devil's Hole is only 36 centimetres long. One potential problem with the record is that ages have to be estimated by linear interpolation between control points determined from the U-Th method. Despite this, the timing of environmental change recorded in Devil's Hole is generally thought to be reliable for the past half a million years. One important observation from this record is that the durations of the last four interglacials were twice as long as thought previously (i.e. ~20 000 years rather than ~10 000 years, Winograd et al., 1997). If the interpretation of the record is correct, it means that the Milankovitch theory either requires significant revision or is wrong for interglacial periods.

On a similar theme, important sea level and palaeoclimate information has been recently obtained from submerged stalagmites in the Bahamas (Richards et al., 1994; Richards et al., submitted). What makes the investigations of speleothems so significant is that in contrast to many palaeoenvironmental records they can be dated directly by U-Th methods at an extremely high resolution. It is widely reported that there is an offset between radiocarbon dates and real calendar years due to variations in the production of ^{14}C. Specifically for the last glaciation, reductions of the Earth's magnetic field in this period meant that ^{14}C production was up to 40 per cent more than at present (Bard et al., 1990; Lao et al., 1992; Stuiver et al., 1986; Stuiver and Reimer, 1993). This causes erroneously young age estimates by the radiocarbon method. However, by coupling radiocarbon dates with U-Th ages from a growth axis of a spelothem, a valuable new calibration for radiocarbon dates has been formulated. The use of speleothems in reconstructing past sea levels and palaeoclimate looks set to increase in the next few years because these carbonate deposits allow accurate dating not currently available in ice cores and sea-floor sediments.

B. RECORDS OF GLOBAL SEA-LEVEL CHANGE

Introduction

Sea-level variation is an important forcing function of glacial cycles (and therefore of numerical ice sheet models), since sea level influences both the area of subaerial bedrock, and the position of the ELA. Given the importance of bedrock elevation and ELA to ice sheet growth and decay, it is highly desirable for an accurate time-dependent sea-level curve to be determined for the last glacial cycle. To do this, it is assumed that the glacial component of eustasy far exceeds other processes affecting sea level, such as ocean temperature and density changes, sedimentary deposition within the ocean and tectonics, so that they can be ignored as significant contributors to the global sea-level signal, over glacial cycle timescales.

Global sea level curve

There are several methods of estimating the ancient global sea-level record: all are subject to some inaccuracies. The first method relies on the fact that the $\delta^{18}O$ concentration within an ice sheet is different to that in the open ocean. Subsequently the $\delta^{18}O$ concentration within the open ocean will be dependent on the global ice volume, and is recorded in deep-sea sediment cores that are rich in benthic foraminifera. Sea-level change can therefore be estimated through benthic oxygen isotope data. However, Shackleton (1987) explains that the $\delta^{18}O$ concentration of deep-sea cores is different to seawater values, and that this difference is related to sea temperature. Thus, as the sea temperature changes through the glacial cycle, the $\delta^{18}O$ value that is recorded will be dependent on two processes. Also, when sea-level change is calculated by oxygen isotope analysis, it is assumed that ice sheets contain a uniform level of isotope concentration. This assumption is invalid since Mix and Ruddiman (1984) suggest that accumulating snow will be isotopically variable because of the variation in surface air temperature of an ice sheet through a glacial cycle. The oxygen isotope method of palaeo sea-level reconstruction is therefore open to a number of errors that are reflected in the confidence of subsequent results.

A second method of eustatic sea-level reconstruction utilises the ^{14}C dating of low latitude raised coral reefs (e.g. Dodge et al., 1983; Chappell and Shackleton, 1986; Fairbanks, 1989). However, these data may also be unreliable for two reasons. First, the effect of local isostasy on the raised beach sequence is invariably unknown. Second, as noted earlier, there is a discrepancy between radiocarbon ages and real calendar years.

Unfortunately, but not surprisingly, low latitude raised coral reef and ^{18}O analysis methods do not provide compatible palaeo sea-level results for the last glacial cycle (Figure 1.4), with the largest difference in the two sea level curves of around 60 metres between 40–28 000 years ago.

The global sea-level curve for the period between 40–18 ka BP is not well-defined (Figure 1.4). However, some clarification on ancient eustasy has been established since Pinter and Gardner (1989) used a LaGrange polynomial interpolation over existing Late Quaternary global sea-level data, to create a polynomial model of sea level with time. The form of the subsequent sea-level function is similar to that shown in raised marine terrace studies from New Guinea (Chappell and Shackleton, 1986). These records indicate that the global sea-level at the LGM was about 120 metres lower than at present.

A third method of predicting global sea-level is by analysing near-coastal sedimentary sequences in tectonically inactive regions. This works because as sea-level changes, so too do the sedimentary characteristics (such as datable faunal assemblages) of shallow marine environments. If such characteristics can be identified, the sea-level can be reconstructed. Analysis of such environments in the Bonaparte Gulf in Australia led to a recent calculation that global sea-level was 135 metres lower than at present (Yokoyama et al., 2000). Importantly, this investigation reveals that global sea-level was held at this low stand for 3000 years between 22 and 19 000 years ago. Deglaciation began after 19 000 years ago with a rapid initial decay of global ice volume, causing sea level to rise by around 10 metres within a few hundred years (Yokoyama et al., 2000).

A well established time-dependent record of sea level for the period of the last global deglaciation (after 19 000 years ago) was calculated by Fairbanks (1989), from ^{14}C dating of a reef-crest coral *Acropora palmata* on the south coast of Barbados. This sea-level curve shows three interesting deglaciation features (Figure 1.4). First, a sea-level rise of 24 metres within 1000 years at around 12 ka (C) BP corresponds to oxygen isotope Termination 1a (14.5–11.5 ka (C) BP). Second, the minimum rate of sea level rise at 11 ka (C) BP reflects the Younger Dryas reversal of deglaciation. Third, the increasing rate of sea-level change just after 10 ka (C) BP, which is centred at 9.5 ka (C) BP, agrees well with oxygen isotope Termination 1b.

A new sea-level record, based on analysis of late glacial sediments on the Sunda Shelf in Southeast Asia, enhances the record established at Barbados (Hanebuth et al., 2000). The Sunda Shelf is the largest region of continental shelf outside of the Arctic, is tectonically stable and would have been exposed subaerially during the last glaciation. Analysis of sediment cores across this shelf reveals mangrove swamps, mud flats, lagoons and delta plains from which relative sea level was able to be identified. AMS radiocarbon dates are available in all these environments, allowing an accurate depiction of the post-glacial sea level at this site (Hanebuth et al., 2000). The record shows that, between 14 600 and 14 300 years ago, sea level rose by as much as 16 metres, reflecting the rapid part-decay of ice sheets at this time.

Sea level variation within the glaciated regions

The global sea-level curves that are described in Figure 1.4 were not necessarily produced within the high latitude seas during the last glacial cycle, because of an anomalous geoidal effect which was caused by the existence of large Northern Hemisphere ice masses (e.g. Chappell, 1974; Farrell and Clark, 1976; Clark et al., 1978). The amount of gravitational attraction that a large ice mass possesses, acts so as to severely distort the sea level near the ice mass, when compared to the sea-level reference spheroid (e.g. Chappell, 1974; Farrell and Clark, 1976; Clark et al., 1978; Fjeldskaar, 1989; Fjeldskaar, 1991). Consequently, although the time dependent pattern of sea-level change during the Quaternary may have been mirrored in high latitudes, the magnitude of sea-level change would not necessarily have been, because of ice sheet presence. This problem has implications for the sea-level function that is used in numerical ice sheet models.

Relative sea level and glacial isostasy

The asthenosphere beneath the thin elastic lithosphere is of a viscous nature, and is denser than the material above it. Therefore, if a load the size of a large ice sheet is placed on the Earth's surface, the crust will act to attain an equilibrium below its original position, according to Archimedes' Principle (Chapter 3). Thus, asthenospheric material is displaced to compensate for the weight of the ice load. This is a form of tectonic loading which causes isostatic depression and, if the reverse happens (unloading), isostatic recovery. Because of the viscous nature of the asthenosphere, it does not respond instantaneously to the isostatic loading of the crust. A lag in the depression and uplift (of the order of a few thousands of years) is therefore observed at the Earth's surface after the tectonic forcing. In contrast, an elastic plate of finite thickness over a viscous medium would act to laterally diffuse a load that is placed on the plate surface; this is the effect of the lithosphere in the process of glacial isostatic loading.

For a marine-based ice sheet isostasy has a number of implications for the reconstruction of ice sheet decay. During deglaciation, when isostatic recovery takes place, uplift features such as raised beaches will be created only once the ice has gone, and also once the surface is at sea level. Therefore, the isostatic response that is observed within sequences of raised beaches will provide only a partial record of isostatic recovery. Sea-level change has a complicating influence on the isostatic signal because it acts as an additional load, positive or negative, to the ice sheet. Thus the isostatic uplift pattern observed within raised beach sequences does not necessarily reflect the total record of postglacial uplift. During deglaciation isostatic recovery will take place while sea levels rise. The uplift recorded in raised beaches is therefore a relative one.

Relative sea-level curves are constructed regularly by dating carbon samples from raised shorelines of known altitudes. When a number of altitudes have been dated within one location, the relative uplift rate can be calculated. It is therefore important to note the factors involved in the formation and development of raised beaches when building relative sea level curves to reconstruct former ice loading.

C. CONCLUDING REMARKS

The identification of global climate and sea-level records for the last glacial cycle is valuable to the understanding of large-scale ice-land-ocean-atmosphere processes. However, these are by no means the only palaeoenvironmental signals that are used to reconstruct the last ice sheets. Proxy climate records that detail the environment of a specific site to a high order of accuracy in time are required to build palaeoclimate parameters for numerical models of climate and ice sheet behaviour.

Numerical ice sheet (Chapter 3), climate and ocean models require proxy records of the local environment for (a) input data and/or (b) comparison with model results. Indicators of global environmental change are of little use to models requiring more regionally specific data. Geological remnants of glacier activity, deposited close to the former ice margin can often be processed to yield regional palaeoenvironmental records. Although such data are useful in their own right, in combination with other proxy records they can contribute to a fuller understanding of the environment at the last ice age, and its variation in time and space. Details of a number of indicators of 'local' ice volume and palaeoclimate are discussed later in relation to specific ice sheet reconstructions.

CHAPTER 3

The Flow of Ice and Ice Sheet Modelling

THE FLOW OF GLACIERS AND ICE SHEETS

Field investigations of glacier motion and laboratory studies of ice rheology (the flow and deformation of ice) have identified three main mechanisms by which glaciers and ice sheets move. These are by internal deformation of the ice itself, and through two processes which take place at the base of the glacier; basal sliding and the deformation of water-saturated sediments (Figure 3.1). Flow by internal ice deformation takes place in all ice masses, and generally accounts for motion of a few metres per year. However, basal motion occurs only in glaciers where the bed is at the pressure melting point and water is present at the bed. Where basal motion takes place glaciers may move at tens to hundreds, and sometimes a few thousand, metres per year.

The flow of ice sheets is organised in general terms as follows. Ice at the centre of an ice sheet flows very slowly (of the order of metres per year), by internal deformation. A particle of ice on the ice sheet surface will be buried by subsequent snowfall and so will possess a relatively significant vertical velocity component downwards into the ice sheet. The flow of ice at the centre of an ice sheet radiates from the ice divide where there is no lateral movement of ice on the surface. The ice sheet interior will be characterised by a series of ice divides that define the margins of ice drainage basins. As one moves away from the ice divide, the ice velocity increases. Ice sheets are effectively 'drained' by fast-flowing rivers of ice, known as ice streams, transporting ice from the interior to the ice margin. The velocity of ice streams is often several hundred metres per year. Ice streams flow quickly because water at their bases causes a reduction in subglacial friction allowing them to effectively slide across the subglacial topography, with ice deformation contributing only a small amount to the total velocity. The transition between the slow-moving ice sheet and the fast-flowing ice streams has been shown recently to occur several hundred kilometres in the ice sheet interior (Bamber et al. 2000a,b). Broadly speaking, ice sheets lose mass by surface melting if they terminate on land, or by iceberg calving if they terminate in the sea. If an ice sheet flows into the sea and becomes afloat in water, it becomes an ice shelf, whereupon subglacial melting may contribute significantly to the mass balance.

In this chapter, the main mechanisms by which glaciers flow are outlined qualitatively and the arrangement of a simple ice sheet model is outlined. Quantitative details of the physics of glacier flow are provided in a number of text books (e.g. Paterson, 1994; Hooke, 1998; van der Veen, 1999).

Internal deformation of ice

From simple mechanical considerations of the effect of gravity on a slab of ice resting on an inclined slope, the shear stress at the base of that ice is given by:

$$\tau_b = \rho_i g H \sin \alpha \qquad (1)$$

where τ_b is the basal shear stress, α is the surface slope of the ice sheet, H is the ice thickness,

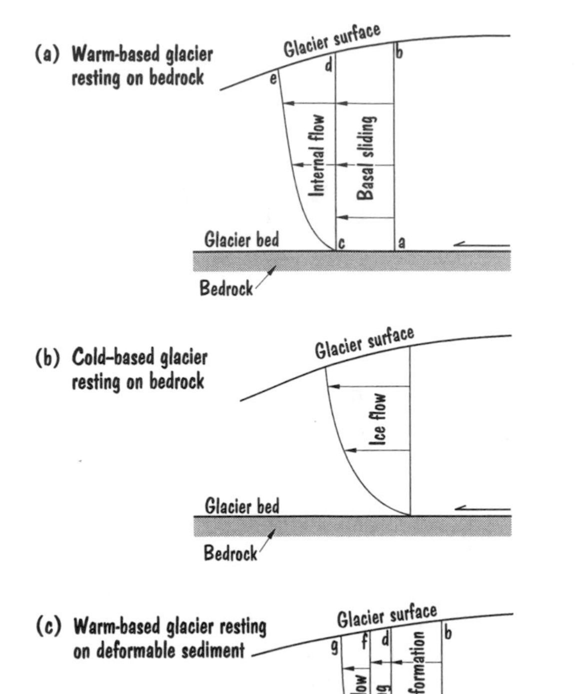

Figure 3.1 Mechanisms of glacier flow. (a) Warm-based glacier resting on bedrock exhibiting ice deformation and basal sliding. (b) A cold-based glacier where ice deformation is the only process contributing to ice flow. (c) A warm-based glacier resting on deformable sediment. In this case ice flow is controlled by ice deformation, basal sliding and the deformation of subglacial material. Adapted from Boulton (1993).

g is the acceleration due to gravity (9.81 m s^{-2}) and ρ_i is the density of glacier ice which varies between 830 and 917 kg m^{-3} and is usually taken as 910 kg m^{-3} (Paterson, 1994). If it is assumed that the ice sheet behaves as a perfectly plastic material, with a yield stress of about 100 kPa, then the shear stress will only reach the yield stress, and hence deformation take place, at the base of the ice sheet. In reality, glacier velocity will vary with depth. Paterson (1994) explains this by considering the relationship between shear strain rate and shear stress:

$$\dot{\varepsilon}_{xy} = A\tau_b^n \tag{2}$$

where $\dot{\varepsilon}_{xy}$ is the shear strain rate, A is a temperature-dependent flow-law parameter, n is a flow-law exponent and τ_b is the effective basal shear stress (Glen, 1955; Nye, 1957). This approximation of glacier flow is known as Glen's flow law, and is a reasonable representation of observed glacier flow through internal deformation. However, it should be noted that this relationship is capable of yielding widely varying results by altering the ice flow exponent n and parameter A. Values of n for ice have been suggested by Weertman (1973) to range between 1.5 and 4.2, with a mean of 3 (which is usually adopted in glacier modelling investigations). The flow parameter A can be thought of as a temperature-dependent quantity relating to ice hardness; the warmer ice gets, the softer it becomes. Usually, when Glen's flow law is applied to calculate velocities within a glacier, the results are well below those that have been measured from the glacier surface. The difference between calculated and measured velocities is normally a result of the occurrence of motion at the glacier bed (Figure 3.1).

The velocity at the ice sheet surface caused by the deformation of ice, u_d, can be calculated by the following equation:

$$u_d = \frac{2A}{n+1}\tau_b^n H = \frac{2A}{n+1}\rho_i g \alpha^n H^{n+1} \tag{3}$$

Thus if $n = 3$, the velocity of ice due to internal deformation is proportional to the surface slope to the power 3 and the ice thickness to the power 4 (Paterson, 1994).

Basal sliding

Many sliding laws assume that the glacier rests upon smooth rigid bedrock, and that ice surface slope, through basal shear stress, controls the glacier velocity. For example, Weertman's (1957) theory of glacier sliding was dependent on two processes; melting and refreezing of ice around bedrock obstacles, and enhanced plastic flow of ice at the bed. Since the publication of Weertman's ideas, the theory has been extended by the utilisation of a more realistic subglacial topography in calculations (Nye, 1970; Kamb, 1970). However, these theoretical sliding relations, when compared with glacier velocity observations, yield dissimilar results. This is because glacier velocity is known to fluctuate on seasonal and diurnal timescales (Willis, 1995). Such velocity variations correlate well with measurements of basal water pressure in the glacier. Because of this, it is clear that subglacial water has an important effect on the motion of ice masses. Field investigations and theoretical studies of sub-glacial hydrology have suggested that relatively high basal water pressures lead to enhanced basal sliding because the shear stress at the glacier base is reduced or eliminated. This is most evident at the beginning of the melt season where surface meltwater reaches the bed, but the network of channels within the ice through which water flows has narrowed or closed during the winter. High basal water pressure results, and enhanced velocity events are observed at this time as water under pressure lubricates extensive areas of the glacier bed (e.g. Iken et al., 1983). If water is produced at the glacier base, there must be a system through which the hydrology is organised. A number of hydrological arrangements have been proposed, from continuous films of water (which would act to reduce the basal shear stress over large portions of the glacier base) to well-defined channels through which water flows. Examples of such channels are where water erodes into ice (known as Röthlisberger or 'R' channels) or where water cuts into bedrock (Nye or 'N' channels). Both channel-types have been observed in the field, and both act to reduce the water contact over large parts of the glacier base. More recently, a system of linked cavities, where water collects on the downstream side of bedrock obstacles, and is connected to other cavities through a channel network, has been proposed (Kamb 1987). Under a linked cavity system, large portions of the glacier bed can remain in contact with water, and water channels can still form in a manner similar to that proposed by Röthlisberger and Nye. The difference between a 'linked cavity system', compared with an organised channel network, is that the stable configuration for the linked cavities results in many small channels, rather than the few large channels of the latter system.

For an ice sheet model, an algorithm that describes basal sliding is required in order to determine the velocity in ice streams. By studying the sliding of ice over different types of known beds in the laboratory, under loads which are representative of glaciers, Budd and Smith (1979) determined a simple relation between the basal shear stress, basal normal stress, and the roughness of the base. This relation was altered after numerical modelling of the West Antarctic Ice Sheet by Budd et al. (1984). The sliding velocity was calculated by Budd et al. (1984) as the difference between the balance velocity and ice deformation velocity, and related to the normal and shear stresses. The relation that was found by Budd et al. (1984) to best describe the ice stream velocities in West Antarctica is:

$$u_s = r\frac{\tau_b}{H_e^2} \qquad (4)$$

where r is a constant relating to the basal roughness and the sliding velocity, u_s, is proportional to the basal shear stress and inversely proportional to the normal load squared. In an ice sheet, the normal load can be approximated by the effective thickness, H_e, of ice (the thickness of ice, unsupported by buoyancy):

$$H_e = H - \frac{\rho_w}{\rho_i}(-B) \qquad B < 0 \qquad (5)$$

$$H_e = H \qquad B > 0 \qquad (6)$$

if one assumes that there is 'free communication between the subglacial water and the sea'

(Bentley, 1987). Here ρ_w is the density of seawater (1025 kg m–3) and B is the elevation of the subglacial topography ($B > 0$) or sea floor ($B < 0$). Thus a simple equation for the ice sliding of an ice sheet is defined. If there is no buoyancy effect on the ice, then the effective ice thickness will be equal to the actual ice thickness.

The action of rising sea level on a marine-based ice sheet theoretically reduces effective thickness, which may cause an increase in the sliding velocity contribution to the ice stream velocity. This process would transfer more ice from the accumulation zone to the ablation area and, in doing so, act to reduce the volume of the ice sheet (i.e. deglaciation). It should be noted that Bentley (1987) detailed a variety of sliding relations that are similar to Budd et al. (1984).

Basal sediment deformation

In the presence of unconsolidated sediments, the assumption of ice resting on rigid bedrock no longer holds (Boulton and Jones, 1979). In this situation, the hydrological system is altered and, as a consequence, so too is the method by which basal motion takes place. In the presence of unconsolidated sediment, water becomes incorporated into the body of the sediment until it is saturated. Once this has occurred, the yield stress of the material is reduced substantially compared with its unsaturated equivalent. Thus, friction at the ice-sediment interface is reduced to very low values over large regions of the glacier base. The subsequent deformation of basal till under the load of overlying ice will induce glacier motion and the transfer of the basal sediments down-glacier. If excess water is generated at the glacier base (perhaps as water originating from an upstream catchment area such as is inferred to take place within Ice Stream B, Antarctica), then transport of water will occur through R channels cut into the ice above the sediment. The quantification of glacier motion by water-saturated sediment deformation has been made possible in recent years by glaciological investigations of ice masses where deforming sediment is known to exist (e.g., Breidamerkurjökull in Iceland, Trapridge Glacier in the Yukon, and Ice Stream B, Antarctica; Paterson, 1994). Such investigations have led to the conclusion that the rate of sediment deformation is related not only to the stress induced in it by the overlying ice mass, but by the geotechnical properties of the subglacial till itself. Algorithms that describe the velocity of ice due to the deformation of subglacial material beneath an ice stream are similar to the sliding equations detailed above in that it is proportional to the basal shear stress and inversely proportional to the effective pressure (e.g. Alley, 1990).

Non-steady and fast glacier motion

Surge-type glaciers exhibit cyclical instabilities in the form of short phases of rapid motion (a few months to a few years), punctuating significantly longer periods of quiescence and stagnation (20–200 years) (e.g., Meier and Post, 1969; Raymond, 1987). During the active phase, mass is transferred rapidly down-glacier in association with heavy surface crevassing and an advancing surge front. Ice velocity may increase by two orders of magnitude. In the quiescent phase, there is net accumulation of mass in an upper 'reservoir area', which thickens and steepens to a critical point, at which fast flow is triggered by a reorganisation of the hydrological system at the ice-bed interface, and associated changes to the geotechnical properties of any soft basal sediments (Clarke et al., 1984; Kamb, 1987). The mechanism which triggers glacier surges is, therefore, independent of any direct climatic control. Prior to the active phase, the glacier surface profile is steep and, conversely, the surface profile is very flat during early quiescence. About 4 per cent of glaciers are known to surge, and these ice masses have a geographical distribution which is markedly non-random. In the Arctic, the archipelago of Svalbard contains many surge-type glaciers

(Hamilton and Dowdeswell, 1996), whereas on the more easterly islands of Russian Franz Josef Land and Severnaya Zemlya very little evidence of past surges has been found (Dowdeswell and Williams, 1997). Several looped moraine systems, characteristic of non-steady surge-type flow, have been reported from East Greenland (Weidick, 1995), and many surge-type glaciers are present in Alaska and the Yukon Territory of Canada (Meier and Post, 1969; Post, 1969). A number of hypotheses regarding the triggering of glacier surges have been proposed. One such theory indicates that the surging of glaciers resting over a rigid bed is dependent on basal hydrology. Quiescent-phase behaviour is typified by a basal hydrological system dominated by tunnels that transport water effectively beneath the glacier to its margin. A surge is initiated by a shift to a less efficient drainage system of linked cavities, which acts to trap basal water, and to increase water pressure (Kamb, 1987). Consequently, the glacier becomes decoupled from the bed, allowing rapid sliding. A second hypothesis, applicable to glaciers resting on deformable sedimentary beds, suggests that as ice thickens the permeability of sediment reduces and high pore-water pressures build up. This transforms the sediment into a non-cohesive 'slurry' with very low yield stress, at which point rapid glacier motion is initiated (Clarke et al., 1984; MacAyeal, 1993). There is also evidence, in the form of buried crevasse fields identified from radar records, that a large West Antarctic ice stream (Ice Stream C) may have switched from fast flow to stagnation a few hundred years ago (Rose, 1979). Satellite radar interferometric methods have also been used to observe smaller-scale motion events, for example a 'mini surge' within the Ryder Glacier, an outlet glacier in North Greenland (Joughin et al., 1996).

GLACIAL ISOSTASY

When a glacial load is placed on the Earth's crust, the weight of ice will act to displace the crust into the viscous asthenosphere (which lies beneath the lithosphere), until isostatic equilibrium is achieved according to Archimedes' Principle. The process by which this is done involves (1) diffusion into the asthenosphere and (2) deflection of the lithosphere. When the ice load is removed (deglaciation), the crust will return to its original position (i.e. isostatic uplift or recovery).

Asthenospheric diffusion

Although there have been a number of models that describe isostasy, such as Budd and Smith's (1979) time dependent studies (where the crustal depression at any time is linked to current and previous ice thicknesses from up to 10 ka ago), the method that is often adopted in ice sheet models is taken from Brotchie and Sylvester (1969). Here, the process of glacial isostasy is treated simply as a mechanics problem. The viscous asthenosphere allows the loaded upper mantle to subside slowly to an equilibrium position after a period of time that is dependent on the load and the viscosity of the asthenosphere. Additionally, as the asthenosphere is displaced by lithospheric material, it moves to find an equilibrium state, in a lateral position away from the ice load. The process can be described easily by a 'diffusion equation', where the viscosity of the asthenosphere determines the rate at which the Earth responds to loading. The maximum amount of glacial isostatic depression is related simply to the ice thickness and the ratio of the densities of ice and the asthenosphere. Making a huge generalisation that the densities of asthenospheric material and ice are 3300 and 910 kg m^{-3}, respectively, suggests that the maximum amount of depression will equate to about one third of ice thickness. The isostatic displacement is damped by the viscous asthenosphere and rigid lithosphere (see below). Thus, it takes several thousand years for the Earth to respond to glacial loading and unloading (see Chapter 5, Figure 5.1b); a fact measurable today as many previously glaciated regions (such as Scandinavia) continue post-glacial uplift (see Chapter 5, Figure 5.3).

Lithospheric displacement

An elastic plate of finite thickness over a viscous medium would act to laterally diffuse a load that is placed on the plate surface; this is the effect of the lithosphere in the process of glacial isostatic loading. It can be modelled in a similar way to the asthenospheric diffusion, by employing a fourth order differential equation (for example Oerlemans and van der Veen (1984)). However, since both the majority of glacio-isostatic features (bedrock depression and forebulge formation) are largely controlled by the asthenosphere, and the reduction in the equilibrium position of the bedrock due to a rigid lithosphere is small compared to the theoretical depression without the lithosphere (van der Veen, 1999), lithospheric rigidity was not accounted for in simple models. However, recently, a simple method by which lithospheric rigidity is accounted for has been developed (Le Meur and Huybrechts, 1996). This model makes the assumption that it takes a known amount of time for the lithosphere to adjust to the 'equilibrium' position under the weight of ice. To do this, a simple exponential decay function can be used to describe this time-dependent modification by the lithosphere. This simple technique compares well with the performance of more sophisticated 'Earth models' (Le Meur and Huybrechts, 1996).

ICE SHEET MODELS

Numerical ice sheet models have been used extensively in the 1990s to aid the reconstruction of Quaternary ice sheets. Results from ice sheet models are only as good as the data used as input to force the model. Model results do not represent unequivocal 'solutions' or 'data'. However, the development of 'inverse' procedures, where model results are made to match geological evidence, means that model results can represent 'plausible quantitative glacial scenarios' to aid the interpretation of the geological evidence. In the 1990s, a series of European-led international workshops and meetings have compared and tested a variety of numerical ice sheet models. This European Ice Sheet Modelling Initiative (or EISMINT) has resulted in a number of 'benchmarks' whereby models can be calibrated (Huybrechts et al. 1996).

The principle behind numerical ice sheet modelling is that an ice sheet can be divided into a number of 'ice columns'. Each of these columns represents a 'cell' in the model's 2-D 'grid'. Ice sheet models are usually arranged in a 'loop' that begins by applying a series of algorithms, determining the flow of ice, mass balance and interaction with the Earth, in each cell. The loop is completed by a final equation (widely known as a 'continuity' equation), used on the full grid, to calculate the interaction and flow of ice between cells. One loop of the model is called an 'iteration'. The accuracy of the model is dependent on the width of the grid cells, and the time period over which a single iteration represents. For continental-scale ice sheet models, where the time-dependent change in ice sheet behaviour is calculated over several thousand years, the grid cell width is usually between 5 and 25 kilometres, and the iteration time step is between 1 and 10 years. Because ice sheet modelling is a valuable tool for Quaternary glaciologists, it is appropriate here to outline briefly how ice sheet models are organised.

Most ice sheet models are centred around the continuity equation for ice (Mahaffy, 1976), where the time-dependent change in ice thickness of a grid cell is associated with the specific net mass budget of a cell:

$$\frac{dH}{dt} = b_s(x,t) - \nabla . F(u,H) \qquad (7)$$

where $F(u,H)$ is the net flux of ice from a grid cell ($m^2\ yr^{-1}$) (the flux of ice being the product

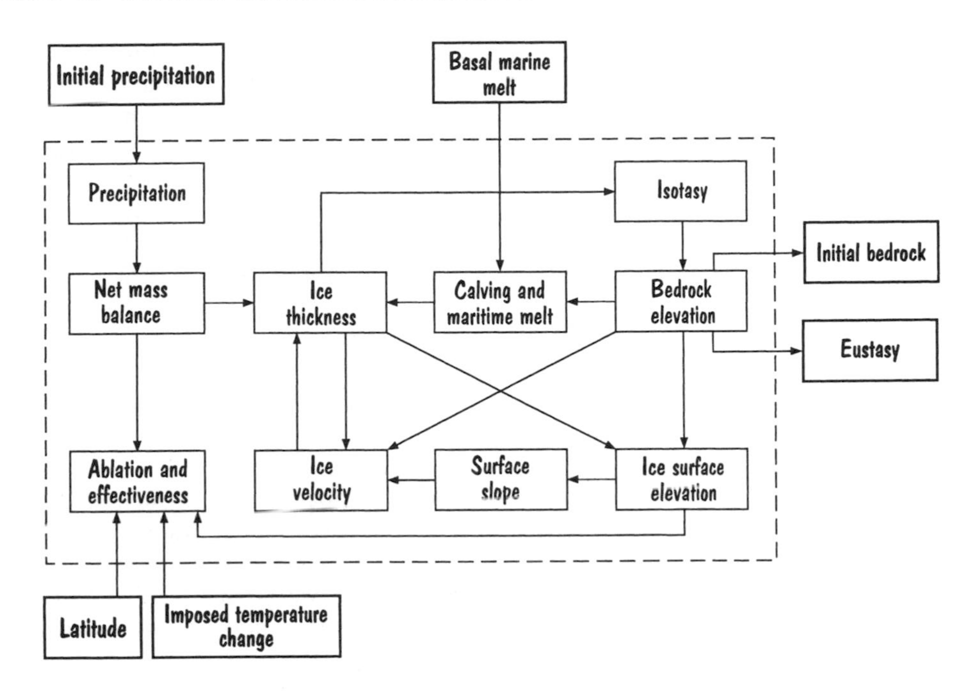

Figure 3.2 Flow chart to show the interaction of parameters that describe large-scale ice-sheet behaviour. Note the feedback process that is focused around the isostasy component. The box denotes the internal physical processes of the ice-sheet system.

of ice velocity, u, and ice thickness, H). The depth-averaged ice velocity, u (m s^{-1}), is calculated by the sum of depth-averaged internal ice deformation and basal motion. The specific mass budget term b_s, is a function of a number of processes including ice sheet surface mass balance, iceberg calving and ice shelf basal melting. The flux between grid cells is determined by the ice thickness and the vertically averaged ice velocity of the grid cell in question, and also of neighbouring grid cells. Numerical methods that can be used to solve this partial differential equation (PDE) are described in, for example, Press et al. (1989), and other standard numerical recipes.

Numerical ice sheet models (detailing the growth and decay of the late Quaternary ice sheet) require the following inputs and boundary conditions. First, the subglacial topography on which the ice sheet forms. As the ice grows this topography will be altered by the action of glacial isostasy. Second, a time and spatially dependent function of climate over a glacial cycle is required to formulate rates of accumulation and ablation of ice as well as air temperature (used in some models to calculate the temperature of the ice sheet). Third, the time-dependent change in sea level. Parameters describing rates of iceberg calving and ice shelf melting are also required.

FEEDBACK WITHIN ICE SHEET MODELS

The relationship between the parameters of an ice sheet system can be summarised within a flow diagram: the mechanism by which the ice sheet responds to external forcing (Figure 3.2). A number of feedback processes can be seen in this diagram that influence the stability of the modelled ice sheet significantly. First, glacial isostasy acts so as to affect the bedrock elevation and hence the ice sheet surface elevation. This process is important for several reasons: primarily that the ice velocity will be changed by the surface elevation effect on basal shear stress; additionally, the accumulation (which is altitude related) will be altered by the change in surface altitude of the ice sheet. The changes to these parameters will alter the isostatic load and, hence, change the surface elevation of the ice sheet. This will cause the ice velocity, accumulation and consequently the ice thickness to be altered (which in turn feeds back to isostasy again). Another feedback loop within the ice sheet system has been examined by Payne et al. (1989), who found a feedback due to the way that precipitation is related to altitude. For the specific case of the ice sheet in Scotland during the Loch Lomond Stadial (or Younger Dryas) at around 11 ka BP, an increase in ice sheet surface elevation forced a higher accumulation rate which led to further ice sheet growth. However, this growth will only occur until a threshold altitude is reached, after which higher ice sheet elevations cause more extreme desert-like conditions, leading to a reduction in the rate of snowfall. This feedback between ice sheet elevation, precipitation and ice growth is complicated further by both the temperature controlled change in the ELA, and the change in sea level, which could both act to alter the elevation at which maximum precipitation occurs.

Another feedback loop acts to change thermal conditions at the base of ice sheets. This is important because the ice velocity due to rapid basal motion (e.g. sliding) is influenced by the thermal regime. In a cold-based ice sheet, ice velocities are generally low and so, theoretically, the ice sheet is allowed to build up with a minimum of ice loss. However, as the ice sheet thickens, the basal temperature is likely to increase. Once the base of the ice sheet starts to melt, rapid basal motion becomes possible. Fast ice flow leads to ice being taken away from the ice sheet and, so, to ice sheet thinning. Thinner ice then causes a reduction in the temperature of the ice sheet base and, possibly to the curtailment of rapid basal motion. This feedback process is complicated because the thermal regime is influenced by the basal heat gradient, which is in part related to the heat due to basal sliding. Thus, rapid basal motion may uphold warm-based conditions even if the ice sheet thins. For ice sheet models, this feedback only applies where ice sheet temperatures are calculated.

The ice sheet model system, as described in Figure 3.2, is influenced heavily by feedback processes. However, the feedback processes do not act as totally independent agents since they are, through the glacial cycle, ultimately controlled by the climatic and physical boundary conditions that are externally forced on to the ice sheet. It should be noted that it is not the purpose of this chapter to try and ascertain causes of additional feedback systems within the external climatic control through the glacial cycle, such as ocean current circulation (e.g. Broecker et al., 1990), or atmospheric CO_2 forcing (Barnola et al., 1987). However, such feedbacks are likely to occur during a glacial cycle.

Processes within the ocean and atmosphere can be modelled, just as ice sheets can. Analysis of the interaction of these systems is becoming possible with the advent of modern computer processors. The next decade is likely to see a significant development of numerical modelling technology, allowing the full coupling of ice sheets, oceans and the atmosphere.

CHAPTER 4

Late Quaternary Geology I Terrestrial: Glacial Morphology and Sedimentology

Reconstructing former ice sheets is possible because of the geological remnants left by them. As one might expect, the subject of glacial geology is diverse, ranging from centimetre-scale scratches on formerly glaciated bedrock to proglacial sedimentary fans that are several million square kilometres by area. It is by understanding the processes responsible for geological features, and correlating these geological datasets from different regions, that a full understanding of ice sheet history can be developed. There are, however, three limitations to the usefulness of using geological evidence alone for reconstructing ice sheets. First, due to the abrasive nature of ice sheet bases, terrestrial geomorphological data for glaciations prior to the last are often eroded by, or covered by sediment from, the latest glaciation. Thus, terrestrial glacial geology from the early stages of glaciation are often missing from records. Secondly, formerly glaciated regions can be so difficult to access that full details of the glacial geology have yet to be acquired in some places. Thirdly, interpretation of geological records is often equivocal. These latter problems have been addressed in recent years by the establishment of multinational research groups, set up to correlate existing records and focus energies on regions where data are absent. A good example of this is the ESF PONAM and QUEEN projects of the 1990s, where the geological history of the Eurasian Arctic has been evaluated (see Chapter 11). Additionally, for North America, the Canadian Geological Survey has been instrumental in organising a focused field campaign, during the last 30 years, to record the geological remnants of the former ice cover.

In this chapter terrestrial glacial geology will be examined by analysing features formed from the centre of the ice sheet first, to the edge of the ice sheet, and then to coastal regions. In the subsequent two chapters, geological evidence from shallow seas and deep ocean floors will be introduced.

A. GLACIAL EROSION FEATURES

Introduction

We will concentrate on determining the causes of some commonly known glacial erosional features. Subsequent identification of such features will, thus, provide information about the nature of the ice sheet which formed them. In order that we can determine a method by which past glaciers and ice sheets actually created glacial geological features, we look at modern glacial erosion processes. A very good evaluation of glacial erosional landforms is available in Hambrey (1994) and Bennett and Glasser (1996) and much of the discussion in this chapter has been developed from these sources.

Glacial (ice) erosion processes

There are four main ways in which glacial environments can sustain erosion:

1. Abrasion (of which there are several forms) and plucking; *subglacial*
2. Rock fracturing; *subglacial*
3. Pressure release (and dilation); *subglacial-subaerial*
4. Frost shattering; *subaerial*

Abrasion

The process of abrasion involves basal ice containing debris sliding over bedrock. The ice-held debris scores the bedrock to produce fine grained material (<100 m) known as rock flour. This scratching of the rock causes a variety of morphological bedforms, many of which will be known well to the lay person. On the basis of laboratory work and field evidence, the principal glaciological factors affecting abrasion have been evaluated as:

- presence and concentration of basal debris – sliding ice without debris abrades less than ice which contains much material;
- sliding velocity of glacier (through regelation or over a planar water film) – high sliding velocities results in greater rates of abrasion;
- transport of debris (to continue abrasion process) – the removal of debris from the base of the glacier needs to be countered by the creation of new material, either directly from subglacial sources, or from supra-glacial regions, which feed material downwards via crevasses to the subglacial region;
- ice thickness – ice thickness is linearly proportional to the effective pressure at the base of the glacier which, in turn, is related to the rate of abrasion;
- presence of water – for sliding to take place, subglacial water is required;
- lithology of basal debris – hard minerals (like quartz) erode more than soft minerals (like feldspar and limestone);
- size and shape of basal debris – angular clasts will scratch more effectively than smooth rounded clasts.

Plucking

Plucking, is where ice is frozen (like being welded) to bedrock. Stresses induced within the overriding ice extenuate weakness within the rock (faults, cleavage, bedding etc.) such that failure of the rock occurs, resulting in a clast attached to the basal ice being removed from the bedrock. Abrasion of bedrock, further down the glacier, can occur by this clast within the basal ice (Figure 4.1).

Regelation

Ice will slide over small obstacles at the base of a glacier by the process of regelation. The upstream side of the obstacle (stoss face) experiences a drop in the pressure melting temperature due to enhanced pressure by the weight of ice pushing against it. This leads to melting and sliding of ice over the obstacle. However, heat required to melt the ice is taken from the downstream end of the obstacle (lee side), via thermal conduction in the rock. This reduces the temperature across the lee face, and re-freezing occurs. Since sliding leads to abrasion and freezing leads to plucking, the stoss face will be smoothed, polished or striated, whilst the lee side will be jagged and rough. This is how 'roches moutonnées' are thought to form (Figure 4.2).

Drewry (1986) showed how, by simple mechanical theory, the dimensions of striations can be used to determine information about the glacier responsible for their creation (Figure 4.3). This analysis showed that the subglacial wear is proportional, in some way, to the ice load, and the sliding velocity (Drewry, 1986, pp. 48–50). Sharp et al. (1989) used this theory to determine information about the past

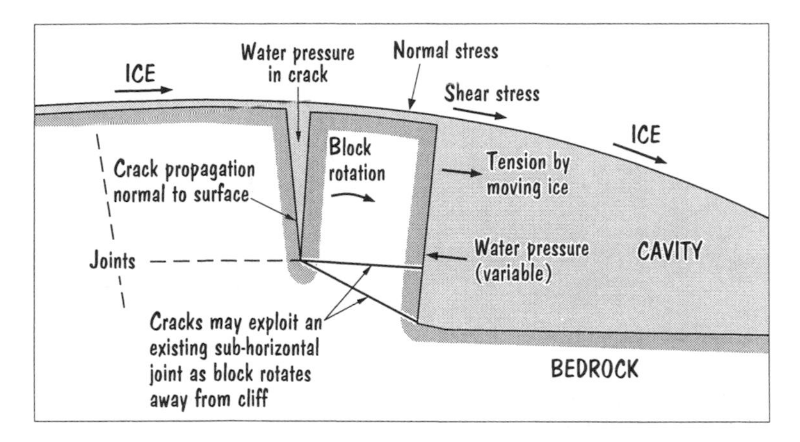

Figure 4.1 Mechanism of glacial 'plucking'. Adapted from Bennett and Glasser (1996).

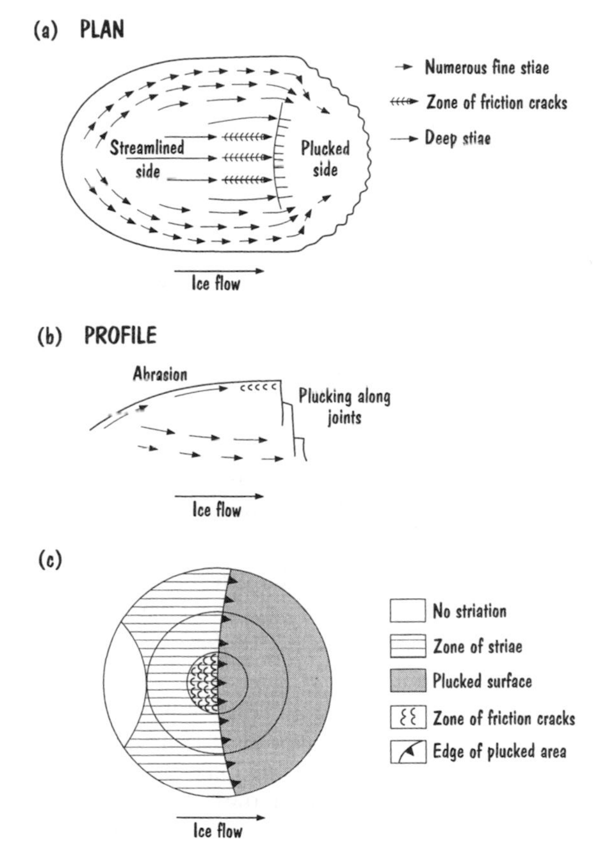

Figure 4.2 Regelation and the formation of Roches Moutonnées. Adapted from Chorley et al. (1984).

glaciation of Snowdonia, based on the size of striations.

There are two schools of thought concerning the actual mechanics of abrasion as follows (Figure 4.4). First, Boulton (1974) determined the velocity of the clast within basal ice, to be controlled by the size of the particle. His theory is that there is a critical particle size (or obstacle size within the regelation theory) at which rock fragments are moved at the greatest speed. Below this critical value, regelation occurs between ice and clast, thus minimising clast velocity. Above this value ice creep occurs instead of basal sliding, thus reducing the particle speed. The normal force exerted by the particle is related to the weight of ice above it (or the ice thickness). Secondly, Hallet (1979, 1981) found that as sliding velocity increased, so did the particle velocity. However, he found that the downward force on the particle was independent of ice thickness. In his theory, Hallet suggested that the ice behaves as a viscous fluid, supporting the particle in a buoyant fashion. Thus, the force exerted by the particle is far less than that determined by Boulton. However the velocity of the rock particles is generally higher in Hallet's theory than in Boulton's.

Most people expect that the actual mechanics of abrasion change from glacier to glacier, and that there is some truth in both of these ideas.

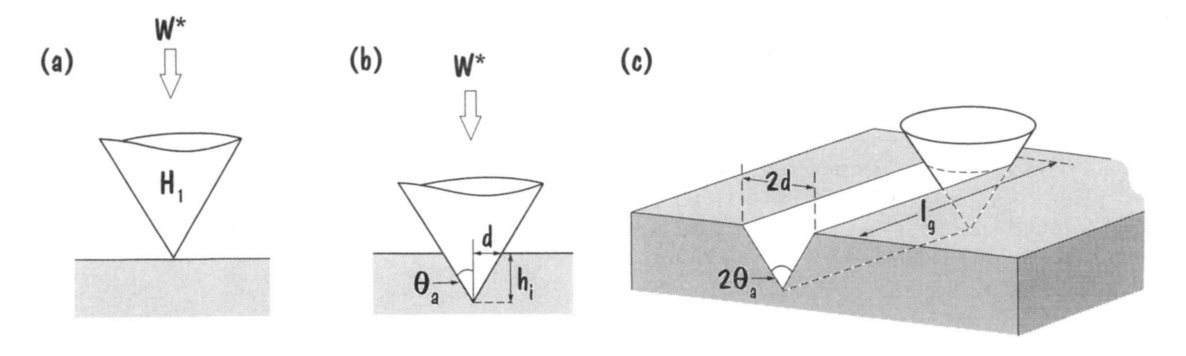

Figure 4.3 A simple model of glacier abrasion assuming a single striation is caused by a conical-shaped clast. If the clast is pressed down by the weight of ice, it will indent the bedrock (i.e. the clast is harder than the rock). The size of this indentation is related to the load on the clast and the hardness of the clast and the bed. The area of the indentation from a conical clast is: $A = \pi d^2 = W^*/\sigma_y$, or rearranged, $d = \sqrt{W^2/\sigma_y \pi}$, where d is the radius of the indentation, W^* is the load and σ_y is the yield strength of the bedrock (above this yield strength the rock deforms, below this value it does not deform). Given that θ_a is half of the angle at the tip of the asperity, the depth of the indentation (h_i) is: $h_i = d \cot\theta_a = \sqrt{W^*/\sigma_y \pi} \times \cot\theta_a$. When the clast is moved across the bedrock by a known amount (l_g), it will plough a track across the bedrock according to: $V_g = (d^2 \cot\theta_a) l_g$. Because only half of the clast makes contact with the bedrock during this abrasion, the radius of the indentation can be redefined as: $d = \sqrt{2W^*/\sigma_y \pi}$. If the clast is moving at a velocity of U_p, the rate of abrasion A_b can be calculated by: $A_b = V_g \times U_p = [2W^*/\theta_a \pi]\cot\theta_a$. According to these ideas the volume of wear is proportional to distance travelled by the clast, load on the clast and bedrock and inversely proportional to the yield stress or hardness of the bedrock. Figure and text adapted from Drewry (1986).

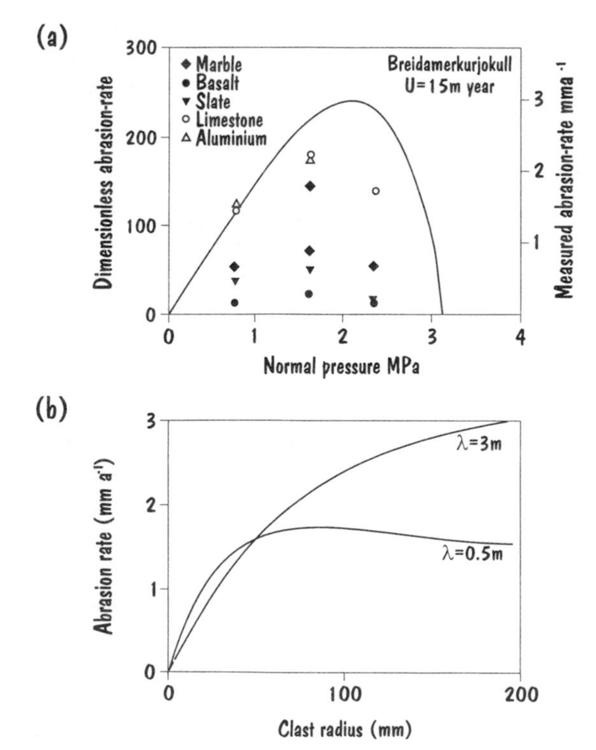

Figure 4.4 Results from two abrasion models. (a) Boulton's mechanism, showing how abrasion changes with normal pressure. Adapted from Boulton (1979). (b) Hallet's mechanism, illustrating the abrasion rate with respect to clast size, for two subglacial undulation wavelengths. Adapted from Hallet (1981).

Glaci-erosional landforms

It is assumed the reader will be aware of the following glacial landforms formed by glacial abrasion (Hambrey, 1994 is a good source for those who require further reading):

- Aerial scouring; low amplitude, irregular relief (e.g., 'knock and lochan' topography). Such morphology is often associated with the north-western Highlands, where 'knock' refers to the small hills often with exposed bedrock and 'lochan' relates to the basins of boggy ground or small lakes.
- Glacial trough; U-shaped valleys and fjords (associated hanging valleys and truncated

spurs). A classic glacial geomorphological landform.

- Domes and whaleback forms; similar to roches moutonnées (Figure 4.2) but with no plucking on down-glacier side.
- Striation; among the most common features of glacial erosion, striations are narrow (sub cm) scratches etched into bedrock by particles held in the basal layers of a sliding glacier or ice sheet.
- Groove; an enlargement of singular striations.
- Polished surface; caused by abrasion by very fine material producing a smooth surface.

Some landforms are created by two or more processes including abrasion, such as:

- Roche Moutonnée; as already mentioned, formed by abrasion plus plucking (Figure 4.2).
- Reigel; a rock barrier perturbation, caused by a band of resistant rock often aligned non-parallel to the direction of ice.
- Cirque; formed by combinations of abrasion on their floor, and rock fracturing (crushing) and plucking along their walls, cirques are commonly found in north-facing slopes (in the Northern Hemisphere) of high mountain regions. They often have over-deepened floors in which lakes or boggy ground exist when the glacier melts.
- Crag and Tail; caused by erosion (abrasion) of upside face, deposition across the downside face. The most famous crag and tail is Castle Rock and the Royal Mile in Edinburgh, Scotland.

Note: relative dating of erosional features can be made through analysis of cross-cutting of these landforms.

Rock crushing

Under the repeated impacting of debris within basal ice, the integrity of bedrock will eventually 'fail'. This process is known as rock crushing.

Four main rock crushing features are (Figure 4.5):

- Lunate features
- Crescentic gouges
- Crescentic fractures
- Chattermarks

These are quite difficult to distinguish in the field. However, they are often associated with striations, which aid their interpretation.

Erosion of ice combined with frost action

The rapid freezing and warming of water within rock fissures can mean the rock expands and contracts at such a rate that it 'shatters'. A number of glacial landforms are derived from this process, the most significant are:

- Arêtes
- Horns
- Nunataks

These are all macro-landforms an order of magnitude larger than most features discussed thus far.

Pressure release and dilation

Repeated loading of valley side walls causes the bedrock in such places to 'peel'. This process is known as exfoliation. Hallet and Boulton have expressed differing views as to how exfoliated valley-sides form. Boulton suggests they form because of the unloading of ice and the release of sideways pressure. Hallet does not agree with this, since he suggests water at the valley side causes the ice to float and, therefore, exert effectively no force on the rock. He suggests alternatively that exfoliation is due to the subglacial water, at high pressure, forcing itself deep into crevasses and cleavage planes within the rock, thus causing the rock to 'peel' along these lines of weakness when the ice is removed.

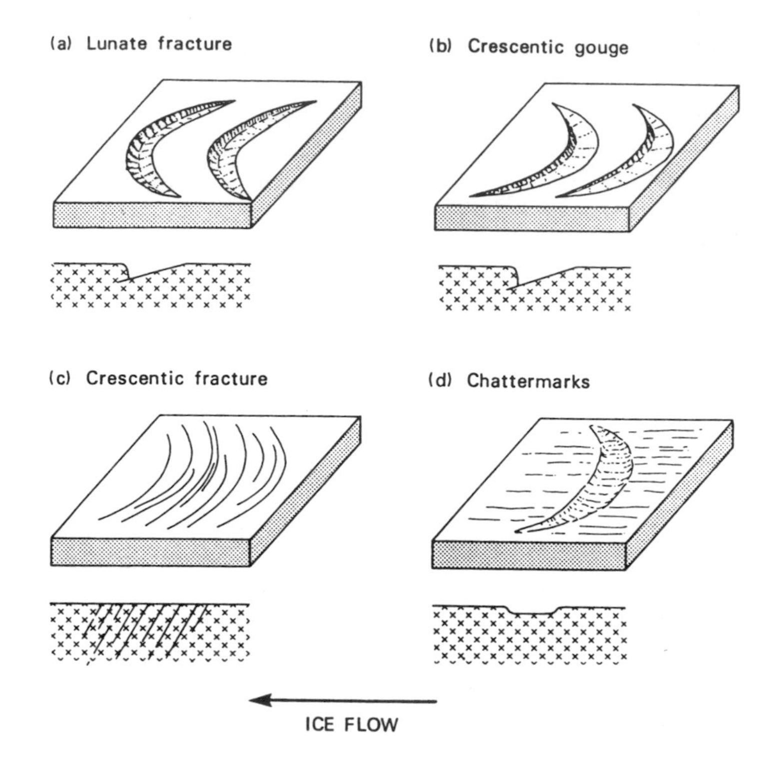

Figure 4.5 An illustration of four small-scale geomorphological features caused by glacial erosion. (a) Lunate fracture. (b) Crescentic gouge. (c) Crescentic fractures. (d) Chattermarks. Taken from Hambrey (1994) and reproduced by permission of UCL Press.

B. SUBGLACIAL WATER ACTIVITY

There are two processes by which subglacial water activity may cause erosion of bedrock; mechanical erosion and chemical erosion. These process are relevant when basal sliding occurs. Identification of such features therefore reveals information about the dynamics of former ice sheets. Mechanical erosion: subglacial water channels (e.g., Nye channels and linked cavities), and proglacial water channels, caused by abrasion of water and rock particles within the water on the substrate. Chemical erosion: dissolving (up glacier) and precipitation (down glacier) of carbonate minerals is indicative of the chemical action of subglacial meltwater, especially over a limestone environment.

A very good example of both mechanical and chemical erosion is the forefield of the Glacier de Tsanfleuron in Switzerland. Recent rapid retreat of this glacier has exposed fresh limestone bedrock formerly beneath the glacier. This bedrock displays an abundance of abrasion features as well as formations resulting from the action of subglacial water. Nye channels (channels incised into the bedrock) link water cavities (as in Kamb, 1987), and two forms of a carbonate precipitate, sparite and micrite, are observed on the stoss and lee faces of obstacles, respectively (Hubbard and Hubbard, 1998; Hubbard et al., 2000). These precipitates are evidence of the dissolving of limestone in the water upstream. Technical matters relating to these mechanical and chemical processes can be found in a number of books (e.g. Drewry, 1986 Chapter 5; Hooke, 1998 Chapter 7; van der Veen, 1999 Chapter 4) and are not elaborated here.

The actual processes involved in subglacial hydrology are not yet known in real detail because of the inaccessibility of the subglacial environment. However, from what we understand of glacial hydrology beneath small glaciers and ice caps, subglacial water has a fundamental control on overriding ice sheet dynamics. There is strong evidence for subglacial water as a major control on former ice sheet dynamics in North America (Chapter 12). Also, a huge amount of subglacial water exists within subglacial lakes beneath Antarctica, detected by ice-penetrating radar, several of which occur at the onset of enhanced ice flow (Siegert and Bamber, in press). It is therefore important to understand that subglacial hydrology has the potential to influence dramatically the dynamics of the world's largest ice sheets.

C. GLACIAL SEDIMENTATION AND DEPOSITION

Glaciers and ice sheets contain debris (e.g. the base of the GRIP and GISP2 ice cores show several tens of metres of sediment laden ice) and, in many cases, water-saturated deformable sediment at their base. Subglacial water emerging from glacier snouts is often observed to contain a high proportion of material ranging from rock flour (where the glacier rests predominantly over bedrock) and the sand and mud proportions of subglacial diamict. It is small wonder that glacier sedimentation is a major process operating in glacial environments. Understanding the process by which modern glacigenic sedimentary features are formed allows us to understand the former process responsible for similar features recorded in glacial geology.

Glacigenic sediment landforms will be left by receding ice masses. The geology therefore tells us about glacial maximum conditions and processes involved in deglaciation. Information relating to the processes of glacial sedimentation is discussed in detail in Hambrey (1994, Chapter 4). Since those reconstructing ice sheets rely on the identification of sediment landforms, we will deal, in a little detail, with some of the more important features of glacial sedimentation, which tell us something about ice flow, and the nature of the ice sheet that left the features.

Landforms formed subglacially, parallel to ice flow

Drumlins

One of the most common, and easy to identify, glacial-depositional landforms. They have an elongated plan, with the long axis parallel to ice flow, and the blunt end facing in an up-stream direction. There are a number of methods by which drumlins may form. However, there are three main methods that have received quite a lot of recent attention (Figure 4.6).

Deformation of sediments

Deforming sediment beneath a glacier may organise itself into regions of competency. Around these areas sediment may flow more easily, thus shaping the lobate appearance of the competent zone (Hindmarsh, 1998). If this process is a major process by which drumlins form, we should expect to see them within the subglacial topography of modern ice sheets. Ice-penetrating radar has provided some evidence for drumlins beneath East Antarctica (e.g. Wilkes subglacial basin, Siegert, in press).

Fluvial action

(i) Sediments deposited within subglacial cavities which were previously eroded by large subglacial floods. (ii) Direct erosion of drift sheets due to catastrophic subglacial fluvial processes. This latter process occurs in the proglacial environment, but its inclusion here is necessary as a method by which drumlins can form.

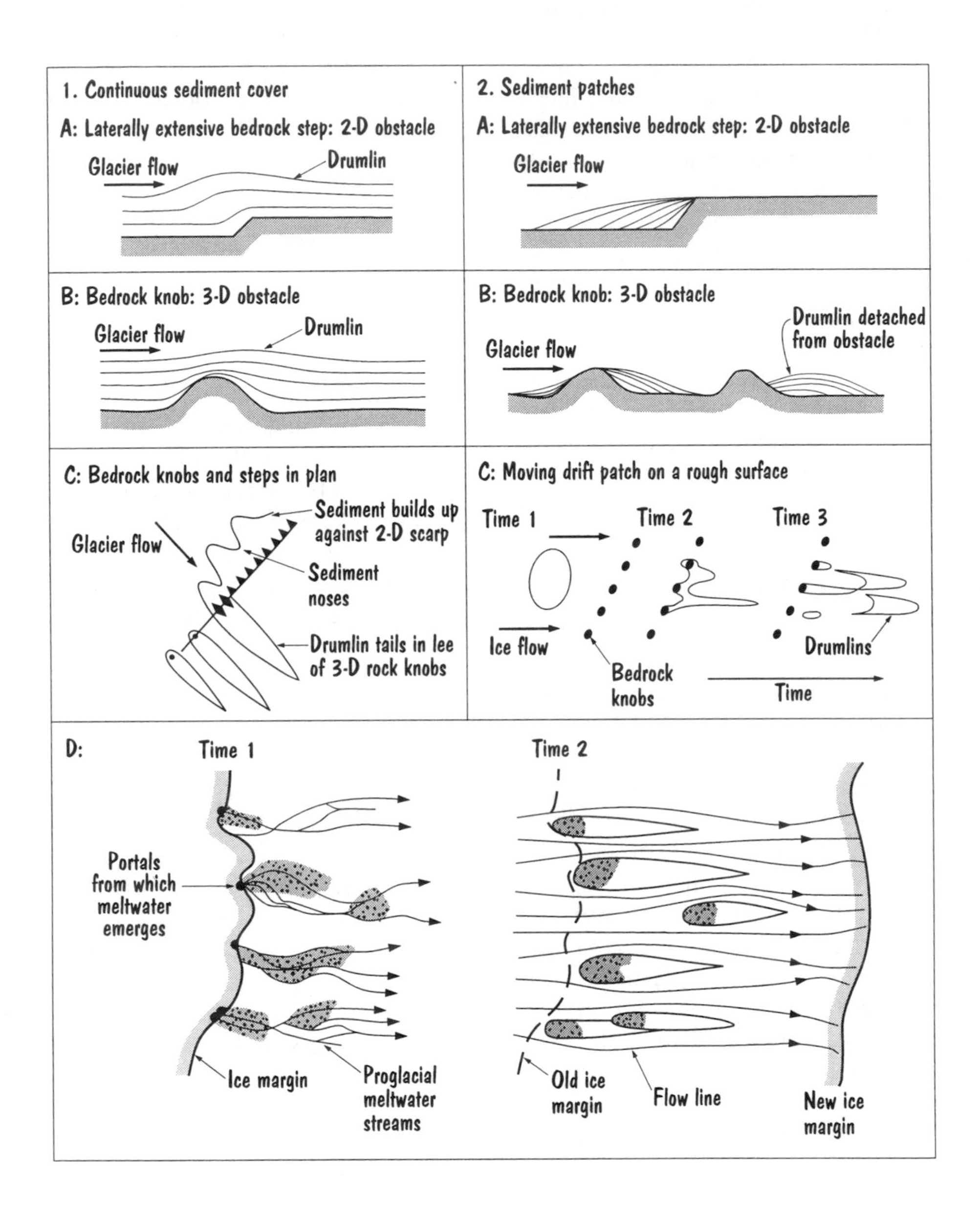

Figure 4.6 The formation of drumlins. (a) Drumlins formed by a deforming sediment layer across a bedrock step. (b) Drumlins formed subglacially due to sediment flow around bedrock perturbation. (c) Reworking of drumlins due to an alteration in the glacial conditions. (d) Drumlins formed subglacially as a result of sediment flow around 'stiff' sediment formed previously across the proglacial area. Adapted from Boulton (1987).

Drumlinoid ridges, or drumlinised ground moraine

Elongated cigar-shaped ridges, formed by subglacial 'streamlining', but under basal conditions unsuitable for the formation of discrete drumlins.

Flutes and fluted moraines

As in drumlins, the formation of flutes has been disputed in recent times. Hambrey (1994) indicates that flutes will form when ice moves over a large boulder imbedded in till. This results in a hollow within the basal ice layer. This hollow is filled by sediment, which leaves a ridge of sediments behind once the ice has melted. Usually they appear as large furrows with a wavelength of a couple of metres (although less common mega furrows 100 metres wide, 25 metres high and 25 kilometres long, have been observed). Rarely, the boulder that caused the flute may be left within the upstream end. This mechanism contrasts significantly with that proposed by Shaw et al. (2000) who suggest that the large-scale fluting within the Athabasca fluting field was caused by the action of subglacial meltwater, which acted to streamline the glacial sediment

Landforms formed subglacially, transverse to ice flow

Ribbed moraines (or rogen moraines)

Large-scale landforms aligned transverse to the direction of ice flow. They are 10–20 metres high, 50–100 metres wide and 1–2 kilometres long. There are six processes by which ribbed moraines may form:

1. Deposited as marginal moraines originally (at the sides of the glacier), which then become incorporated into subglacial moraines as a complex sequence of sediments.

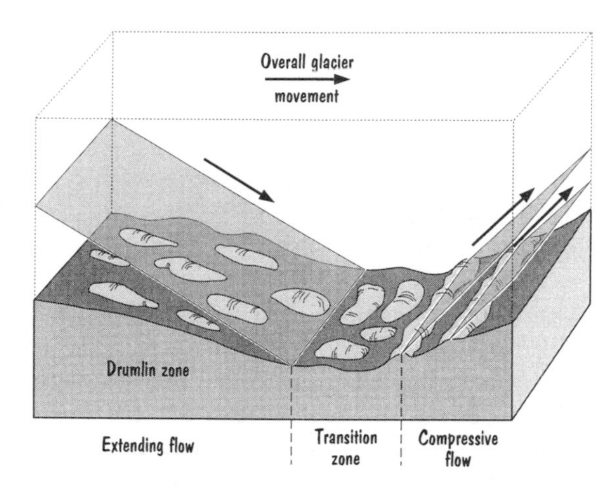

Figure 4.7 An example of how Rogen Moraines may form beneath the compressive flow of a glacier riding over a bedrock perturbation. Adapted from Sugden and John (1976).

2. Deposited as proper subglacial moraines, under thick ice away from the ice front, where a transition occurs between the cold and warm based glacier, where the ice is under compression. There are many examples of glaciers which are thought to display such subglacial conditions ranging from the Trapridge Glacier in the Yukon (Jarvis and Clarke, 1975) to the East Antarctic Ice Sheet (Huybrechts, 1992).
3. Formation by active basal tectonic processes within the ice (e.g. folding of basal sedimentary layers or stacking of debris rich basal layers against obstacles to glacier flow) followed by melt-out leaving the sediments behind.
4. Filling of open crevasses by supraglacial debris.
5. Filling of basal crevasses by subglacial debris.
6. Formed by influence of ice flow and basal topography, where the compression of ice across stoss faces leaves behind a transverse ridge (Figure 4.7).

It should be noted that there are many other glacial sedimentary landforms, and the reader is encouraged to become acquainted with them (e.g. Hambrey, 1994, Chapters 3, 4 and 5).

D. TERRESTRIAL GLACIAL GEOLOGY AND ICE SHEET RECONSTRUCTIONS

Although ice sheet reconstructions will be dealt with in some detail in later chapters, it is an opportune moment to provide a brief example of how terrestrial geology can aid the interpretation of former ice sheets. The extent, ice flow direction, centre of the ice sheet, glacial dynamics and age of the ice sheet can all be found by analysis of the following terrestrial landforms:

- *Ice extent* – end moraines, proglacial deposits, aerial scouring;
- *Flow direction* – striations, drumlins, flutes, roches moutonnées, crag and tails, lateral moraines, erratics;
- *Ice sheet centre* – cirques and other glacial landforms found in mountain regions;
- *Ice dynamics* – landforms from subglacial erosion and subglacial water activity (mechanical and chemical);
- *Age of the ice sheet* – the date ice sheet maximum, and the temporal pattern of ice sheet decay can be calculated; (a) relatively, from cross cutting landforms, (b) absolutely, from geochronological dating of sediment deposits (e.g., carbon dating, thermo-luminescence, chlorine dating).

A good example of how terrestrial geology can allow the determination of past ice sheet behaviour is in the British Isles. Ice marginal moraines mark the maximum extent of the ice sheet well. However, it is the former subglacial landforms that indicate the dynamics of the ice sheet. In the British Isles, extensive drumlin fields indicate (a) the direction of ice flow and (b) that the ice sheet was warm-based in many places. This provided the necessary evidence for Boulton et al. (1977, 1985) to reconstruct the last British Isles Ice Sheet (Figure 4.8).

Background reading

Space in this book prevents a detailed discussion of glacial geology. Very good chapters on glacial geology can be found in the following textbooks:

Benn, D. and Evans, D. 1998. *Glaciers and Glaciation*. Arnold, London.

Bennett, M.R., and Glasser, N.F. 1996. *Glacial Geology: ice sheets and landforms*. John Wiley and Sons.

Hambrey, M.J. 1994. *Glacial Environments*. UCL Press.

Figure 4.8 Reconstructions of the British Isles Ice Sheet and glacial geological evidence. (a) Drumlin fields and the direction of former ice flow that they indicate. (b) Directions of former ice flow from full interpretation of glacial geology. (c) Maximum ice-sheet reconstruction for the British Isles Ice Sheet. (d) Minimum reconstruction for the British Isles Ice Sheet. Adapted from Boulton et al. (1977 and 1985).

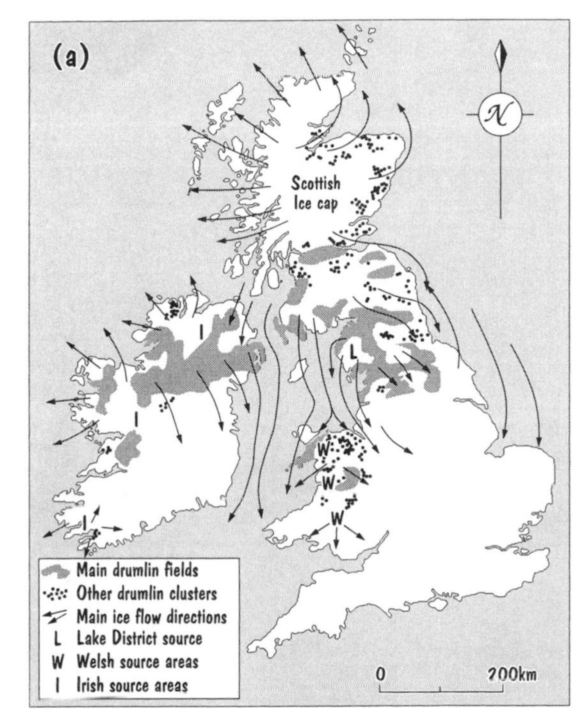
(a)
Scottish
Ice cap
L
W
W
W
W
I
I
I
Main drumlin fields
Other drumlin clusters
Main ice flow directions
L Lake District source
W Welsh source areas
I Irish source areas
0
200km

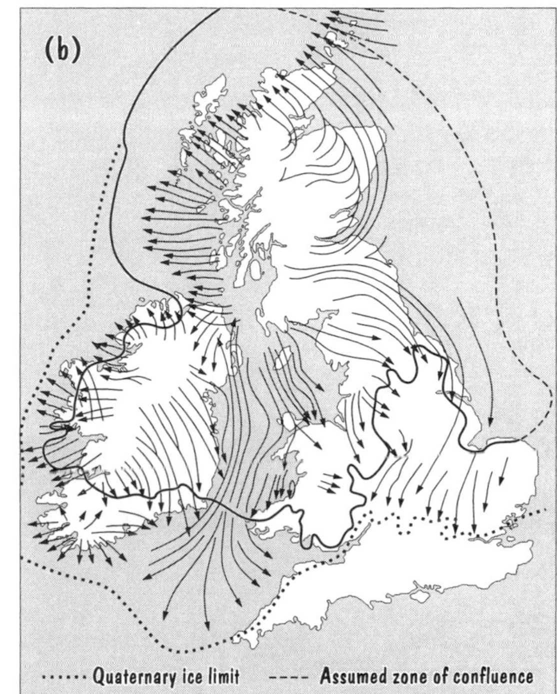
(b)
Quaternary ice limit
Assumed zone of confluence

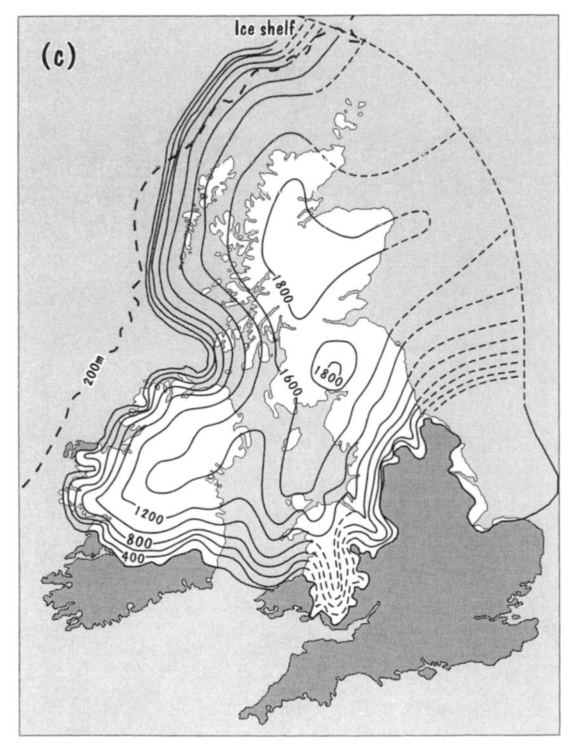
(c)
Ice shelf
1800
1800
1600
200m
1200
800
400

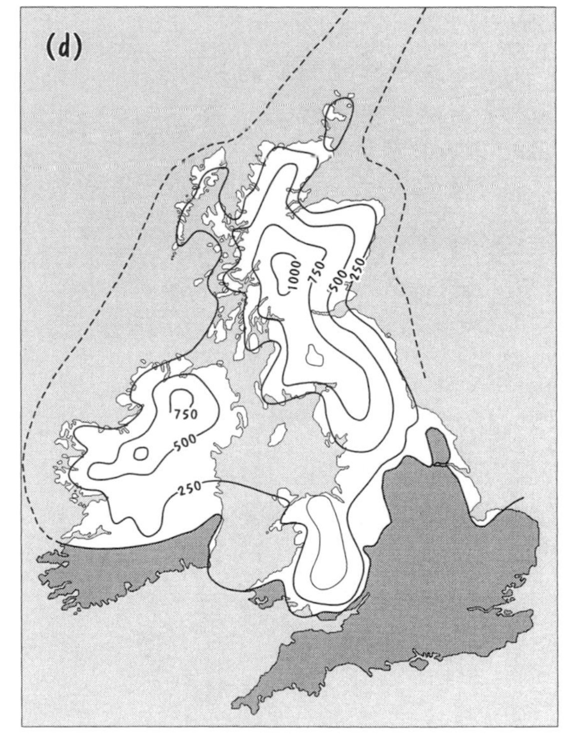
(d)
1000
750
500
250
750
500
250

CHAPTER 5

Late Quaternary Geology II
Raised shorelines and Continental shelf

This chapter discusses the glacial geology that occurs across coastal regions and within marine environments. Such geological evidence is important for the reconstruction of ice sheets because many terminated in what is now shallow sea, or were even marine-based at the LGM.

RAISED SHORELINES AND ISOSTATIC UPLIFT

When a load is exerted on the lithosphere, it depresses into the asthenosphere in accordance with Archimedes' Principle (see Chapter 3). The asthenospheric material (which we can assume is a very viscous fluid) is displaced to accommodate the depression (isostatic subsidence). This action does not happen quickly because of the viscous nature of the deforming substance. When the ice is removed (and remember, this happens very quickly compared with glacial loading), the lithosphere and asthenosphere respond by the process of isostatic uplift. The ice will be removed long before isostatic equilibrium is reached and, thus, the margins of land masses which experienced glacial loading may also have experienced shoreline uplift. The raised beaches that form as a result of shoreline uplift provide information about past relative sea levels of formerly glaciated regions. However, it should be remembered that relative sea level has two components: glacial isostasy and sea-level (or eustatic) change (Figure 5.1). 'Relatively uplift' can be correlated across a number of raised beaches to yield maps of postglacial emergence. The rates and patterns of uplift interpreted from these maps allow us to determine information about:

1. the centre of ice loading – indicated by the pattern of uplift isolines;
2. the timing of deglaciation – by carbon dating of organic material within the beaches (e.g. whalebones);
3. the maximum ice thickness – by assuming that isostatic depression is equivalent to the ice thickness multiplied by the ratio of ice density and asthenosphere density (i.e. about 1/3 of ice thickness). Numerical modelling of the Earth-system can be used to further quantify such information.

A good example of how raised beach analysis has been used in an ice sheet reconstruction is where former shorelines across Svalbard (Norwegian High Arctic) have been correlated to yield the Holocene uplift due to the decay of the Eurasian ice sheet (Figure 5.2: Forman, 1990; Forman et al. 1997; Chapter 11).

In addition to raised beaches, accurate surface altimetry can determine the present rate of postglacial uplift (Figure 5.3). This method has been used to reconstruct the present uplift of Scandinavia. Since there has been negligible change in ice loading since 10 000 years ago, it can be concluded that the pattern of present day uplift reflects that at the beginning of the Holocene. Because we can estimate how isostatic uplift decays exponentially over time, the rate of uplift at 10 000 years ago can also be estimated.

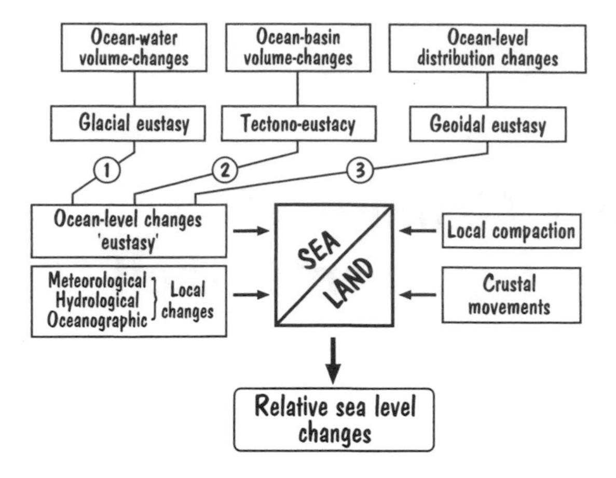

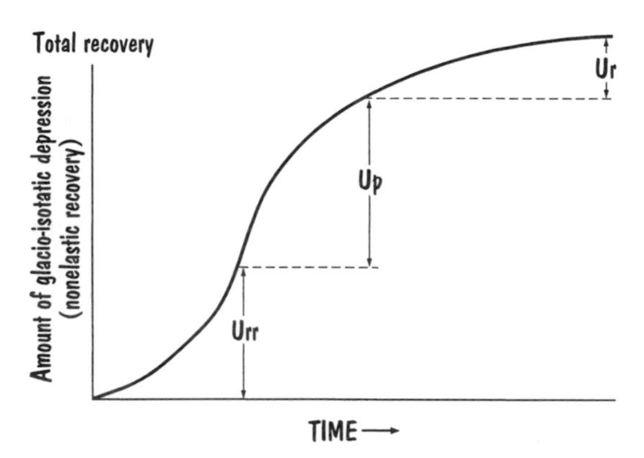

Figure 5.1 (a) Flow diagram of the components involved in relative sea-level changes. Adapted from Mørner (1980). (b) Graph showing the time-dependent change in isostatic uplift during ice-sheet decay. Urr = restrained rebound; Up = postglacial uplift; Ur = residual uplift. Adapted from Andrews (1970).

Numerical models of the Asthenosphere and Lithosphere use uplift information as inputs to establish the size and extent of ice responsible for their formation (e.g. Lambeck, 1995; Tushingham and Peltier, 1991). Although these models have provided information on ice sheet extent and ice volume, this numerical approach is hindered by the fact that uplift patterns are not necessarily characteristic of single, unique ice sheet configurations. This can be exemplified by considering that a large ice sheet that decays early in the last deglaciation may result in a similar uplift pattern at the start of the Holocene as a small ice sheet decaying at the end of the deglacial phase. In addition, as in all numerical models, poor parameterisation of key model components (for example in this case the viscosity of the asthenosphere or, in ice sheet modelling, the rheology of ice) means that results should be viewed with caution (see Chapter 3).

GLACI-MARINE SEDIMENTATION

During the last ice age, many ice sheets that grew originally across land-masses flowed over, and terminated within, the sea. This is true of the Laurentide Ice Sheet (with an iceberg calving margin in the Hudson Strait), the Scandinavian and British ice sheets (terminating in the NE Atlantic), the Greenland and Antarctic ice sheets (as at present) and the ice sheets within the Eurasian High Arctic (terminating in the Norwegian Sea and Arctic Ocean). In addition, smaller ice caps around Iceland, NW Canada, Alaska and the Aleutian Islands (and several others) all had marine margins at the LGM.

The transfer of sediments within and beneath marine ice sheets may occur in a similar way to that in the terrestrial portion of the ice sheet. However, the methods of deposition within marine regions are different to that over land. We can separate glaci-marine sedimentation into four main areas as follows:

1. Lakes
2. Fjords (not covered in this book, see Hambrey 1994, Chapter 7, for a good review on this subject)
3. Continental shelves
4. Deep sea (ocean) environments

If we look at active glacial sedimentation within these regions today, we can apply our knowledge of contemporary processes to sediment features formed in similar environments during the last ice age. In this way we can gain

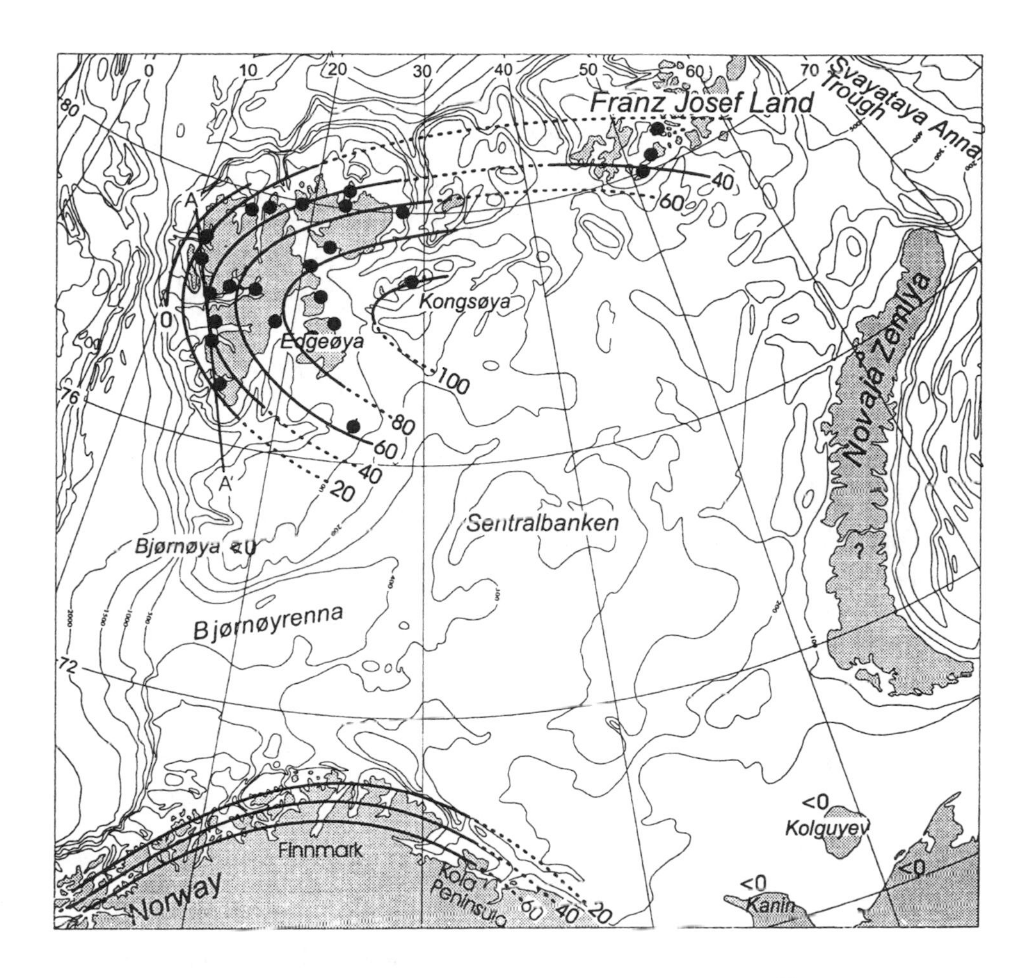

Figure 5.2 Rate of isostatic uplift across the Eurasian High Arctic at 10 000 ^{14}C years ago. Reprinted from *Quaternary Science Reviews*, 17, J. Landvik et al., Last glacial maximum of Svalbard and the Barents Sea area: ice sheet extent and configuration, pp. 43–76, copyright (1998), with permission from Elsevier Science.

information about the physical nature of the ice sheet responsible for transporting and depositing the sediments.

LAKES (GLACI-LACUSTRINE SEDIMENTATION)

During the last ice age, many glaciers and ice sheets lay over the paths of rivers (which may either have existed prior to glaciation, or have origins from glacial meltwater). The lakes that built up as a result are referred to as 'ice-dammed lakes'. Sediments transported to these lakes will be *predominantly* from the ice sheet. When the ice sheet is removed, so too may the lake. However, the glacial geology within the lake will be preserved. We therefore know such lakes existed in the Late Quaternary because of the sedimentary sequences that were deposited within them. Analysing the sediments within the lakes provides evidence of (a) glacier extent – since these lakes form at the margins of the ice sheet, (b) the rate of sediment supply from the glacier, and (c) the dynamics of the ice sheet at its margin.

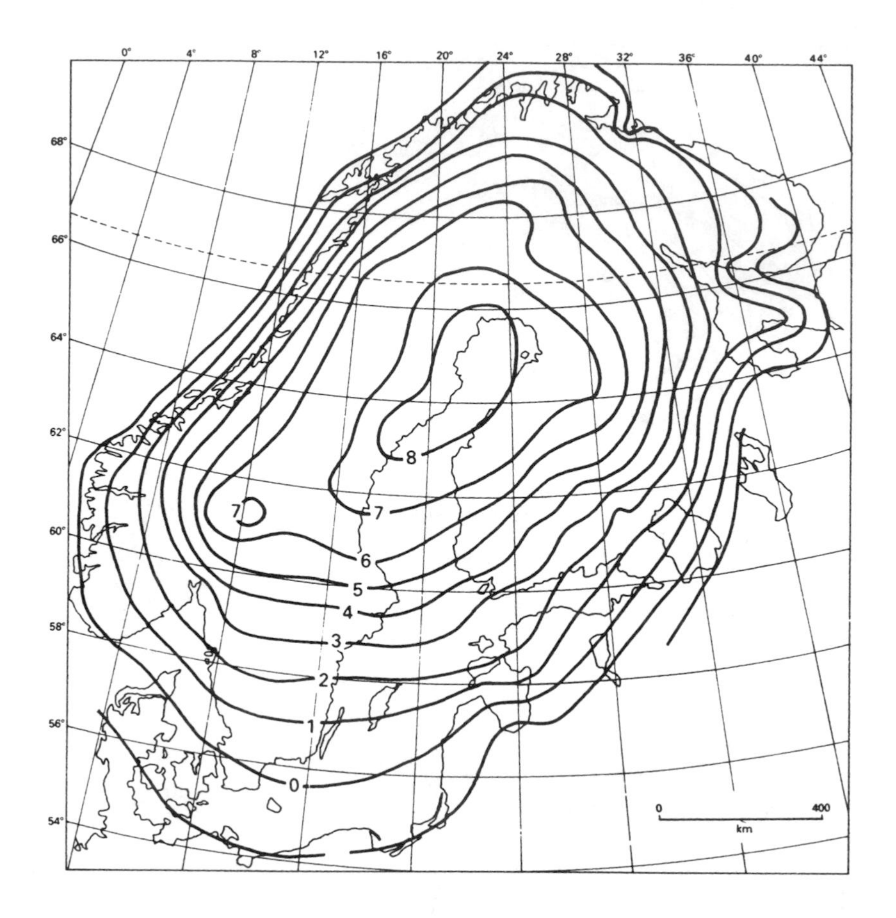

Figure 5.3 Present-day uplift of Scandinavia, in millimetres per year. Taken from Sjöberg (1991) and reprinted from Terra Nova vol 3, pp. 356–357 with permission from Blackwell Science Ltd.

The physical character of glacially-fed or ice-dammed lakes

The pattern of glaci-lacustrine sedimentation is controlled by density differences within the water mass. Density of the lake water is controlled by (a) temperature (greatest density at +4°C), (b) the concentration of dissolved salts, and (c) the amount of suspended sediment. In the summer, warm water develops at the surface of the lake, and cool water at the base. However, in the autumn, cold water forms at the surface, which then sinks (due to density contrast) resulting in complete overturning of the water column. A number of sedimentological possibilities result as a consequence of water density as follows:

- if the meltwater is much less dense than the lake water, the glacial water may remain integrated as a plume;
- however, it is more common for meltwater to be more dense than lake water, such that an underflow forms;
- from underflows, the descending water can behave as a turbidity current, producing graded, rhythmically stratified sediments spread over the whole lake floor;

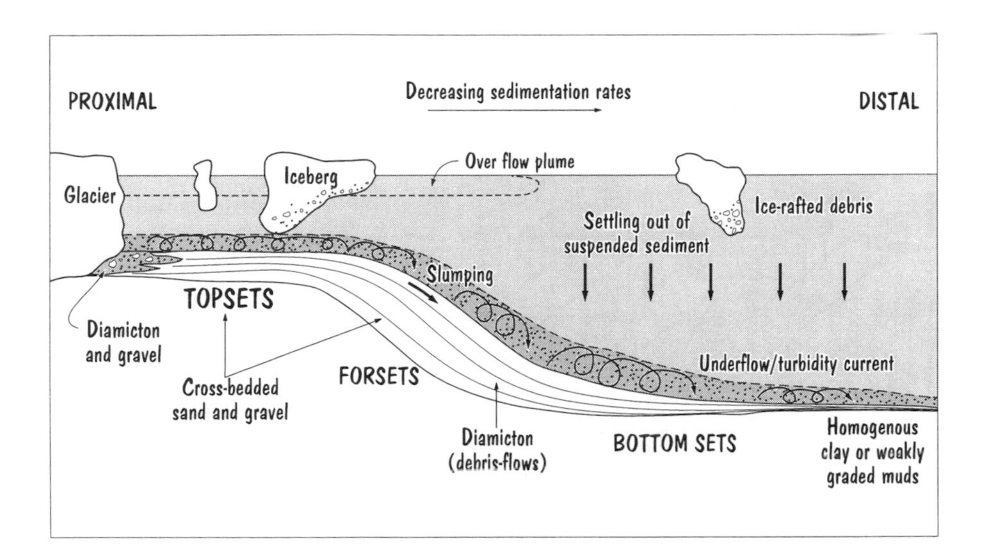

Figure 5.4 Processes and sediments in glaci-lacustrine environments. Adapted from Hambrey (1994).

- if the glacial meltwater has no sediment, and the lake has a high concentration of suspended sediment, a fresh water overflow plume may develop;
- if the lake water and glacial melt water are of the same density, then an interflow can develop.

Glaci-lacustrine sedimentary processes

For glaciers that are directly connected to the water mass, factors affecting the deposition of sediments are (Figure 5.4):

- the direct deposition from glacier ice at the ice margin;
- fluvial deposition from subglacial rivers (usually enter the lake below the surface water level);
- sedimentation from suspension of material in water;
- sedimentation from gravity-driven flows from the margin of the sub-glacial region, where sediment transport is at a maximum;
- iceberg rafted debris (IRD);
- three relatively minor non glacial processes, (1) lake-shore sedimentation, (2) biogenic sedimentation and (3) evaporitic mineral sedimentation.

Glaci-lacustrine sedimentation occurs primarily within a delta environment, from topsets through steeply dipping foresets and on to bottom-sets. Foresets are formed by gravity-driven slumping due to sediment build-up within the topsets. Bottom sets can be generally characterised as rhythmites, or varves. Each layer representing a single slumping event (often annually).

Landforms from glaci-lacustrine deposition

There are a number of tell-tale signs of former lacustrine environments that are useful for ice sheet reconstruction purposes:

- glacigenic deltas;
- delta moraines where the ice front remains stationary within a lake or sea (e.g. Figure 11.6);

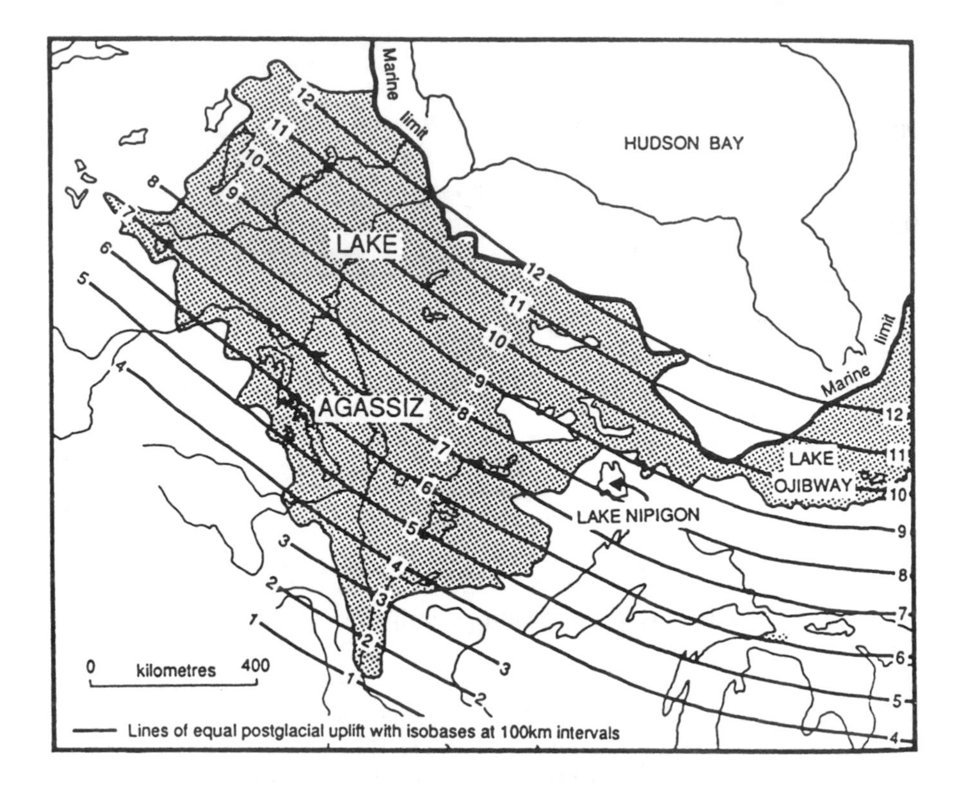

Figure 5.5 Isobases across the former Lake Agassiz, from analysis of glacio-isostatic shorelines. Taken from Dawson (1992).

- former shorelines, or strandlines (e.g., Parallel Roads of Glen Roy, in Scotland).

Identification of these features enables us to locate previous glacially-dammed lakes. Probably the best example of a proglacial lake is 'Lake Agassiz' (and other glacially dammed lakes of the Laurentide Ice Sheet) that existed in North America at the last ice age (Figure 5.5). Late Quaternary proglacial lakes in North America are discussed in Chapter 12.

If ice, which acts as the lake dam, is broken (i.e., during deglaciation, when the ice is thinner), the water within the lakes may be subject to outburst events (or Jökulhlaups). Glacial-geologic consequences of these events allow us to determine the pattern of deglaciation for the Laurentide ice sheet, and the Scandinavian ice sheet (where a large ice dammed lake occurred over the Baltic Sea).

Other information from lacustrine environments

Lakes which are not directly influenced by glaciers will have a very different sedimentary signature. Although they may contain sediments from rivers, the rates of sedimentation will be much lower, and the sediments are likely to be consistently finer grained and well sorted compared with their glacial counterparts. Identification of lake-floor sediments that do not have a glacier origin is direct evidence for the non-glaciation of a region. This fact has been fundamental in the recent reconstruction of the last Eurasian High Arctic Ice Sheet, since non-glaciated lake sediments are found on the Taymyr Peninsula, marking a maximum location for the eastern limit of the ice sheet (a point returned to in Chapter 11).

CONTINENTAL SHELF (SHALLOW MARINE SEDIMENTATION)

Introduction

Modern glaci-marine processes occur in several places around the world including Antarctica, Greenland and Svalbard (e.g. Nordaustlandet). These are all relatively large ice masses, and the sediments derived from them, and deposited in the nearby seas, are subject to processes different to those mentioned thus far.

Transport of sediment out of the ice sheet system

The amount of debris held within the base of glaciers and ice sheets is controlled largely by the thermal character of the ice mass. This debris is thought to derive from basal erosion and entrainment by the regelation of ice. However, Cuffey et al. (2000) showed that entrainment beneath Meserve Glacier in the Wright Valley, Antarctica, occurred at temperatures of −17°C without alteration to the glacier ice. Thus, entrainment in this case occurs without melting-freezing (regelation) processes. Entrainment without regelation probably occurs solely as a result of the flow of ice. If this is the case, then cold-based glaciers may be far more important as agents of erosion than thought traditionally. Cold glaciers have a thicker basal debris layer than warm based glaciers. From coring and seismic investigations, the basal layer tends to be about 1 per cent of total ice thickness. At the grounding line, depending on the form of the ice sheet, ice can either break off to form icebergs (e.g. Nordaustlandet in the Arctic), or maintain its integrity within a floating ice shelf (around several places in Antarctica). Sediments within icebergs and ice shelves will have originated from entrainment processes when the ice was grounded over bedrock. In addition to basal debris held within the ice, if a warm-based glacier is moving over water saturated basal sediments, then a 'conveyor' of sediment may flow beneath the glacier towards the grounding line, providing an additional process by which sediment can be transported by ice.

Icebergs from grounded margins of ice sheets

Icebergs calved from grounded margins will possess (initially) the same thickness of basal debris as exists within the grounded ice from which they were formed. As the icebergs escape from the ice sheet system, they will be subject to marine melting and, hence, the sediments will be released from the ice.

Ice shelves and icebergs

The basal sediment layer is largely maintained when grounded ice becomes afloat, within the ice shelf. Two scenarios are possible:

1. Basal melting of the ice shelf occurs: sediment begins to drop out of the ice shelf. Thus, the thickness of the sediment layer within the base of ice shelves reduces with increased distance from the grounding line. However, when icebergs form at the edge of the ice shelf, they usually still possess some debris.
2. Basal freezing of ice shelf: no (or little) sediment deposition. Icebergs retain full thickness of debris from grounded ice sheet.

Sedimentation rates

Since the greatest rate of sedimentation from icebergs will occur immediately after calving, it is the ice proximal region that will receive the largest volume of glacially derived sediments.

This rate decreases with distance from the ice margin. *In addition, the supply of sediments from the deforming till conveyor may add to the rate of sediment supply at the ice margin.*

Thus, we can use our knowledge of modern day glacial processes to compare with the marine sedimentological record from the last ice age. This may tell us about (i) the marine extent of the last ice sheets, and (ii) the rate of iceberg sedimentation from icebergs issued from the ice sheets (Dowdeswell, 1987).

GLACIAL EROSION AND DEPOSITION OVER THE MARINE-BASED CONTINENTAL SHELF

Just as terrestrial parts of ice sheets will erode and manipulate the substrate beneath them, so will grounded parts of marine ice sheets. A good place to analyse such glacial geology is the Barents Sea, north of Scandinavia (Chapter 11), where the north-western sector of the Eurasian ice sheet was located. Although this former ice sheet is classified as marine-based (its base is below sea level) since the ice sheet was grounded to the sea-floor, similar sedimentary and depositional processes occur at its margins and base as terrestrial ice sheets (i.e. formation of moraines and glacial sedimentary landforms). The problem is identifying such features beneath 2–300 metres of water, and a thin covering of fine-grained sea-ice-derived Holocene sediments. However, ship-borne shallow seismic and sonar (side-scan and point source) surveys can be used to remotely sense the floor of the sea.

The western Barents Sea contains a complex series of troughs and banks that are associated with large-scale glacial erosion. On a smaller scale, glacial sedimentary features have been observed within many areas of the sea floor (e.g., moraines, flutes, ice proximal sediments). This information has been used to determine where the ice front was during the LGM (or afterwards during deglaciation) (e.g. Landvik et al., 1998) (Figure 5.6).

Background reading

There are excellent reviews of glaci-marine sedimentation in the following books and papers:

Ashley, G.M. and Smith, N.D. 2000. Marine sedimentation at a calving glacier margin. *Geological Society of America Bulletin*, **112**, 657–667.

Bennett, M.R., and Glasser, N.F. 1996. *Glacial Geology: ice sheets and landforms*. Ch 2. John Wiley.

Dowdeswell, J.A. 1987. Processes of glaci-marine sedimentation. *Progress in Physical Geography*, **11**, 52–90.

Hambrey, M.J. 1994. *Glacial Environments*. Chapters 3 and 4. pp. 93–107. UCL Press.

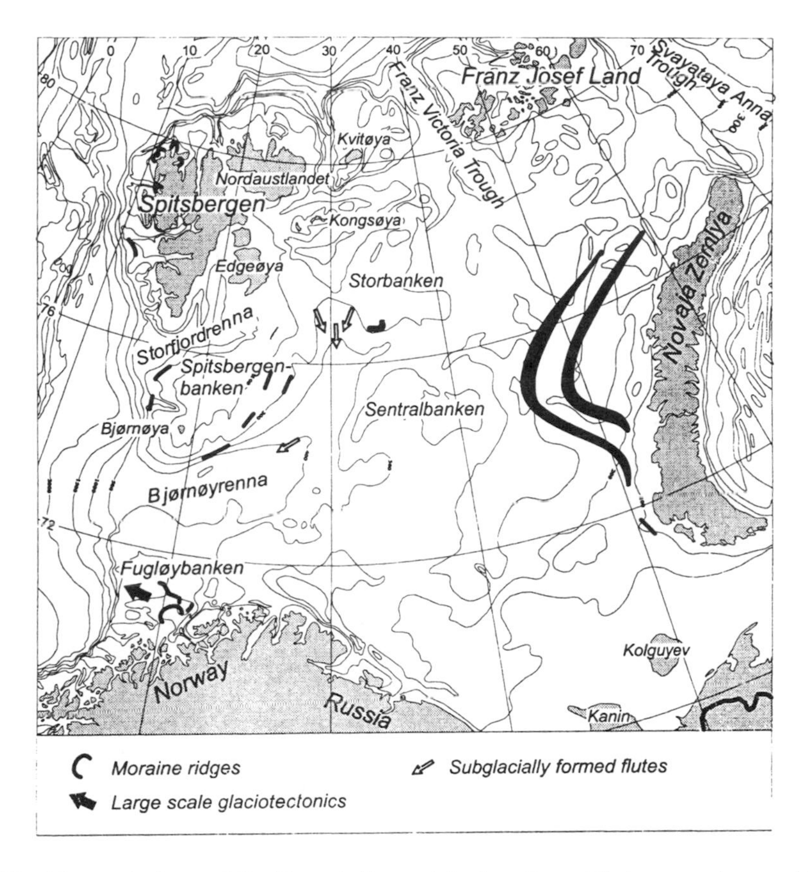

Figure 5.6 Glacial morphology of the Barents Sea and the location of flutes, moraines and large-scale glacial tectonic features. The moraine ridges were measured from sparker seismic profiles and glacial flutes and iceberg ploughmarks observed from side-scan sonar. Reprinted from *Quaternary Science Reviews*, 17, J. Landvik et al., Last glacial maximum of Svalbard and the Barents Sea area: ice sheet extent and configuration, pp. 43–76, copyright (1998), with permission from Elsevier Science.

CHAPTER 6

Late Quaternary Geology III Continental Shelf-break Sediments and Deep Sea Environments and Iceberg Debris

Glacimarine sedimentary environments can be surveyed using ship-borne experiments such as seismic surveys, side-scan sonar surveys and sediment coring. Through side-scan sonar, a map of the surface morphology of the sea floor can be obtained, whilst ice coring and seismics provide information about the subsurface sediment in one and two dimensions respectively. When put together, the data from these techniques yield information on the 3-D architecture of marine sediments.

TROUGH-FAN COMPLEXES

Glacially carved marine troughs are, in a number of cases, associated with large sedimentary fans at their mouths, which are located on the continental shelf break. The best examples of trough mouth fans can be observed within the shelf-break off the western Barents Sea (Figure 6.1). One such fan, located at the mouth of the Bear Island Trough (often named Bjørnøyrenna), is of a size comparable with many large contemporary alluvial fan systems. In addition, there are several other fans in this area of a smaller size. There is no modern deposition of any significance over these troughs. Shallow coring (gravity- and box-) of the uppermost levels of the troughs, sonar and seismic surveying indicate that these sediments are glacigenic. Glacial fan systems provide information on:

- the extent of the ice masses – since they are evidence of the ice sheet at the continental margin
- the dynamics of the ice sheets – since the sedimentation often reflects ice stream activity within the trough
- the timing of ice sheet growth and decay – since the thickness of sediment accumulated over the fan will be a function of the time the glacier margin spent at the trough mouth
- information on past glaciations not available from terrestrial geology – since there is likely to be little erosion of material deposited by ice sheets prior to the last, Late Quaternary ice sheet.

Sediment transport to the trough mouth is due mainly to (Figure 6.2):

- Subglacial till deformation conveyor (e.g. Alley, 1990). Recent evidence from Ice Stream B indicates that such processes operate in modern Antarctic ice streams. It may well be the case, therefore, that glacigenic sedimentary fans are located at the mouths of Antarctic bathymetric troughs, formed when the ice sheet was at its LGM position (see Chapter 9).
- Entrainment of basal ice – as recorded in deep ice cores.

Deposition of sediment material over the fan system is due to the following nine processes (Figures 6.3 and 6.4):

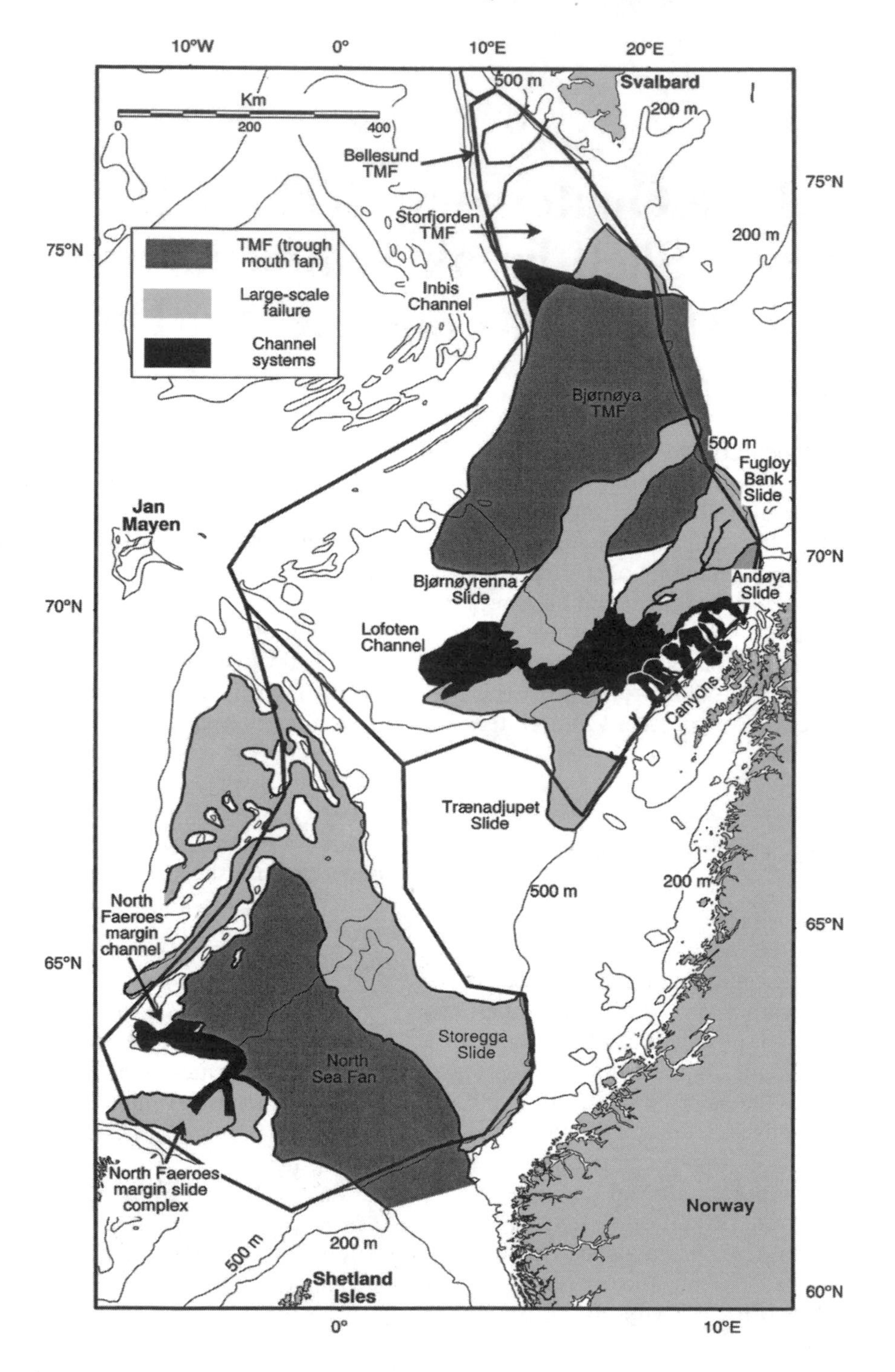

Figure 6.1 Locations of Arctic trough mouth fans, slides and channels. Taken from Taylor (1999), and reproduced by permission of the author.

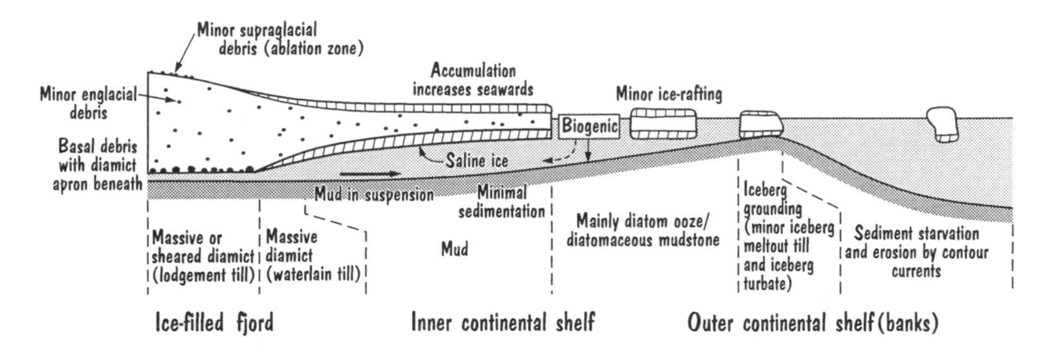

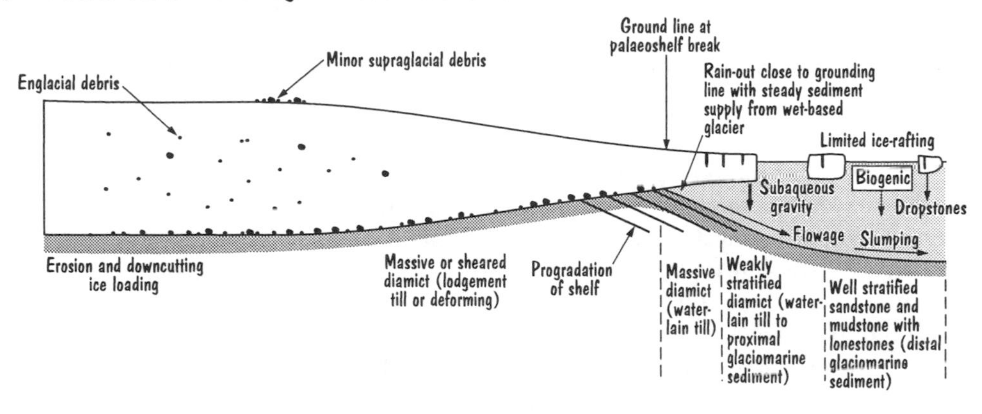

Figure 6.2 Ice-sheet dynamics and glaci-marine sedimentation over (a) continental seas and (b) continental margins. Adapted from Hambrey (1994).

1. meltwater plumes from the ice sheet
2. ice rafting from icebergs calved off the ice sheet
3. sliding/slumping of material, driven by gravity (some of which can be extremely large)
4. sediment push and squeeze at the base of the ice sheet
5. debris flows (comprising the majority of landforms seen across fans along the western margin of the Eurasian Arctic)
6. turbidity currents
7. shallow bank winnowing
8. outflow of cold, dense shelf water
9. upper slope winnowing

These processes will occur at different stages of the sediment-supplying ice sheet's evolution. The trough-fan systems mark the maximum extent of the ice sheet. Geochronological dating, of the moraines and fan can yield information about the exact timing of glaciation, and rates of sedimentation.

RATES OF SEDIMENTATION OVER GLACI-GENIC FANS

It is common knowledge that: (a) glaci-genic fans are supplied with glacial sediments; (b) the fans are of comparable size to large river-fed

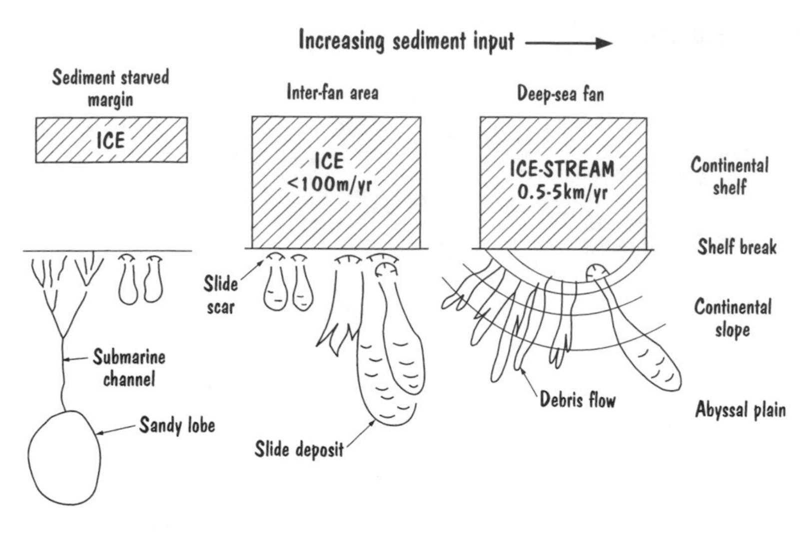

Figure 6.3 Conceptual model of sedimentation across glaciated passive margins. Adapted from Dowdeswell et al., (1996).

fans, (c) glacial sedimentation occurs for a few thousand years during the maximum phase of glaciation, followed by much longer periods of no activity, (d) river fans are being fed constantly. Therefore, we can conclude that glacial sedimentation on to continental fan systems occurs at a much higher rate than that on modern river-fan systems. Geochronological dating of individual layers within the fan complex have yielded active sedimentation rates of up to 120 cm per thousand years (i.e. enormously high). The dates on which these sedimentation rates are based also provide information on the period in which the ice sheet was feeding glacial sediments to the trough (i.e. when it was at its maximum size) (Laberg, 1996). The measurement of large-scale glacigenic material in trough mouth fans is discussed later, in Chapter 11.

MARINE GEOLOGICAL DATA FROM THE EURASIAN ARCTIC: FAN DISTRIBUTION AND STRUCTURE

A series of large, glacier-influenced submarine fans is present on the passive continental margins of the Eurasian Arctic. The extent and thickness of these fans have been investigated using a combination of geophysical tools, including reflection seismic methods and long-range side-scan sonar systems (Vorren et al., 1998). Each of the fans has formed at the mouth of a cross-shelf trough, some of which are offshore of major fjord systems. The largest is the Bear Island Fan, located west of the epicontinental Barents Sea, with an area of approximately 280 000 square kilometres (Figure 6.2). A number of other fans are present on the northern Barents and Svalbard margin (Vorren et al., 1998); Storfjorden Fan (40 000 square kilometres), Bellsund Fan (6000 square kilometres), Isfjorden Fan (3700 square kilometres) and Kongsfjorden Fan (2700 square kilometres). North of Eurasia, on the Arctic Ocean margin, there is also less detailed evidence of large fans off the Franz-Victoria, St. Anna and Voronin troughs, between the archipelagos of Svalbard, Franz Josef Land and Severnaya Zemlya (e.g. Knies et al., 2000).

These fans appear to have a similar internal architecture, and are made up of a series of stacked debris flows. The uppermost debris flows are imaged clearly on long-range side-

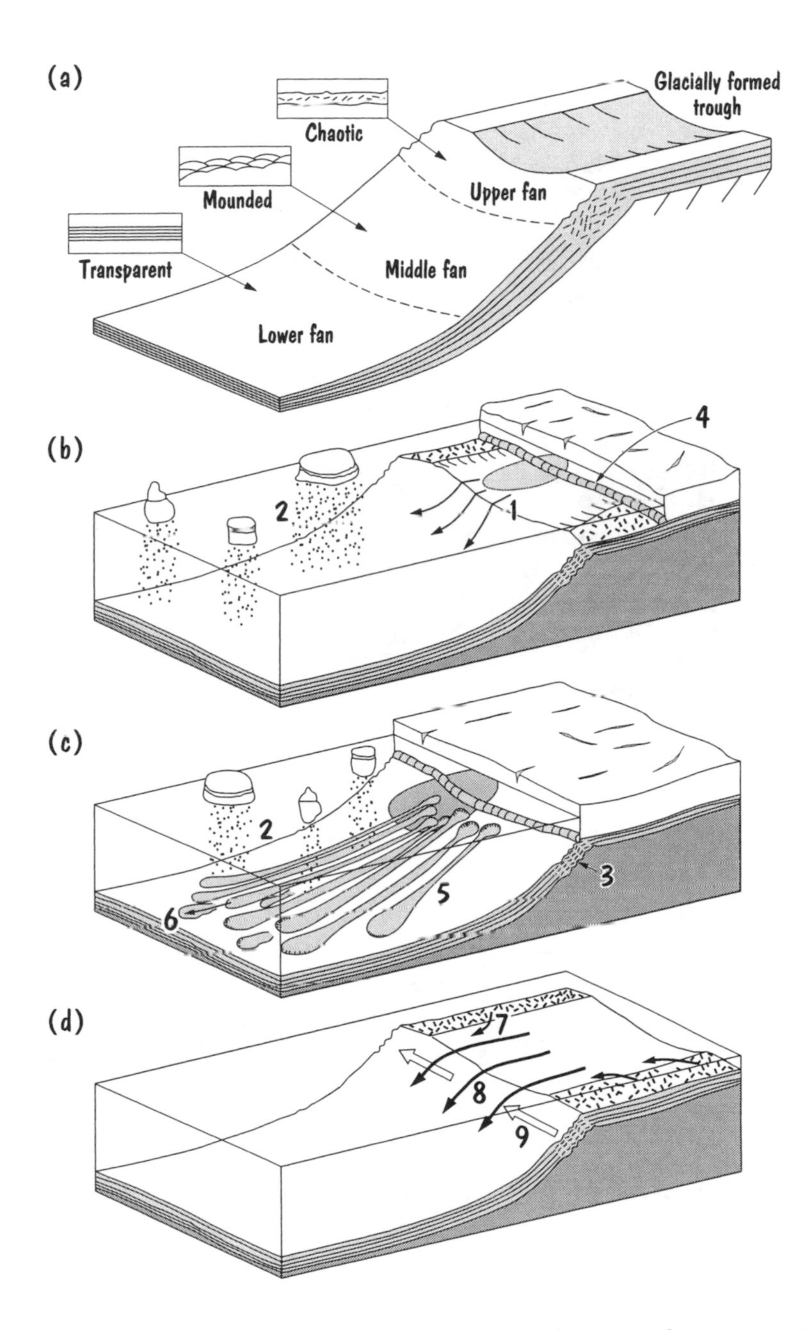

Figure 6.4 Morphology and processes within glacigenic trough mouth fan systems. Numbers refer to processes detailed in the text. (a) generalised diagram showing the seismic signature of sediments. Fan sedimentary processes operating under (b) glacial advance, (c) glacial maximum and (d) glacial retreat. Adapted from Laberg (1996).

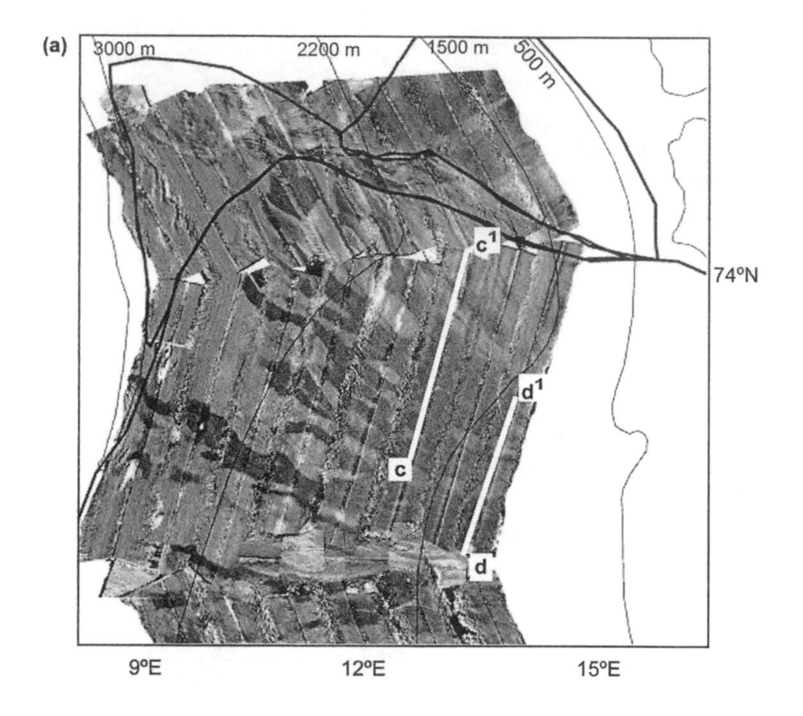

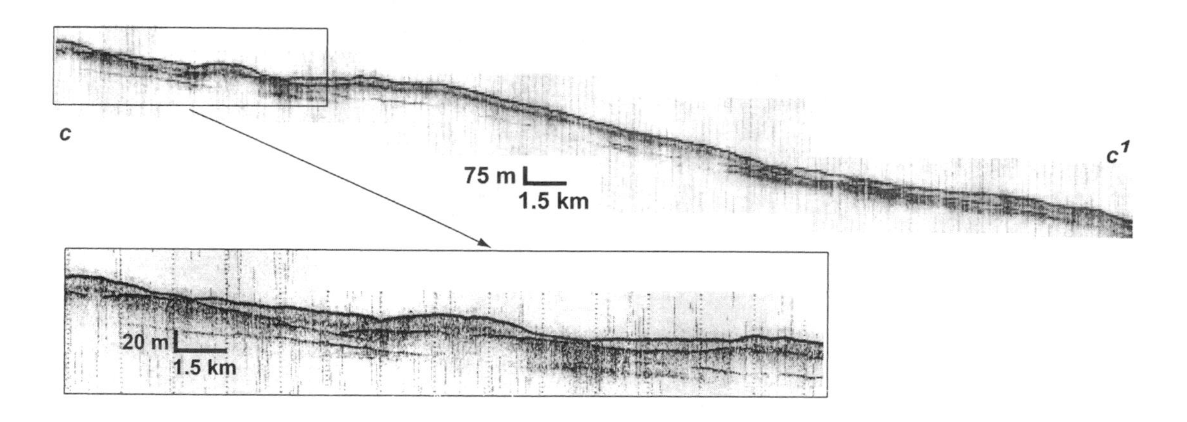

Figure 6.5 The Bear Island Trough Mouth Fan, western Barents Sea. (a) GLORIA side-scan sonar data collected by J.A. Dowdeswell in 1994. (b) interpretation of the GLORIA data. (c,d) 3.5 kHz records taken from transects indicated in (a). Taken from Taylor (1999), with permission of the author.

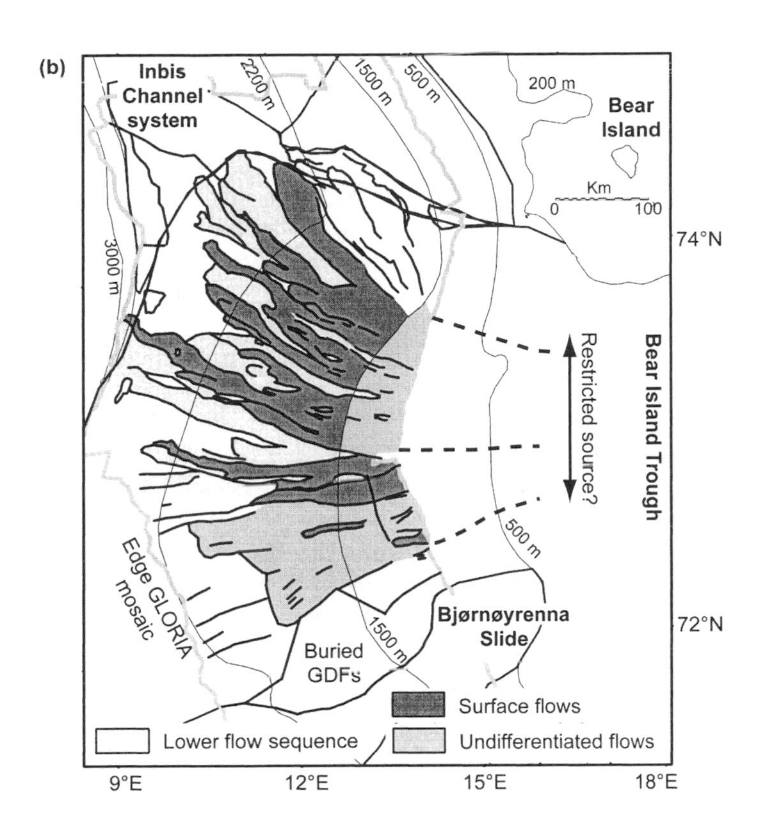

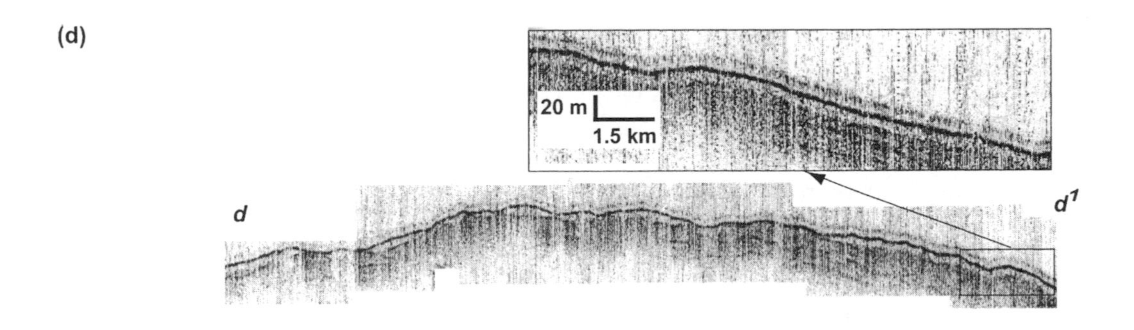

scan sonar records and high resolution 3.5 kHz profiles (e.g. Vogt et al., 1993). An example of GLORIA 6.5 kHz side-scan sonar imagery and a 3.5 kHz profile from the Bear Island Fan illustrates the character of these debris flows in plan and cross-section (Figure 6.5, Taylor, 1999). The acoustically transparent debris flows range between about 2–10 kilometres in width, 10–50 metres in thickness, and are 30–200 kilometres in length (Dowdeswell et al., 1996a; Elverhøi et al., 1997). In seismic lines over several fans a number of acoustic packages are identified, each made up of a group of stacked debris flows and separated from the underlying packages by a major reflector (e.g. Vorren et al., 1989; Laberg and Vorren, 1995). Each acoustic unit is interpreted to represent debris flows from a single glacial-interglacial cycle (Laberg and Vorren, 1995). The debris flows are assumed to be derived from the intermittent failure of clay-rich glacier-derived sediments, deposited on the upper slope during full glacial conditions when the ice sheet margin was at the shelf break (e.g. Laberg and Vorren, 1995; Dowdeswell et al., 1996a; Elverhøi et al., 1997).

Age-constrained seismic data have been used to construct isopach maps of sediments making up the fans off the Barents Sea continental margin. From these studies, estimates of volume and sedimentation rate for several time intervals have been provided (e.g. Fiedler and Faleide, 1996; Hjelstuen et al., 1996; Laberg and Vorren, 1995). The thickest parts of the glacial sequence, on the Bear Island and Storfjorden fans, measure between 3.5 and 4 kilometres, and about 2 kilometres on the Isfjorden Fan to the north (Faleide et al., 1996). For the Bear Island Fan, Laberg and Vorren (1996a) estimate sedimentation rates of 124 cm 1000 yr^{-1} for the Late Weichselian glacial maximum, as compared with only 10–15 cm 1000 yr^{-1} for the Holocene. Over longer periods, Fiedler and Faleide (1996) calculate a total sediment volume of about 340 000 cubic kilometres for the Bear Island Fan since about 2.3 million years ago, shortly after the onset of Pleistocene glaciation.

The long-term sedimentation rates for the Bear Island and Storfjorden fans can be regarded as measures of full-glacial sediment delivery, which operated over approximately 10 per cent of the 440 000 year time interval. This increases the mean and maximum sedimentation rates on each of the fans during full glacials to the following values: a mean of 13 cm 1000 yr^{-1} and maximum of 124 cm 1000 yr^{-1} for the Bear Island Fan (Laberg and Vorren, 1996a); and a mean of 29 cm 1000 yr^{-1} and a maximum of 172 cm 1000 yr^{-1} for the Storfjorden Fan (Laberg and Vorren, 1996b). Thus, for the Late Weichselian glaciation, 4200 cubic kilometres of sediments were deposited in 12 000 years over the 280 000 square kilometre area of the Bear Island TMF (Laberg and Vorren, 1996a), and 700 cubic kilometres was deposited in 10 000 years over the 40 000 square kilometre Storfjorden TMF (Laberg and Vorren, 1996b).

However, it should be noted that Taylor (1999) recently recalculated the volume of sediment deposited over the Bear Island Fan during the last glaciation to be around 2000 cubic kilometres (about half of that calculated by Laberg and Vorren, 1996a). This difference is mainly because Taylor found that debris flows were limited to stacks of about three deep for the last glaciation. Since the volume of a debris flow was calculated, Taylor was able to calculate the total volume of sediment comprising Late Weichselian debris flows.

A QUALITATIVE MODEL FOR SEDIMENTATION ON GLACIER-INFLUENCED MARGINS

Dowdeswell et al. (1996a) developed a conceptual model for glacier-influenced sedimentation on high-latitude continental margins (Figure 6.3) summarising the role of ice sheet extent and dynamics on the geographical location and rate of sediment build-up. The model is based on a large volume of long-range side-scan sonar and seismic data from the Polar North Atlantic. During interglacial and interstadial periods, ice is far from the continental shelf edge, and sedimentation on the outer shelf and slope is slow and hemipelagic in character. However, in full glacial and some stadial periods, ice advances on to the shelf and across it to the shelf break. Ice stream activity is most likely within bathymetric troughs because, firstly, basal temperatures are more likely to reach the melting point in these regions of relatively thick ice and, secondly, a reduction in basal drag may occur as a result of low buoyancy-induced effective basal pressures in such areas (Bentley, 1987). Thus, the maximum-sized ice sheet was characterised by a series of ice streams situated within troughs which fed ice to the shelf break (Figure 6.5). It should be noted that modern Arctic ice streams are often, although not always, located in subglacial troughs (e.g. Dowdeswell and Collin, 1990).

The delivery of sediments in icebergs, meltwater and, in particular, from deforming subglacial sediment (Alley et al., 1989), is enhanced at the margins of ice streams relative to slower moving ice. The development of fast-flowing ice streams is, therefore, proposed to be fundamental to the growth of prograding fans, and the debris flows which are their building blocks, during full glacials (Dowdeswell et al., 1996a; Laberg and Vorren, 1996a, b; King et al., 1996).

Between ice streams, although ice may still reach the shelf break, the rate of ice sheet flow is one to two orders of magnitude slower. In a number of the proglacial inter-fan regions, very large but infrequent slide events have been observed (Bugge et al., 1988; Dowdeswell et al., 1996a). In some high latitude areas, ice fails to reach the shelf break even during full glacials, and here sedimentation rates will remain low throughout a glacial-interglacial cycle. The margins and deep-ocean basins beyond them are dominated by large submarine channel systems (Mienert et al., 1993).

DEEP SEA ENVIRONMENTS

The flux of sediments derived from icebergs (ice-rafted debris or IRD), will diminish with distance from the ice sheet margin. However, IRD is found within the deep-sea regions bordering glaciated continents. Because of the low-energy environment that exists at the deep-sea floor, sediments deposited in these regions will be preserved well. Methods of deposition of sediments within deep sea environments include (Figures 6.2 and 6.6):

- iceberg rafting and meltout of material (i.e. IRD)
- wind-blown material over the ocean
- material incorporated within sea ice
- sea floor processes such as slides, slumps and turbidity currents from more elevated regions bordering the abyssal plains (e.g. material from glacially-fed fan systems)
- biogenic processes such as calcium carbonate precipitation from foraminifera (discussed in Chapter 1 on global indicators of ice volume)

IRD is easily identifiable compared with sediments derived from these other sources since it has relatively large clasts within a finer surrounding matrix which are glacigenic. Deep-sea coring (from the Ocean Drilling Program (ODP)) allows marine geologists to sample such material. We can use IRD, and other sediment from deep sea regions, to provide information on (a) rates of iceberg production (especially in the North Atlantic where large thicknesses have been found), and (b) insights into ocean water conditions (especially local variations from the global record).

SPATIAL DISTRIBUTION OF IRD

Assuming that the drift of icebergs is controlled by surface ocean currents, one would expect that the thickness of IRD deposits will decrease with increase in distance (along the line of iceberg flow) from the ice margin. Subsequently, if/when an anomalous concentration of iceberg debris is located far from the reconstructed ice front, there must be a glaciological reason. There are two possibilities. First, a significant increase in ice-rafted material is indicative of an increase in iceberg production and/or an increase in the volume of sediment carried by icebergs. Secondly, however, a decrease in ice-rafted material may represent a pause in iceberg production and/or a decrease in the volume of sediment carried by icebergs.

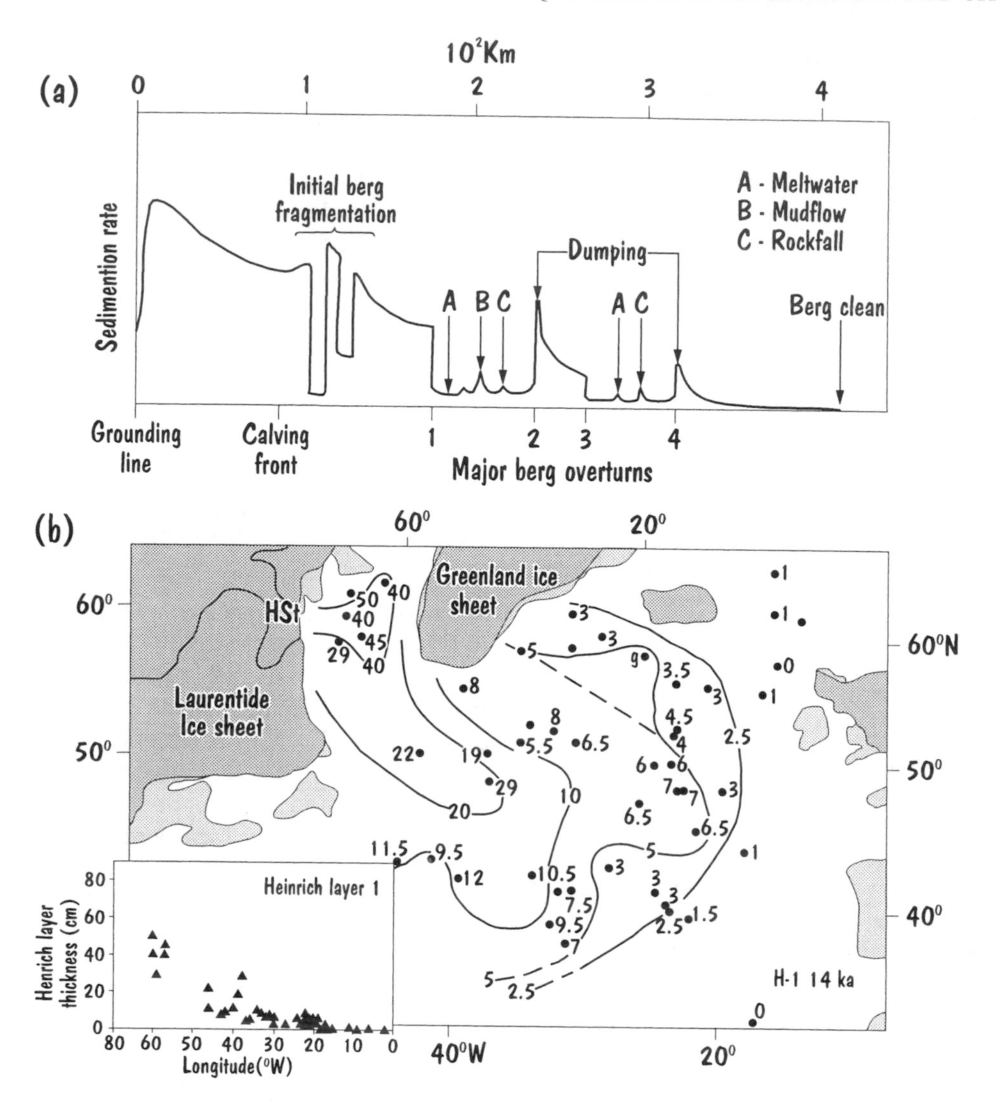

Figure 6.6 Iceberg rafted debris: model and measurements. (a) Conceptual model of sedimentation from icebergs. (b) A Map of Heinrich Layer 1 across the floor of the North Atlantic. Adapted from Drewry (1986) (a) and Dowdeswell et al. (1995) (b).

HEINRICH LAYERS

ODP investigations within the North Atlantic during the 1980s, discovered a significant coverage of IRD. These sediments exist within a number of unique layers. This sediment complex has been interpreted by many to represent the increase in drop-out due to increases in the amount of icebergs derived from the Laurentide Ice Sheet (Figure 6.6). Laurentide Ice Sheet reconstructions in recent years have addressed this issue, and most now account, in some way, for discharges of icebergs at the dates calculated for the respective layers (this matter is returned to later in Chapter 12). The most likely explanation is that Heinrich layers represent an increase in iceberg production, emanating from an ice stream within the Hudson Strait, which acted to drain the ice dome over central North America. However, there are two possibilities to explain this increase in iceberg calving:

1. The iceberg production was due to climate controlled deglaciation. If this was the case, then we should see similar IRD deposits from other ice sheets. Such sediments have not been found yet, suggesting this climate-controlled ice sheet behaviour is an unlikely cause (Dowdeswell et al., 1999).
2. The iceberg production was due to dynamic surging (which may be periodic) of the ice sheet (See Chapter 12 – Binge Purge theory).

Other explanations for Heinrich layers include:

- The Laurentide ice sheet response to changes in the European ice sheets. The influence of the European ice sheets on IRD in the North Atlantic is not well known. However, European-derived IRD has been reported at the beginning of several H-events. This implies that the European ice sheets may have been influential in the dynamics of the Laurentide ice sheet by surging (or responding to climate) first, causing sea-level rise which then caused surging of the Hudson Strait sector of the Laurentide ice sheet (Grousset et al., 2000).
- An increase in debris content within an otherwise steady flux of icebergs. The debris content of icebergs is clearly an important factor for the interpretation of Heinrich layers, and one that will be difficult to address.
- A sudden break up of an ice shelf within the Labrador Sea (Hulbe, 1997). However, foraminifera from sea floor sediments in this region suggests that the sea was seasonally open around the LGM.
- Earthquake-induced decay of the ice sheet, due to ice sheet loading of the crust (Hunt and Malin, 1998).

To summarise, although many people believe in Heinrich layers, their exact cause is not known with any certainty.

LOCAL VARIATIONS IN THE PALAEO-ENVIRONMENTAL SIGNALS

Later in Chapter 11 an ice sheet reconstruction across the marine regions of the Eurasian ice sheet will be presented. This reconstruction has profited in recent years from detailed analysis of the sea floor sediments surrounding the Barents Shelf, the interpretation of which have implications for ice sheet growth and decay. For example:

- local oxygen isotope signal: indications of the timing of local ice sheet melting;
- local planktonic foraminifera with sea floor sediments: information about local sea surface temperatures;
- local IRD deposits: details about ocean surface conditions, and regional iceberg production.

CHAPTER 7

Late Quaternary Palaeoclimate

INTRODUCTION

Numerical ice sheet models require climate forcing from which ice accumulation and temperatures inputs can be determined. We can gain information on palaeoclimate from (a) geological information (including ice cores) and (b) numerical modelling of the climate system.

The process of obtaining a palaeoclimate reconstruction has a direct analogy to the reconstruction of ice sheets. In climatology, geological information is used to determine a basic palaeoclimate where information is available, then a numerical model is used to determine information about the climate in regions where there are no data. Existing climate information is used to calibrate the model in a 'present day' situation. Sophisticated models used to describe the climate are called Atmospheric General Circulation Models (GCMs). Important to any model of climate is the account of atmospheric heat and moisture derived from the ocean. Therefore, Oceanic GCMs are often related to Atmospheric GCMs. In glaciology, geological information is used to determine where ice sheets once existed, and what the ice flow patterns and basal conditions were. Numerical ice sheet models are then used to determine further information about the ice masses (where no geological information exists). Present day ice sheets are used to calibrate the models under modern-type settings (e.g. EISMINT block experiments and Greenland programmes).

Before we go any further, we must note a basic problem about reconstructing palaeoclimates from GCM models.

- We noted in the first chapter that ice sheet size controls the Earth's surface albedo.
- Note also that albedo will be related to sea ice extent.
- Surface albedo is an important input to GCM models.
- Therefore, we need information about ice sheets and sea ice to get the albedo information.
- So we are in danger of having a circular argument, where an ice sheet reconstruction is used to model the palaeoclimate which is then used to model the ice sheet evolution.

Currently, models are being developed which 'couple' GCMs with ice sheet models. However, these studies are in an early stage of development. Most GCMs utilise the ice sheet distribution determined by the CLIMAP project members (1976; 1981) or as published in the two special volumes of *Quaternary Science Reviews* in 1986 (vol. 5, Quaternary Glaciations in the Northern Hemisphere) and in 1990 (vol. 9, Quaternary Glaciations in the Southern Hemisphere). The reconstruction of the world's ice sheets by these studies will be returned to in later chapters on specific ice sheets. The discussion here will rest with global climate information and will not involve regional (local) climate variations at this stage.

GEOLOGICAL INFORMATION ON PALAEOCLIMATE

The most important palaeoclimate indicator comes from ice core studies from Greenland

and Antarctica, deep sea sediment cores and a few other records, such as calcite veins. From ice cores, CO_2 (that translates to air temperature) and snow accumulation rates can be established.

Important ice cores are located in Canada (Devon Island Ice Cap); Greenland (Camp Century, Renland, Dye 3, GISP2, GRIP); and Antarctica (Dome C, Byrd, Vostok). Other important geological climate information comes from sea floor sediments, indicating SSTs from planktonic forams, and calcite veins such as the Devil's Hole, Nevada, providing oxygen isotope values from outside of the marine/ice sheet environment. By collating these sorts of data together, we can obtain a picture of global climate change (Figure 7.1). Recent advances in palaeoclimate reconstructions from ice cores have provided a detailed temperature record for the last deglaciation in Greenland (Dahl-Jenssen et al., 1998) (Figure 2.3), and a palaeoclimate dating back over 400 000 years in Antarctica (Petit et al., 1999) (Figure 2.4).

EXAMPLES OF ICE-CORE-GEOLOGICAL DATA INTER-COMPARISON

Figure 7.1 shows information gained from a variety of geological records, plotted along the same linear timescale. Here we can see how global climate varies, for the large part, in a synchronous manner. This is an important feature, since it demonstrates that Antarctic and Arctic glaciations occur near-simultaneously. However, this finding is relatively hard to understand, given an assumption that climate is forced by Milankovitch oscillations, dominated by solar insolation reduction in the Northern Hemisphere's summer. *One* hypothesis is, therefore, that Northern Hemisphere ice masses are forced by Milankovitch (plus amplification), which yields global sea level fall which, in turn, forces ice advance of the largely marine Antarctic Ice Sheet. Consequently, Antarctic ice advances lag behind, and are forced, by Arctic ice advances. This may be also similarly true of deglaciation. However, there is little evidence to support this hypothesis. As pointed out below, there is currently a major debate concerning the timing of climate events in each hemisphere.

The Northern Hemisphere ice core record is far more 'spiky' than that from Antarctic cores. For example, predicted temperatures in Greenland during deglaciation are observed to vary by ±5°C in 30 years (Dahl-Jenssen et al., 1998); oscillations which are largely absent from records from the Antarctic interior (Figure 2.3). A number of authors cite variations in neighbouring ocean conditions as an explanation for such variation (see Chapter 8). However, for now, it is important to note that the North Atlantic Ocean provides a large amount of heat to high latitudes. Consequently, the presence/absence of this important heat source needs to be accounted for when modelling climate conditions during the last ice age.

The main problem in attempting to correlate cross-hemisphere ice core and sea floor sediment records is that it relies on the chronology of the record being exact. Unfortunately, most ice cores use numerical models to guide the ice core chronology. Consequently, as these models are refined, so the chronology of an ice core may be adjusted. For example, the Vostok ice core in Antarctica has had several adjustments made to its chronology over the past decade or so. Sea floor sediment cores suffer similar problems with geochronological dating techniques. Bearing this dating problem in mind, several authors have attempted to correlate Northern and Southern Hemisphere palaeoclimate records. Opinions are currently divided as to whether this correlation indicates that Antarctic records lag or lead those from Greenland. Blunier et al. (1998) suggested that 'warm' peaks in the $\delta^{18}O$ records from the Byrd and Vostok ice cores at about 36 000 and 45 000 years ago occur about 1000 years earlier than the comparable records in Greenland (Figure 7.2). However,

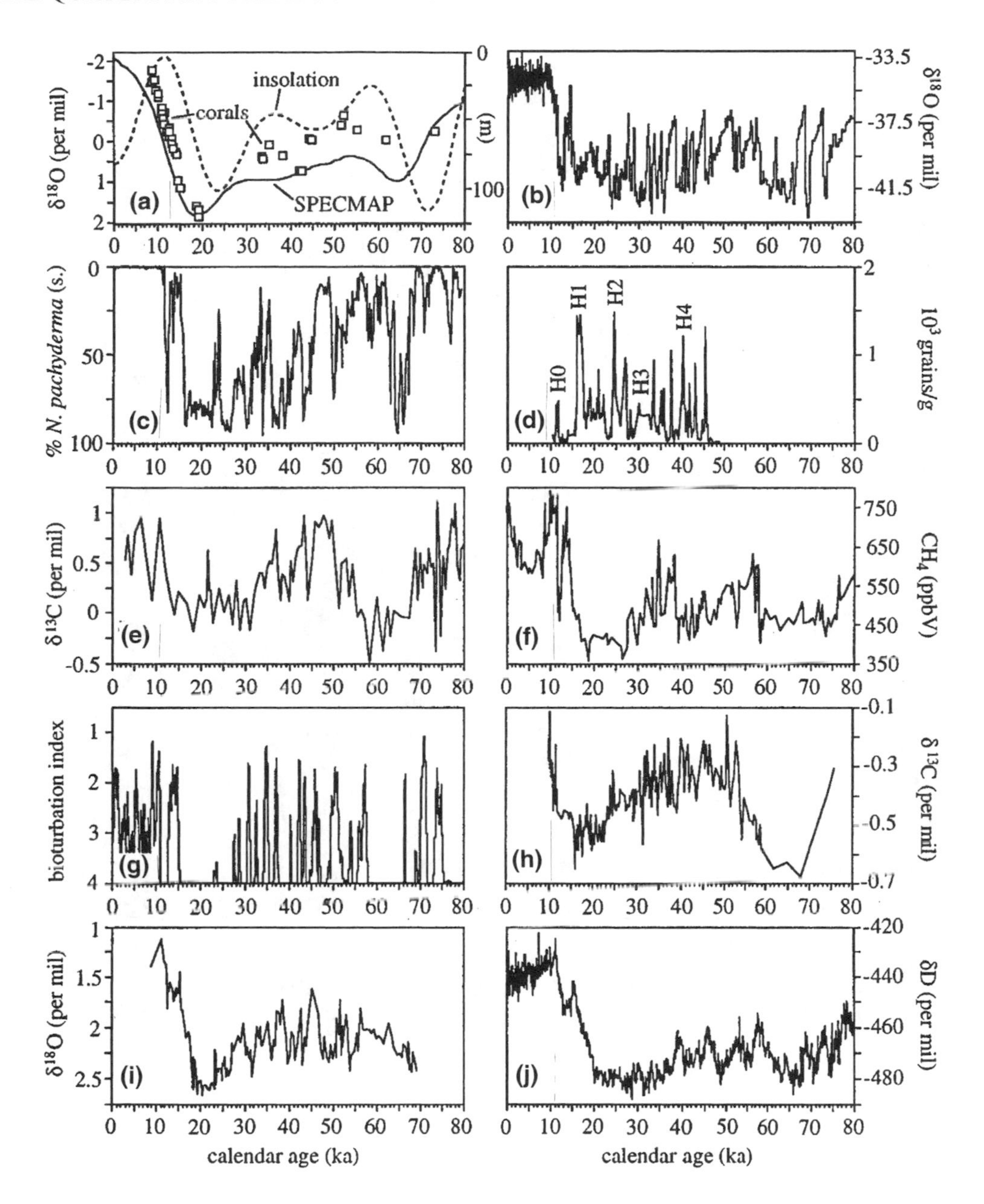

Figure 7.1 Proxy climate records for the last 80 000 years from several independent datasets around the world. (a) Summer insolation at 60°N and the SPECMAP $\delta^{18}O$ signal. (b) GISP2 $\delta^{18}O$ signal. (c) Changes in the percentage of *neogloboquadrina pachyderma* (s.) from a core in the North Atlantic. (d) Concentration of grains within a core from the North Atlantic. (e) Record of $\delta^{13}C$ from an equatorial sediment core. (f) GISP2 record of atmospheric methane. (g) Changes in the bioturbation index from an ODP site in the Santa Barbara basin (1=laminated sediments, 4=massive sediments). (h) Record of $\delta^{13}C$ from a north-eastern Pacific sediment core. (i) Record of $\delta^{18}O$ from a southern ocean sediment core. (j) The δD record from the Vostok ice core. Taken from Alley and Clark (1999). With permission, from the *Annual Reviews of Earth and Planetary Sciences*, Volume 27 ©1999 by Annual Reviews www.AnualReviews.org

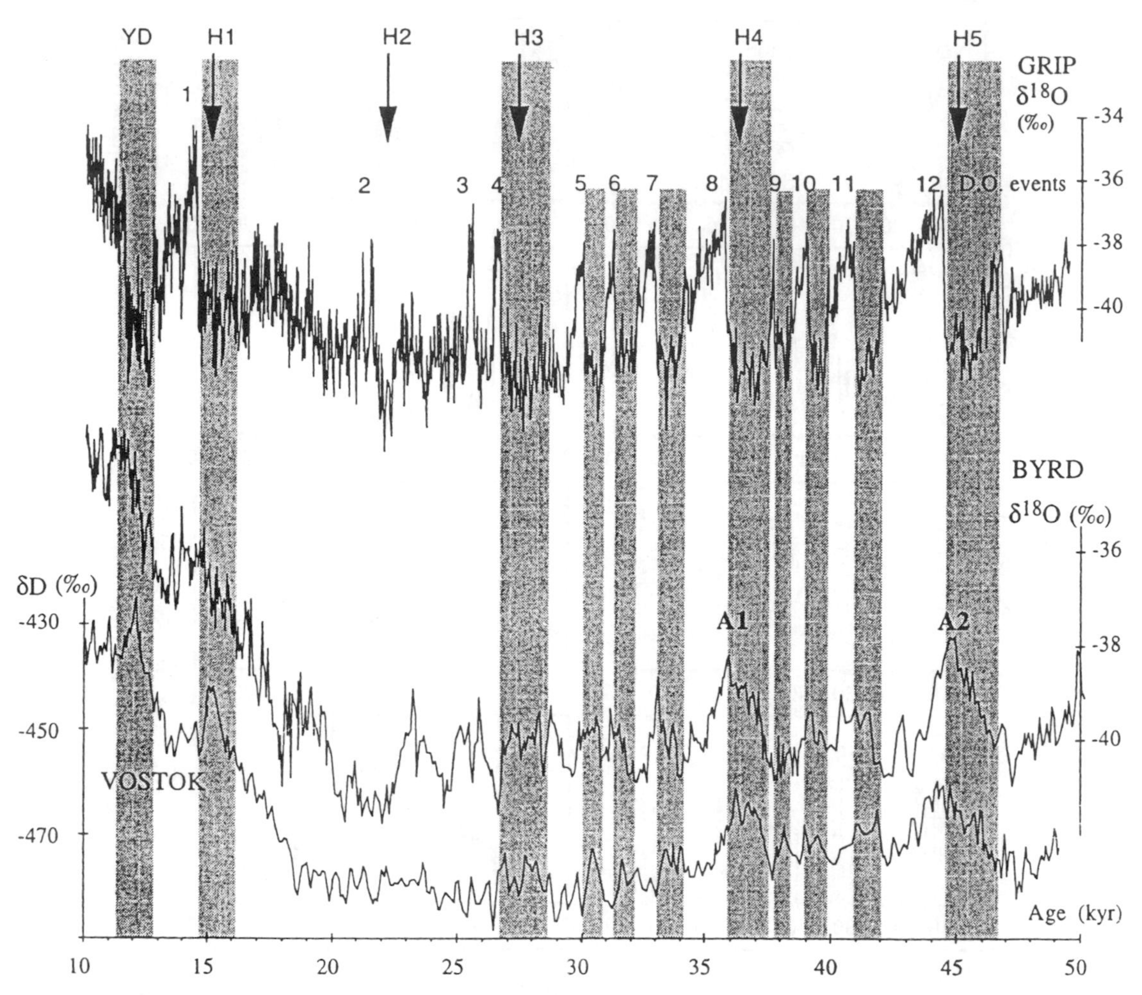

Figure 7.2 A correlation between Greenland (GRIP) and Antarctic (Byrd and Vostok) ice core records. H (Heinrich) and D–O (Dansgaard–Oeschger) events are indicated in grey shade. Reprinted from *Earth and Planetary Science Letters*, 1777, A. Mazaud et al., Short fluctuations in Antarctic isotope records: a link with cold events in the North Atlantic? pp. 219–225, copyright (2000) with permission from Elsevier Science.

Mazaud et al. (2000) suggest that short-term changes in Antarctic palaeoclimate may be forced by alterations to North Atlantic conditions, suggesting that the Northern Hemisphere drives short-term climate change in the Southern Hemisphere. Whichever theory is correct, it is clear that climate change in a particular hemisphere will result in alterations to climate in the other. The identification of teleconnections between oceans, climate and ice sheets from palaeoclimate records is becoming a major research topic, and one that will help to establish a more complete understanding of the Earth's climate system.

MODERN CLIMATE CONTROLS (POLAR ATMOSPHERIC CIRCULATION)

To model atmospheric conditions during the last ice age we examine the modern climate

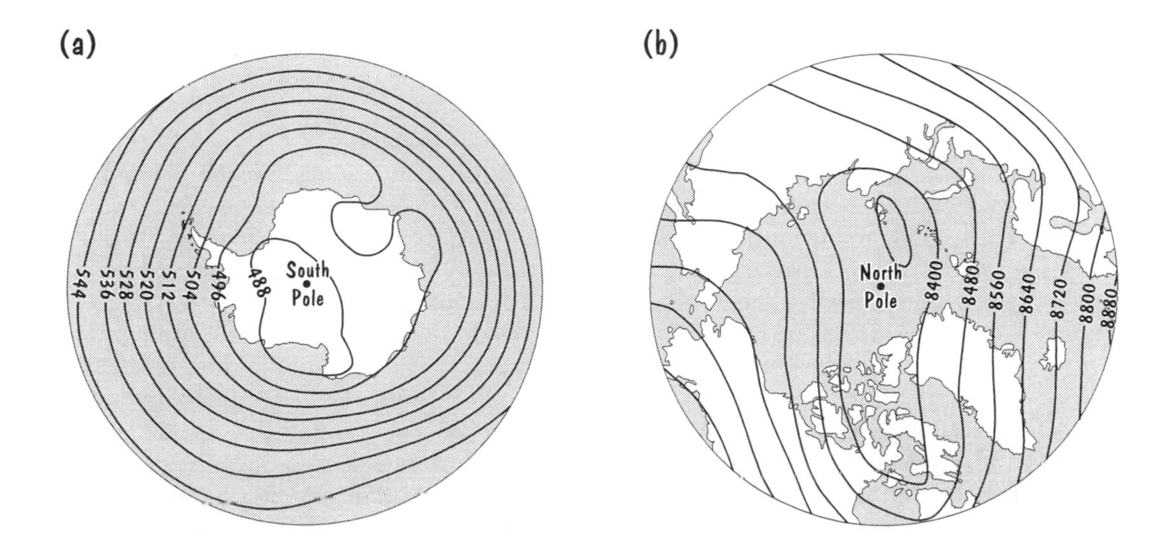

Figure 7.3 Present day atmospheric configuration over the Antarctic and the Arctic. (a) Mean height of the 500 mbar level across the Antarctic. (b) Mean height of the 300 mbar level over the Arctic. Adapted from Sugden (1982).

regime within the Northern Hemisphere, and observe how this varies when one of several climate forcing operators (such as solar insolation changes) take effect. If the Earth had uniform surface, each pole would have an atmospheric vortex above it, surrounded by a zone of permanent westerly winds. This is due to the difference of solar heat to the poles compared with that from the equator, and the rotation of the planet.

- The temperature difference results in denser air over the poles, causing a 'bowl' formation within the lines of equal density.
- As the air flows down gradient it is deflected by the rotation of the Earth so that it flows roughly parallel to the contours.
- Thus air circulates anti-clockwise in the Northern Hemisphere and clockwise in the Southern.

Figure 7.3 indicates the vortex at each pole. Note that the Southern Hemisphere vortex is relatively symmetrical, whilst the Northern, due to the surrounding irregular land masses, is asymmetrical. Note also that the height of the isobar is sufficiently high as to eliminate effects from the ground topography. Thus, the distribution of land influences the modern atmospheric circulation at and around the poles. Subsequently, we can speculate that an increase in Antarctic ice volume would not result in much change to the atmospheric circulation in this region. However, distribution of ice masses in the Northern Hemisphere would have a large influence on the atmosphere of the Northern Hemisphere. In the Arctic the alternation between ocean and landmasses breaks up the continuity and strength of the westerlies, such that a more meridianal circulation occurs. Cooling of the continents in the winter introduce high pressure regions. These act to channel relatively warm North Atlantic air toward the pole, away from the continents.

In the Antarctic, the temperatures are low all year around. However, in the Arctic, temperatures fluctuate significantly between locations and seasons. To model the palaeoclimate, we simply have to be able to model accurately the modern climate over the poles, and then adjust in accordance with the model's forcing functions. However, although several climate models exist, they can yield dissimilar answers. The problem can be traced to two areas. The first is that models can behave differently to each other. The second, and the source for most error, is that there is no consensus as to the true nature of glacial-interglacial climate forcing functions that climate models can use.

CLIMATE MODELS OF THE LAST GLACIAL MAXIMUM WITHIN THE NORTHERN HEMISPHERE

CLIMAP

To illustrate the type of influence large-scale Northern Hemisphere glaciation would have on atmospheric circulation, and therefore climate, we will examine a simple GCM model of the glacial climate. This model is based on two assumptions: (a) distribution of Northern Hemisphere ice-masses is taken as that suggested by CLIMAP (1976); (b) sea surface temperatures (needed for the model input) are as those detailed by CLIMAP (from measurements of planktonic forams with sea cores). When these assumptions are used as model input, together with the appropriate radiative conditions, at the LGM, several important features are calculated (Gates, 1976a,b) as follows:

1. Ice-age evaporation and precipitation were, on average, 15 per cent lower than present July values.
2. July air temperatures were 10–15°C lower over large ice masses than over the southern limits of the ice sheets.
3. Atmospheric circulation was dominated by enhanced west–east air flow and a reduction of the meridianal north–south circulation (i.e. lack of warm air transported northwards over the Atlantic).
4. Climatology of North America was dominated by high-pressure anti-cyclone flow around the Laurentide Ice Sheet, the surface temperature being 20–30°C lower than at present.
5. Similar anticyclones existed around Eurasian and Greenland ice masses.
6. Upper tropospheric airflow was spilt around the Laurentide Ice Sheet and Eurasian Ice Sheet.
7. A dramatic increase in precipitation occurred across the mid-latitudes, from south-western United States to western Europe. This rainfall did not act to raise the level of lakes in these regions (from geological information), implying that a high rate of run-off occurred in these regions.
8. Very dry anticyclones existed over the Arctic Ocean; dry because of the permanent sea ice cover which did not allow evaporation of sea water.

In order to identify the seeding ground of former ice sheets, it is important to know where the source of moisture for snowfall comes from. For example, the Laurentide Ice Sheet in North America was supplied by moisture from the north-west Atlantic, and the Cordilleran Ice Sheet over the northern Rocky Mountains was 'fed' by precipitation from the Pacific. The Greenland and Eurasian ice sheets both had a supply of snow originating from the North Atlantic. In the Southern Hemisphere, the southern Pacific maintained a moisture supply over ice caps across Patagonia and South Island, New Zealand. One thing to note at this point is that the growth of ice sheets is also affected by the prevailing wind directions (usually westerly) that carry the snow-laden clouds across glacierised regions. Across Antarctica, the supply of precipitation comes from the Pacific, Indian and Atlantic oceans that surround the continent. Precipitation will decrease with increase in distance from the source. Thus, regions that are distal to these moisture sources are likely to be 'dry'. Because

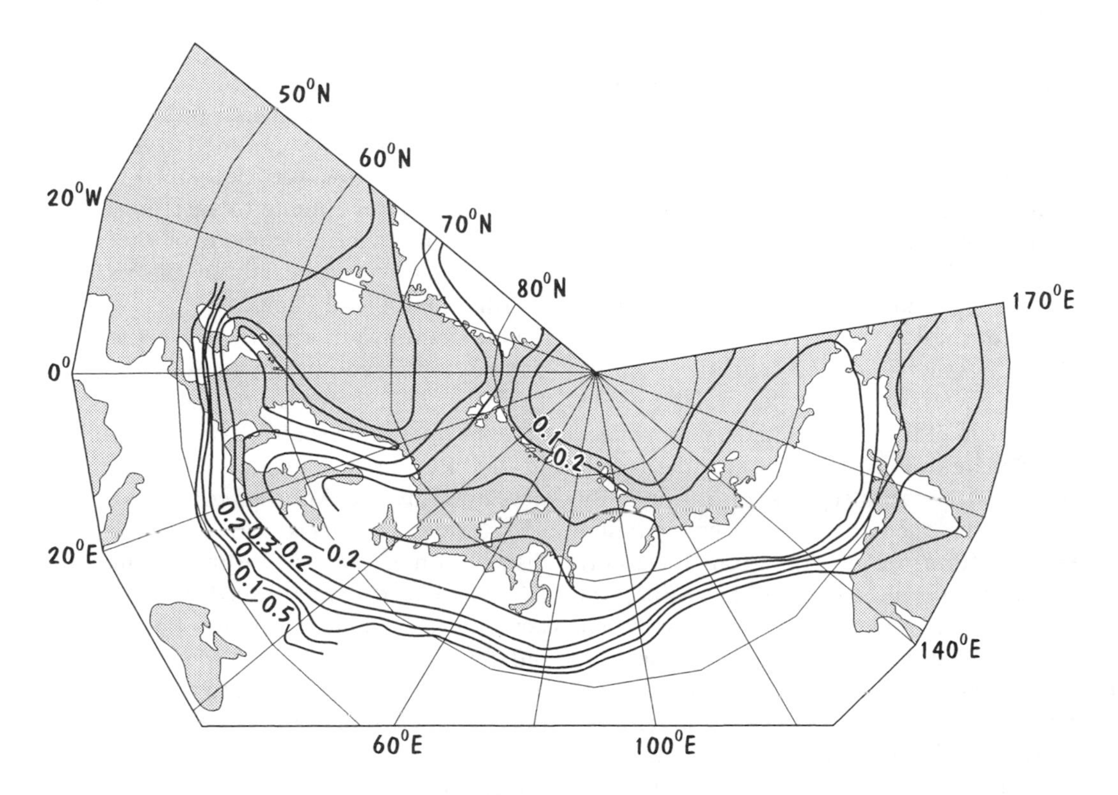

Figure 7.4 Reconstruction of the LGM accumulation of ice across the Eurasian High Arctic according to Hughes (1985). Adapted from Lindstrom (1989).

of this, as we shall see in later chapters, ice sheet growth is critically affected by geographical position.

GCM modelling

Eemian interglacial

GCM models of the last interglacial (the Eemian at 125 000 years ago) have shown that, in general, air temperature conditions were warmer than at present (Montoya et al., 1998; Kukla, 2000). Specifically, this warming has been calculated in the summer time at a global level. However, GCM models have indicated that winter cooling, compared with present day conditions, may have occurred over northern Africa and across continental Asia. As will be seen later, summer warming may be responsible for an Eemian sea level greater than that at present (Chapters 9 and 10).

Last glacial

Since the CLIMAP reconstructions of LGM palaeoclimate (Figure 7.4), GCM models have become more sophisticated. They can now be coupled with (at least) simple ocean models, and there is a lot more climate information from geological sources against which to test model results.

The Laurentide Ice Sheet

Kageyama and Valdes (2000) calculated the effect of the orography of the Late Wisconsin ice sheet over North America (the Laurentide ice sheet – Chapter 12), on the climate at the LGM. They used an Atmospheric GCM with ice sheet dimensions described in Peltier (1994) (Chapter 12). This experiment was referred to as their 'LGM' run. An additional experiment was undertaken where full LGM conditions were accounted for but without any Laurentide ice sheet; named 'NL' for 'no Laurentide'. By comparing the results of these two experiments, the effect of the Laurentide Ice Sheet topography on atmospheric flow and climate could be evaluated. Kageyama and Valdes (2000) found that the height of the ice sheet had a major effect on the climate over North America, the North Atlantic and western regions of Europe. By subtracting the calculated rates of precipitation determined from the NL run from the LGM run, the influence of the Laurentide Ice Sheet on precipitation was identified. This showed that an ice sheet that was 'flatter' than described in Peltier (1994) would cause an increase in precipitation over (i) northern Canada, (ii) the northern North Atlantic and, (iii) western Europe. They suggest that the adaptation of precipitation due to the shape of the Laurentide Ice Sheet means that it influences the growth rate of the Fennoscandian Ice Sheet and, in this way, the Laurentide and European ice sheets are linked (Kageyama and Valdes, 2000).

Effect of sea-surface temperatures on the Eurasian Ice Sheet

Marsiat and Valdes (in press) recently employed an Atmospheric GCM to investigate the role of sea surface temperatures (SSTs) on the LGM climate. Their work was necessary because of the apparent discrepancy between recently established SSTs for the northern Atlantic and Pacific oceans, and SSTs derived from the CLIMAP experiments in 1981. Specifically, there is a large marine geological database to show that the North Atlantic was several degrees warmer than that depicted in the CLIMAP reconstruction. Conversely, the northern Pacific is most likely to be colder than the CLIMAP analysis indicates.

The Atmospheric GCM used by Marsiat and Valdes (in press) was based on the European Medium Range Weather Forecasting model coupled with a slab-ocean model. A number of modelling scenarios were developed to examine the variability of climate to variations in SSTs. They found that the climate of the Eurasian Arctic is influenced heavily by the SSTs in the North Atlantic and Pacific.

One significant feature was that, under oceanographic conditions representative of the LGM, a cold temperature anomaly appeared across western Siberia from the Ural Mountains to the Putorana Plateau. The palaeoclimate developed by the Atmospheric GCM (Marsiat and Valdes, in press) is characterised by a distribution of temperature and accumulation of ice similar to that established from an independent ice sheet model (Figure 7.5). The major difference is that the AGCM model indicates even colder conditions across the Kara Sea (mean annual temperature less than –30°C). The model of Marsiat and Valdes (in press) provides meteorological information on the LGM palaeoclimate. For example, the cooling of surface air temperatures across western Siberia extends up into the atmosphere, causing a temperature inversion during winter. There is a reduction in the atmospheric water content and a decrease in winter precipitation. The model indicates very little variations in the mean–annual cloud cover over Siberia. However, as cooling extends into the atmosphere, changes occur to the structure of the cloudiness due to an alteration in the advection of heat, associated with a reorganisation of atmospheric circulation. The traditional large-scale atmospheric circulation across the region is characterised by a band of strong westerlies which cross the Atlantic. However, Marsiat and Valdes (in press) show that the low-level winds have a stronger northward component across the western margin of Eurasia, and a split in the wind direction occurs. One arm of this warm jet is directed north and a weaker branch flows on to the continent. This latter flow of air causes west Siberia to be influenced by cold winds from

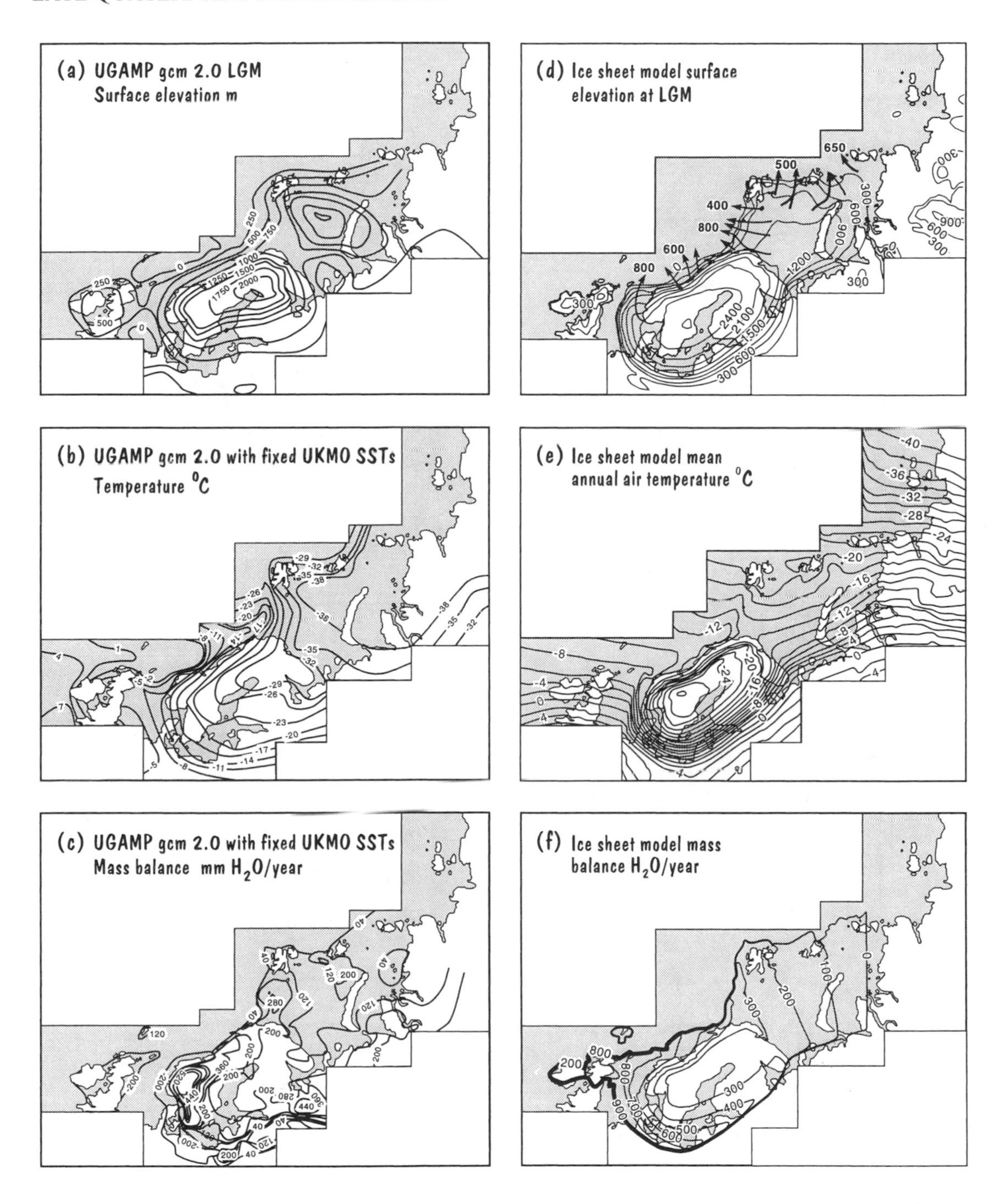

Figure 7.5 (a–c) AGCM palaeoclimate results. (a) Ice sheet topography (contours every 250 m). (b) Mean annual temperature (°C). (c) Annual rate of ice accumulation (mm yr^{-1}). These results are extracted from the output of a global AGCM model, adapted from Marsiat and Valdes, (*in press*). (d–f) Ice sheet model results. (d) Ice sheet topography (contours every 250 m). (e) Mean annual temperature (°C). (f) Annual rate of ice accumulation (mm yr^{-1}), adapted from Siegert et al. (1999).

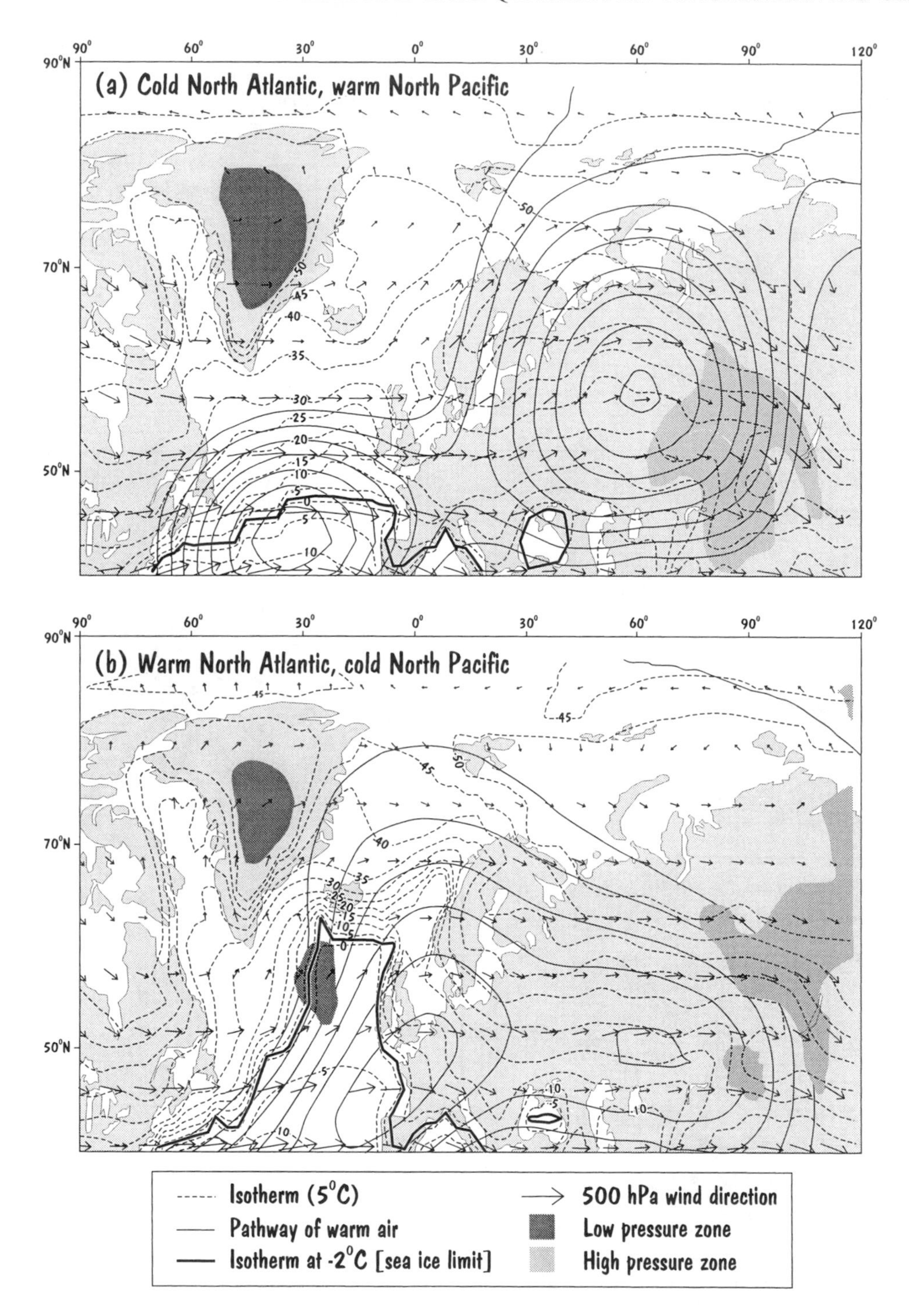

Figure 7.6 Atmospheric conditions across the Eurasian Arctic at the LGM for (a) the CLIMAP reconstruction and (b) recent AGCM results. From information supplied by I. Marsiat (University of Reading, England).

the Fennoscandian ice sheet. Marsiat and Valdes (in press) showed that variations to the SSTs in the northern Atlantic influenced only western Europe. However, variations in the Pacific Ocean SSTs caused climate change as far as 10°E. Therefore, the effect of the Pacific Ocean appears to dominate the climate across the Eurasian Arctic. Details of the global palaeoclimate established by Marsiat and Valdes is provided in Figure 7.6. These results show how recent advances in climate modelling technology and proxy climate records have allowed the CLIMAP LGM climate to become updated.

LGM GLOBAL CLIMATE TRANSECT

A recent international programme organised to establish past climate scenarios (called PAGES: PAst Global changES) has resulted in the collection and collation of vast quantities of information about the LGM climate. One of the aims of PAGES is to produce a number of transects across the planet showing how the climate varied at the LGM. One of these transects is aligned parallel to a previous study of LGM ELA from the North to South poles via the west coast of America (e.g. Broecker and Denton, 1990; Markgraf et al., 2000). The ELA (a function of mean annual air temperature) is shown in Figure 7.7 to be depressed by several hundred metres at the LGM, exposing mountain ranges across North and South America to the ice accumulation zone. Hence, glaciers are likely to build up in these regions. However, glacier growth will not only be affected by the ELA, but also by the supply of moisture to form snow and, hence, ice. Markgraf et al. (2000) have integrated a variety of palaeoclimate records (such as limnological records and lacustrine conditions) that indicate information on past atmospheric moisture conditions, between the LGM and early Holocene. The palaeoclimate records show that the Arctic and Antarctic climates were characterised by dry cold conditions at the LGM, with more moisture available during the Holocene. Western North America experienced potentially higher moisture conditions at the LGM than in the Holocene; a pattern that is reversed across the northern part of South America, where the LGM is quite dry. Across the southern part of South America, the LGM is typified by moist conditions in contrast to the Holocene when a dry climate prevailed. These palaeoclimate and ELA reconstructions provide the environmental background to understanding the glaciation of North and South America, and are referred to later in Chapters 12 and 13.

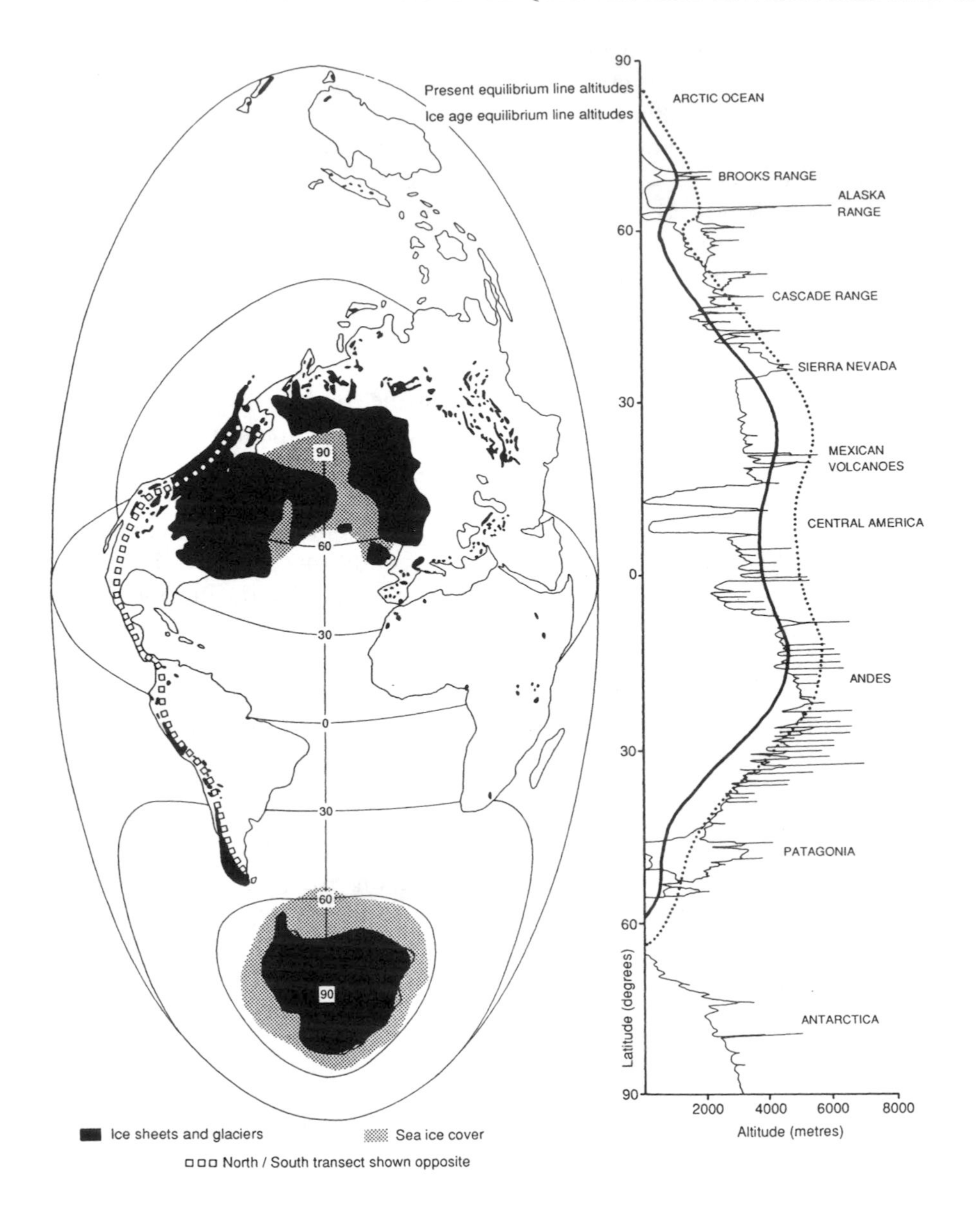

Figure 7.7 Position of the ELA across a transect from the Arctic to the Antarctic at the LGM (dotted line) and present day (dashed line). Also shown is the location of ice sheets according to CLIMAP. Taken from Dawson (1992) after Broecker and Denton (1990).

CHAPTER 8

Late Quaternary Palaeoceanography

INTRODUCTION

Information regarding the ice age climate has thus far focused on atmospheric circulation problems with little consideration of how the ocean may be involved (except for the recent work by Marsiat and Valdes where an ocean model and atmospheric model have been coupled). Knowledge about the oceans is important for a number of reasons:

1. Oceans are capable of transporting heat-from mid-Atlantic to the North Atlantic.
2. Ocean thermal conditions are one of several factors affecting the growth of sea ice.
3. Ocean thermal conditions affect the evaporation rate, and thus are a control of moisture availability to ice sheets.
4. Ocean temperatures are a control on the rate of ice shelf melting, and possibly iceberg calving. Therefore, ocean conditions may play an important role in terms of ice sheet mass balance.
5. Eustatic sea level change is directly involved in forcing the growth and decay of marine ice sheets.

It is, therefore, essential to note that palaeoclimate and palaeoceanography are inextricably linked (Pinot et al., 1999, Marsiat and Valdes, in press).

OCEAN INFLUENCE ON CLIMATE: PRESENT DAY

Ocean flow has two components: surface currents and thermohaline circulation.

Surface currents

Ocean surface currents are controlled by the direction of the prevailing winds. Therefore, they are driven by: (i) Earth rotation. The prevailing direction of ocean currents is west to east. Thus, the ocean basins have higher water levels on their eastern sides than their western sides. A good example of this is the Panama Canal problem, where the sea on the Pacific Ocean side of the canal is several metres higher than on the Atlantic Ocean side. However, since ocean currents cannot penetrate through land masses their distribution is also governed by: (ii) Distribution of land masses. Because of this boundary condition, the dominant pattern of ocean surface currents is a gyre system. The effect of Earth rotation is to create fast flowing currents from west–east, and slow diffuse currents in return. A good example of a fast west–east current is the Gulf Stream in the North Atlantic. Because the currents are driven by wind directions, changes in the wind motion will effect changes in surface water currents.

Oceanic conveyor belt

In recent years there have been several advances in the understanding of the global flow of oceans. The relevant points of the ocean circulation system are as follows.

- Oceanic convection is driven from the polar regions.
- Cold, salty water sinks to depth and moves toward the opposite poles. There are two main locations in the North Atlantic where

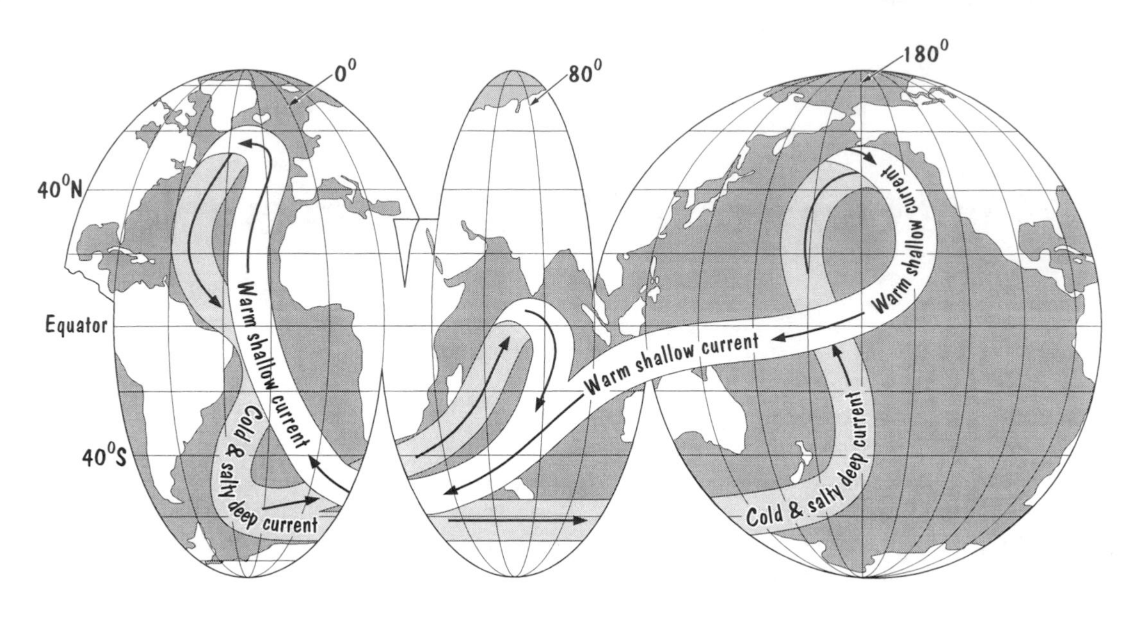

Figure 8.1 The global ocean conveyor system. Adapted from Schmitz (1995), reprinted from *Reviews of Geophysics*, with permission of American Geophysical Union.

this happens: the Labrador Sea between Greenland and America (known as the Boreal region) and in the Norwegian–Greenland Sea (the Nordic region) (Imbrie et al., 1992).

- The processes involved are evaporation of sea water, and formation of sea ice (e.g., the formation of the Odden Ice Tongue in the Norwegian–Greenland Sea), both causing dense salty water to develop at the upper boundary, which then sinks to form 'deep water'.
- At present, North Atlantic Deep Water (NADW) overcomes the opposing influence of Antarctic Bottom Water (AABW) because a large proportion of the AABW is held within the clockwise rotation around the Antarctic continent.
- NADW is replaced by warm mid latitude surface water, which flows northwards.
- The result is known as the ocean conveyor belt: which possesses a flux of water greater than 20 times that of all the rivers combined (Figure 8.1).
- The pattern of ocean circulation is self sustaining: warm water in the North Atlantic will preferentially evaporate, causing dense salty water to sink and flow back southwards.
- Up-welling of NADW eventually occurs in the Indian and North Pacific oceans.
- This system provides the northern latitudes of Europe with approximately one-third of the heat derived from the Sun.
- However, the ocean conveyor could conceivably exist in an opposite mode, resulting in cold water to the North Atlantic and the denial of this huge amount of heat (the land around the North Atlantic would be cooler by 6°C).
- In addition, a sudden influx of fresh water to the North Atlantic may cause the circulation to halt (albeit for a limited period). The Greenland ice cores showed rapid and significant fluctuations in the local climate during the last deglaciation, which may be associated with the release of melt-water and the ocean circulation.

OCEAN INFLUENCE ON CLIMATE DURING THE LAST GLACIATION

Surface ocean currents under a CLIMAP-type glacial scenario

The most important current during the present day is the Gulf Stream. However, from the CLIMAP reconstruction of ice masses, the LGM Polar front existed somewhere to the south of its present location. This configuration would have eliminated North Atlantic Drift flow into the Norwegian Sea. Thus, under the CLIMAP reconstruction, the glacial North Atlantic was characterised by cold water, overlain with multi-year sea ice, with no Gulf Stream input (Figure 8.2). However, recent analysis of sea floor sediments from the north-east Atlantic and Norwegian Sea show this scenario to be incorrect north of the British Isles.

Surface currents under a non-CLIMAP-type scenario

The CLIMAP programme was performed over 20 years ago. Since then, much more information has been gathered concerning the palaeoceanography of the Atlantic. To reiterate, the most important region of the oceans during the LGM was the North Atlantic. If the CLIMAP situation were correct, deep sea cores should indicate permanent cold surface water conditions within the lack of coccolithic forams in sea floor sediments. However, cores from Fram Strait indicate that coccoliths (warm surface water forams) were present within two distinct periods (27–22.5 and 19–14.5 ka ago). During these intervals, warm waters existed within the Greenland–Norwegian seas (Figure 8.3). But, what was the origin of this warm water? The answer comes from analysis of sea floor sediments in the Norwegian Sea. IRD chalk fragments have been found within the Fram Strait cores (e.g. Hebbeln et al., 1994. Figure 3). This chalk originates from south of 59°N. Thus, these fragments are transported north into Fram Strait, by icebergs calved off the Scandinavian Ice Sheet. Iceberg drift is controlled mainly by ocean currents. Therefore, during the two periods depicted by coccolith concentrations in the Fram Strait record, a warm northward travelling current was active, providing relatively warm, seasonally open conditions to the Glacial North Atlantic (Figure 8.4). Lassen et al. (1999) added to the idea of an active, warm north-east Atlantic during the LGM, through analysis of sea floor sediment cores from the Faeroe–Shetland gateway. They showed that warm Atlantic surface water was channelled northwards through this gateway to the eastern margin of the Norwegian Sea (Figure 8.5).

So, what are the consequences of this oceanic action? A local thermohaline circulation may have been set up within the GIN seas during two phases in the last glaciation. Seasonally open waters may have provided a local moisture source for precipitation over Scandinavia and the Barents Sea. From this work, and the discussion at the end of the last Chapter, there are at least two major problems with the CLIMAP ocean reconstruction. First, the SSTs in the Norwegian Sea are too cold and secondly, the SSTs in the Pacific are too warm.

OCEAN CONVEYOR CIRCULATION: A GLACIAL CYCLE PREDICTION

There are clear records showing that ocean conditions are related to the general patterns of global climate change (Figure 8.6). However, it is only relatively recently that the mechanisms that control ocean circulation and, in turn, the climate have been established. In order to determine how ocean conditions changed during the last glaciation palaeoceanographic records are required, allowing an assessment of how these mechanisms altered over time. At present, the NADW dominates the Atlantic Ocean circulation. During the last ice age the

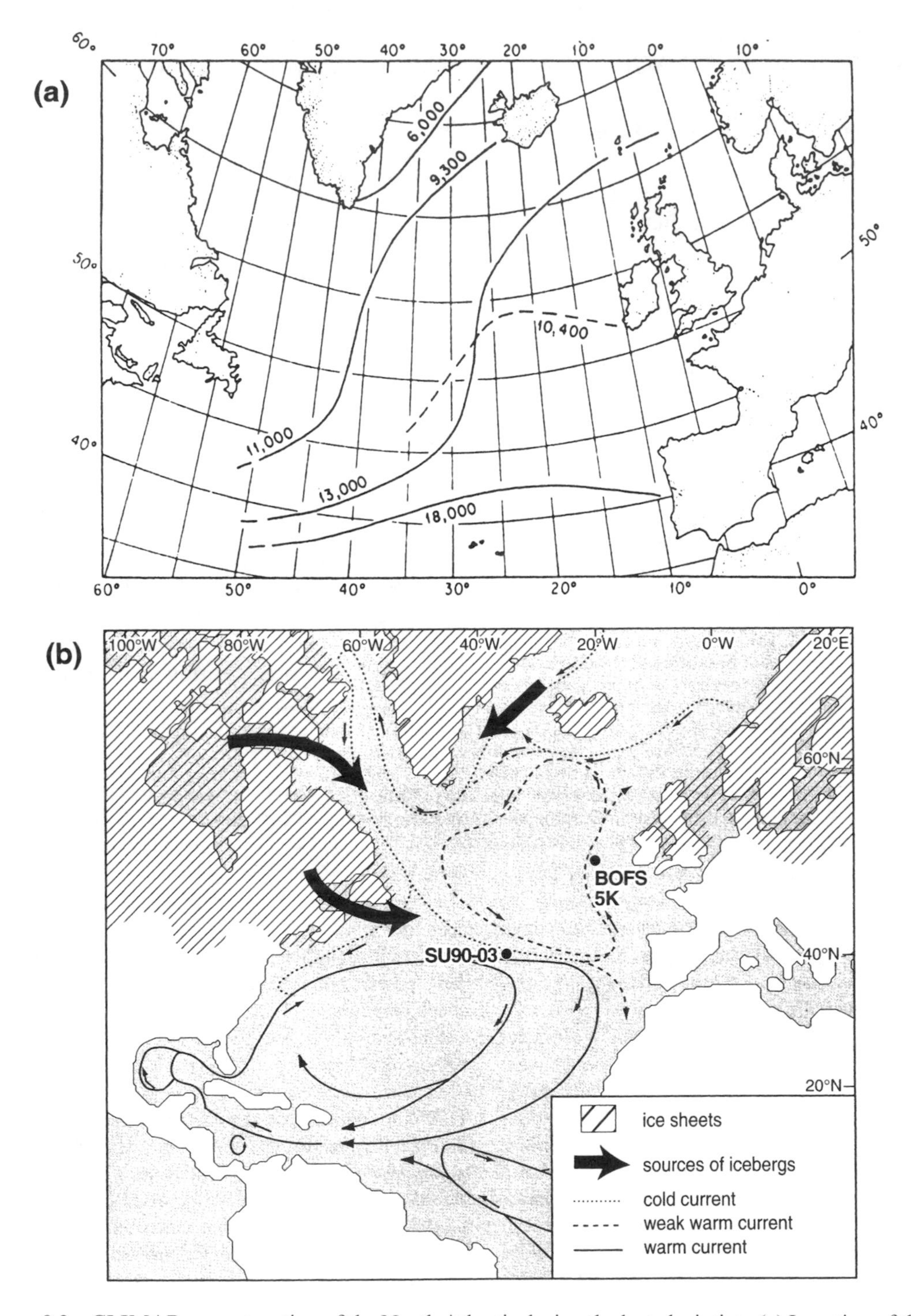

Figure 8.2 CLIMAP reconstruction of the North Atlantic during the last glaciation. (a) Location of the North Atlantic Polar Front between the LGM and mid-Holocene. (b) North Atlantic surface currents at the LGM. Taken from Chapman and Maslin (1999), reproduced with permission of Geological Society of America.

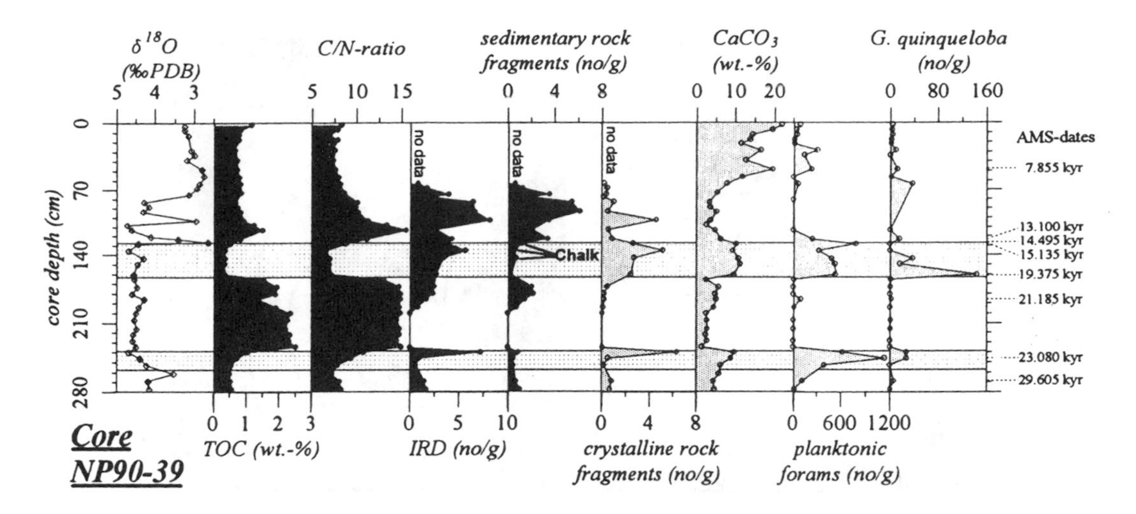

Figure 8.3 Palaeoceanographic information from a sea floor sediment core (NP90-39) taken from the Fram Strait (Hebbeln et al., 1994). Locations of this core site (and two other comparable cores cited in Hebbeln et al., 1994) are provided in Figure 8.4. Reprinted with permission from *Nature* vol 370, pp. 357–360, copyright (1994) Macmillan Magazines Limited.

overall pattern of the ocean conveying system was similar to that at present, but driven by different mechanisms. The differences occur within the method of attaining warm to cold water conversion in the North Atlantic. There are two similar mechanisms by which warm-cold water conversion may occur: (i) the open ocean convection within the Boreal North Atlantic/Labrador Sea (Boreal Heat Pump), and (ii) overflow of water from the Atlantic into the GIN seas (Nordic Heat Pump).

Imbrie et al. (1992) identified four end-member states, in which the heat transport to the North Atlantic is enhanced or restricted due to the turning on and off of these two heat pumps across a glacial–interglacial cycle (Figure 8.7). These states are detailed below, with some additional oceanography relating to the LGM conditions.

Interglacial state

It is assumed that the Holocene is a good example of this condition, although atmospheric conditions may have been even warmer in previous occasions (e.g. Montoya et al. 1998). Northern ice sheets are at their minimum extent and the climate is at a humid extreme. The terrestrial biomass is at a maximum and the northern Atlantic sea ice field is at a minimum. The salinity of surface waters in the GIN seas is also at a maximum. The warm to cold water conversion in the North Atlantic has two branches (the Nordic and Boreal heat pumps). However, the Nordic heat pump is more effective than the Boreal heat pump at this stage. Warm water passed to the southern Atlantic causes the Antarctic Ice Sheet to remain at its modern limits. The sea ice field and resulting latitudinal temperature gradient interact with the atmosphere and ocean to produce the modern pattern of winds, ocean fronts and plankton communities.

Preglacial state

In response to the decrease in Northern Hemisphere solar radiation, the atmosphere and surface ocean in the Arctic cool, evaporation decreases and the fields of snow and sea ice

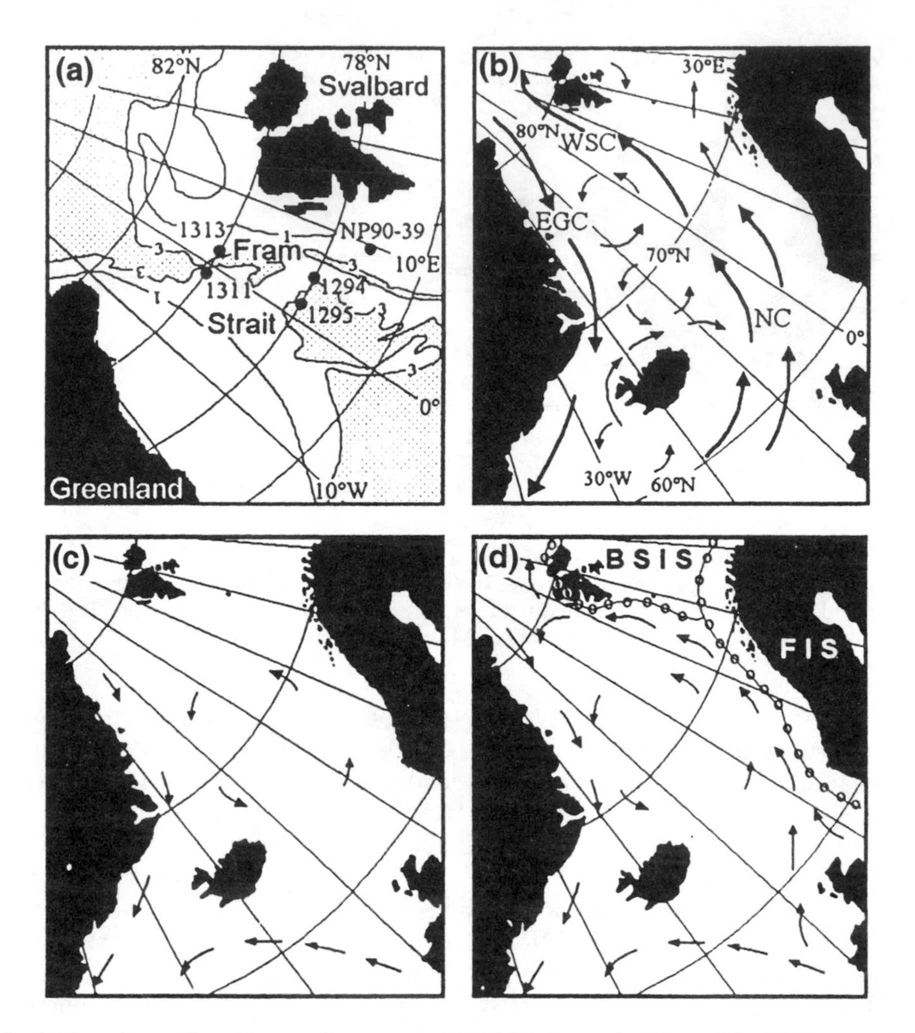

Figure 8.4 (a) Location of sea floor sediment core NP90-39 shown in Figure 8.3. (b) Present-day pattern of surface currents in the Norwegian–Greenland Sea. (c) CLIMAP reconstruction of surface currents at the LGM. (d) Reconstructions of surface currents adapted from (c) by information provided in Figure 8.3. Reprinted with permission from *Nature* vol 370, pp. 357–360, copyright (1994) Macmillan Magazines Limited.

expand. Because the ice sheets have not grown yet, no significant sea level reduction occurs. The lack of evaporation results in a freshening of the Arctic waters which stops the Nordic heat pump and, in turn, halts the relatively warm waters from flowing into the Antarctic. The boreal heat pump and associated warm water convection to the Antarctic is unaffected. There is an increase, in relative terms, of northward transport of AABW.

Figure 8.5 Surface currents along the Faeroe–Shetland Channel at (a) Present-day and (b) the LGM. Taken from Lassen et al. (1999), reprinted from *Paleoceanography*, with permission of American Geophysical Union.

(a)

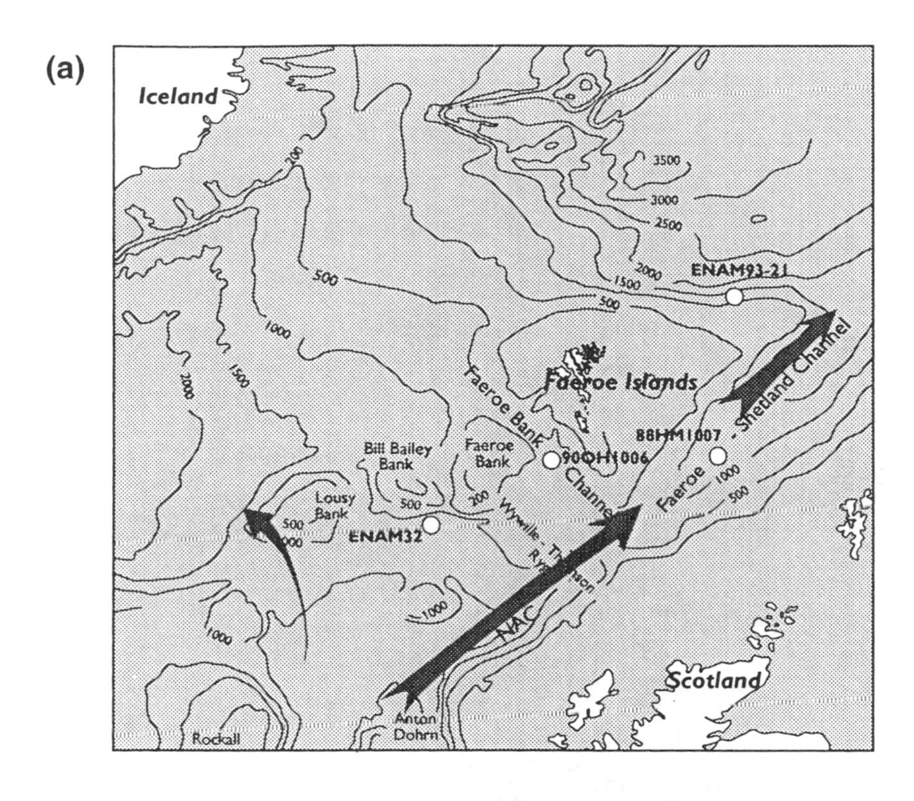
Iceland
Faeroe Islands
Faeroe Bank Channel
Faeroe - Shetland Channel
Bill Bailey Bank
Faeroe Bank
Lousy Bank
Wyville - Thomson Ridge
NAC
Anton Dohrn
Rockall
Scotland
ENAM93-21
88HM1007
90OH1006
ENAM32

(b)

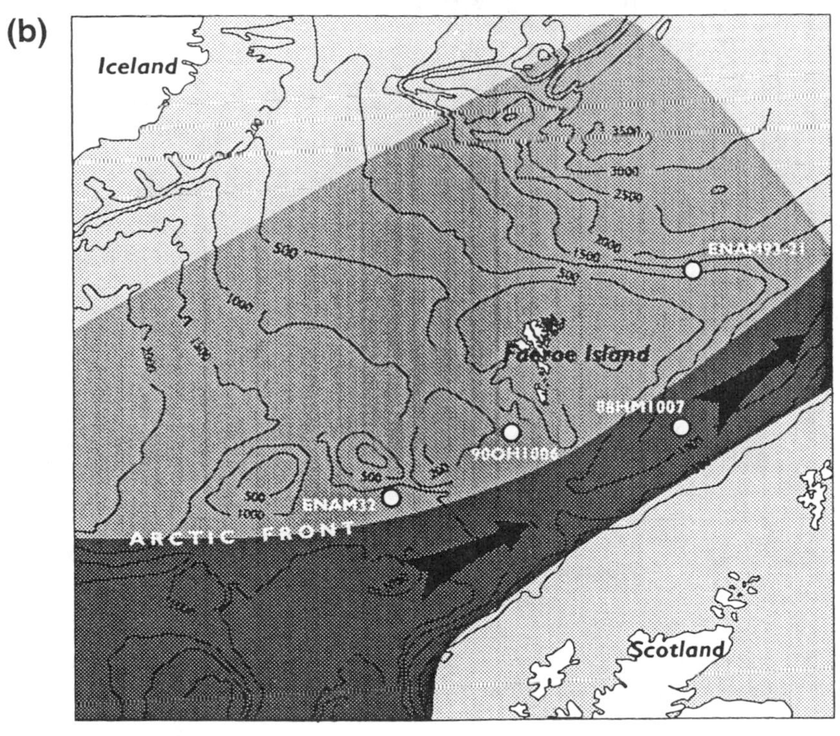
Iceland
Faeroe Island
Scotland
ARCTIC FRONT
ENAM93-21
88HM1007
90OH1006
ENAM32

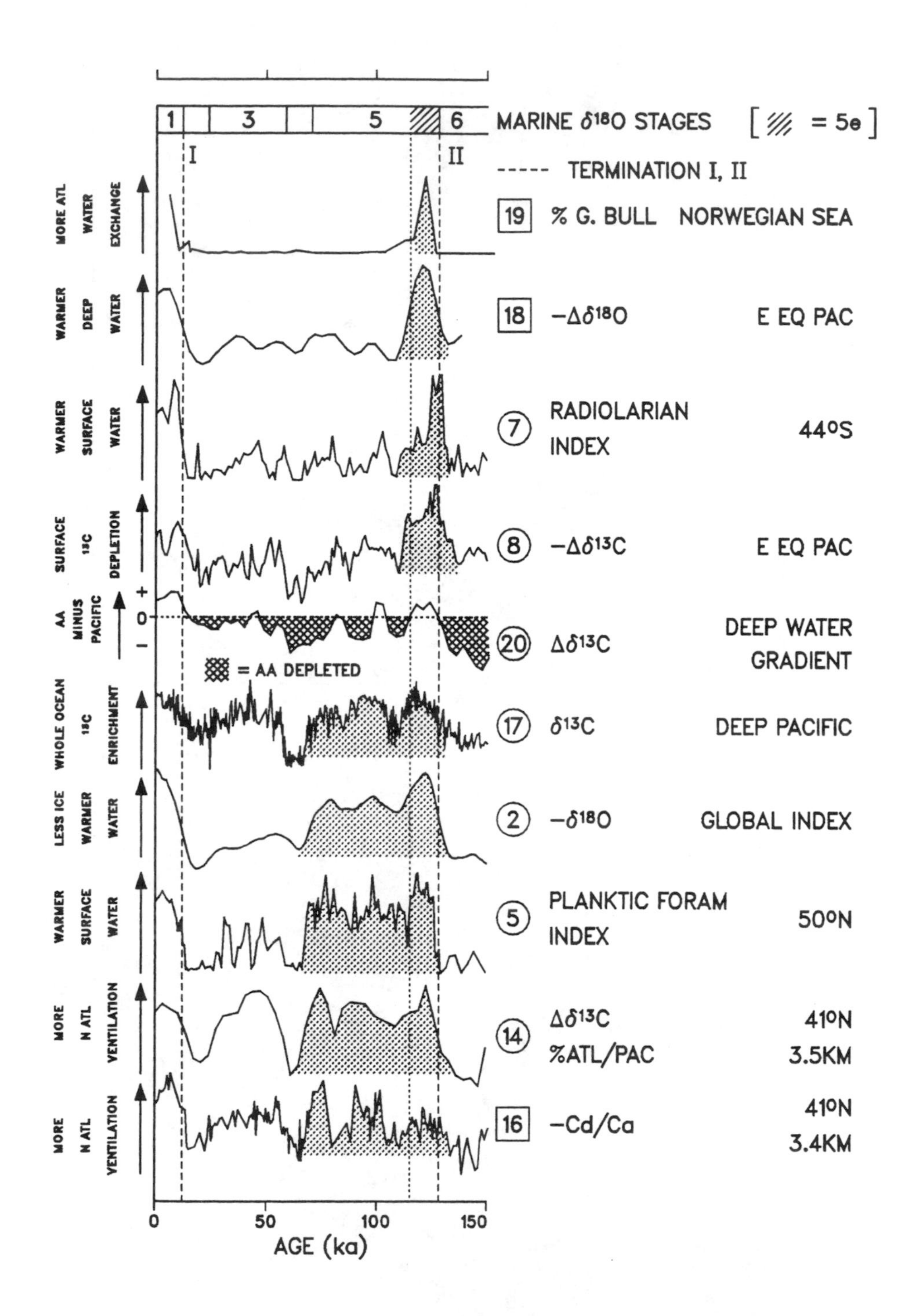

Figure 8.6 Global climate change recorded in several oceanic records. The type and location of the record is highlighted on the right of the figure. Taken from Imbrie et al. (1992), reprinted from *Paleoceanography*, with permission of American Geophysical Union.

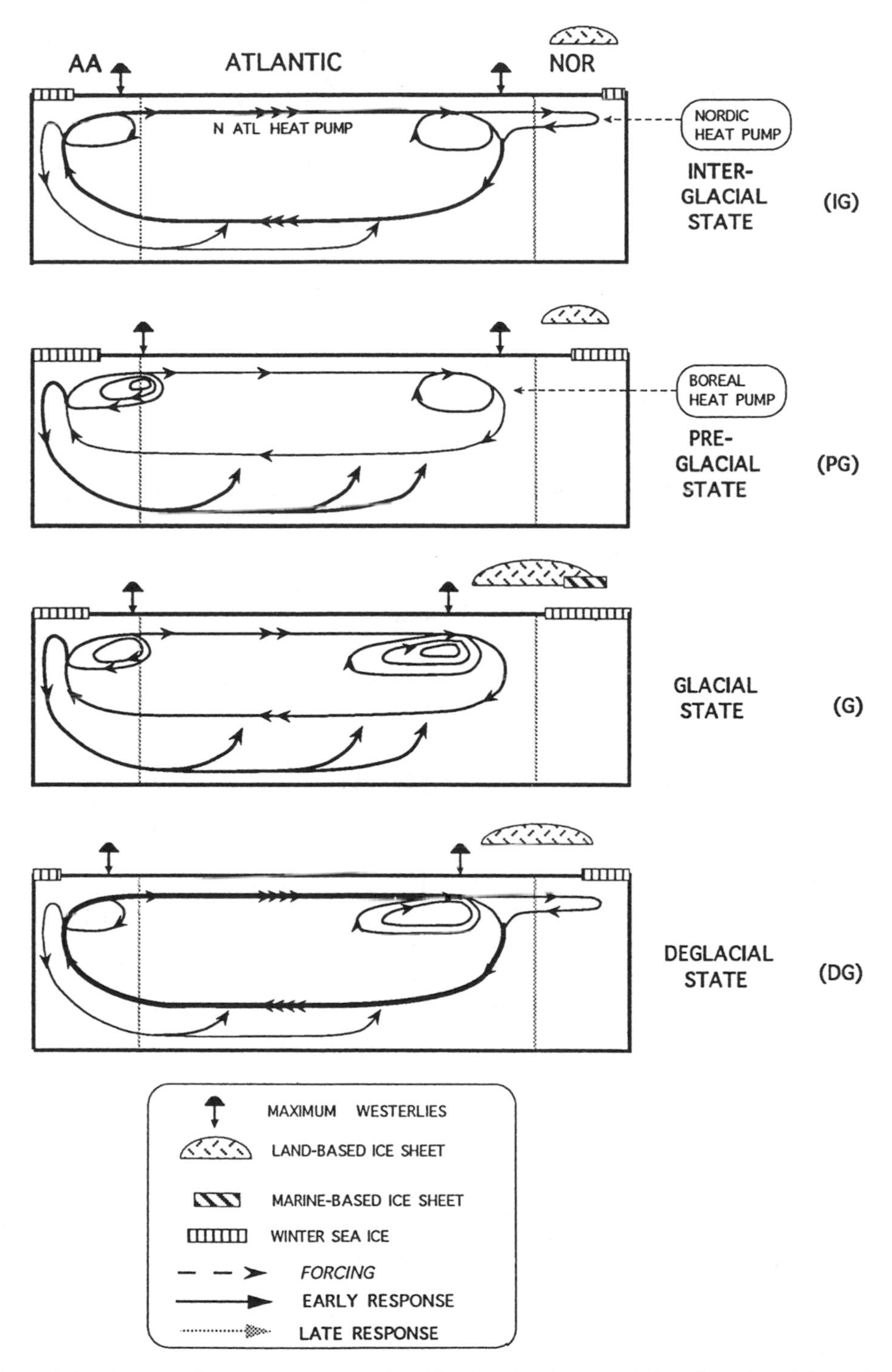

Figure 8.7 Atlantic oceanic processes operating during the four periods of a glacial-interglacial cycle (Interglacial, preglacial, glacial and deglacial states). Taken from Imbrie et al. (1992), reprinted from *Paleoceanography*, with permission of American Geophysical Union.

Glacial state

As the Northern Hemisphere ice sheets grow they exert an ever-stronger control on the position, speed, temperature and water content of winds. When the ice sheets attain their maximum size, the area that is influenced by cold dry winds also reaches a maximum. Eustatic sea level lowering of around 120–135 metres is attained. In the Boreal Atlantic, the ocean's response is to expand the area where winter time convection occurs due to the annual growth of sea ice conditions. This yields a modest increase in the heat export by NADW compared with the preglacial condition. This, in turn, leads to a temperature increase in Antarctic waters and, hence a reduction in Antarctic sea ice compared with the preglacial state. However, there are two recent pieces of evidence that call into question the increase in NADW proposed by Imbrie et al. (1992). The first is that geochemical investigations of sea floor sediments reveal that NADW formation may have been reduced at the LGM (Rutberg et al., 2000). The second is evidence from South Atlantic sediment cores to suggest that the LGM Antarctic sea-ice extent was greater than at present. Ancient diatoms in sea floor sediments provide detailed information about the ocean environment at the LGM. Because diatomaceous accumulation on the ocean floor alters by rate and species when open ocean conditions are replaced with a sea ice cover, sea ice extent can be reconstructed. Compilation of diatom data in recent years confirms that the LGM sea ice extent was much greater than at present, especially in the winter period (Armand, 2000). This would have had major implications for the availability of a moisture source for precipitation over the Antarctic Ice Sheet at this time. Unfortunately at present there is no consensus on how an expanded sea ice cover in the North Atlantic is compatible with a reduced NADW formation.

Deglacial state

Increase in summer insolation causes warming of atmosphere and oceans, and the snow fields to shrink. In the Nordic Sea, the permanent sea ice recedes. In the Boreal latitudes, the ELA increases, causing the Laurentide Ice Sheet to melt in the south. The effect on the wind field is to re-establish the Gulf Stream, and transport warm saline waters northward, increasing the exchange of waters within the Nordic regions. This heat helps to warm the atmosphere that, in turn, leads to the melting of ice across Scandinavia and Eurasia. This deglaciation induces sea level rise and, hence, the decay of marine ice sheets and then terrestrial ice sheets in the extreme north. The largest ice sheet for decay was the Laurentide ice sheet, however, the most significant decay would have been due to the marine ice sheets (e.g., Barents Sea Ice Sheet). The extra heat in the ocean increases evaporation and thus induces the Nordic Heat Pump (in addition to the boreal heat pump). Thus heat export to the south reaches a maximum. This induces further melting of sea ice in the Antarctic, and a return to modern-type conditions.

SHORT-TERM CHANGES IN OCEAN CONDITIONS AND CLIMATE

The GRIP and GISP2 cores, through over 3000 metres of the Greenland Ice Sheet, cover time intervals back to the last, or Eemian, interglacial. The two cores, drilled about 30 kilometres apart close to the summit of the Greenland Ice Sheet, yield a very similar palaeoenvironmental picture of the last 90 000 years or so (GRIP Project Members, 1993; Grootes et al., 1993). A series of relatively colder stadials and warmer interstadials are seen during the last glacial period (Johnsen et al., 1992, 1995) (Figure 7.2). A number of major fluctuations in oxygen isotope ratios and electrical conductivity suggest mean annual temperature variations of 10 to 15°C, with shifts taking place rapidly over time periods on the order of decades. A doubling of accumulation at the beginning of the present interglacial also appears to have taken place over as little as three years (Alley et al., 1993). The ice-core records, with their very high temporal resolution, therefore show that climate during the last deglaciation has been particularly variable and that alterations between warmer and colder, and wetter and drier conditions, have been very rapid. By contrast, the isotopic records from the Greenland ice cores demonstrate that climate variability during the present interglacial has been subdued by comparison, with changes of only a few degrees at most.

The detailed and rapid climatic variability recorded in the GISP2 Greenland ice core has been compared with isotopic records of climatic and ocean circulation changes in marine sediment cores from deep North Atlantic waters (Alley and Clark, 1999). There is a marked similarity in the timing and rapidity of change in these records (Bond et al., 1993). It has been proposed that the changes recorded in the Greenland ice cores are related to shifts in the pattern of thermohaline circulation in the oceans and the changing nature of deep bottom water formation and of northward heat transfer in near-surface currents such as the North Atlantic Drift.

Recently, Clark et al (1999) collated a number of palaeoclimate and palaeoceanic records to identify the interaction between ice, ocean and atmosphere during the last deglaciation. They found that the ocean circulation system as described by Imbrie et al. (1992) (Figure 8.7) probably oversimplifies the situation. In particular, Clark et al. (1999) focused on the causes of two periodic oceanic phenomena: Dansgaard-Oeschger and Heinrich events.

Heinrich events are (probably) the periodic release of huge armadas of icebergs from the Laurentide ice sheet (Chapter 12). The events were detected from measurements of the IRD that these icebergs left over the North Atlantic (Heinrich, 1988). They have a periodicity of about 7000 years and may be linked with the internal oscillatory dynamics of the Laurentide ice sheet (see Chapter 12). The icebergs within a Heinrich event will be responsible for issuing a significant volume of fresh water across the surface of the North Atlantic. Evidence in support of this comes from sea floor sediments in the North Atlantic which indicate a decrease in sea surface salinities during periods of IRD formation (e.g. Chapman and Maslin, 1999) (Figure 8.8).

In contrast to the 7000 year periodicity of Heinrich events, Dansgaard-Oeschger events are millennial-scale periods of cooling, probably associated with meltwater and iceberg production (Bond and Lotti, 1995; Bond et al., 1997). Although D-O and H events are distinct, they both affect ocean circulation and, because of this, climate (Alley and Clark, 1999). The meltwater released by iceberg discharges and surface run off is thought to reduce the thermohaline-driven formation of North Atlantic Deep Water, the force behind the modern ocean circulation system. The sudden loss of warm Atlantic waters that usually replace NADW at mid-high latitudes will be reflected in cooling of the atmosphere.

It is also possible that the ocean system may oscillate freely at a 1500-year time scale, triggered initially by a meltwater event. Under this idea, according to Alley and Clark (1999), the Northern Hemisphere is able to cool at the same time as warming in certain parts of the Southern Hemisphere. This is because if melting events cause the shut down of NADW, it may be that the ocean demand for deep water formation must come from elsewhere such as the Southern Hemisphere. This then may initiate warm surface waters southwards, warming the climate in these regions. However, once the system in the Northern Hemisphere has stabilised, NADW formation could be reintroduced. The influence of iceberg and surface meltwaters on ocean circulation can be thought of as punctuating the deglacial phase of Imbrie et al.'s (1992) ocean system (Alley and Clark, 1999).

Alley and Clark (1999) also suggest that changes in the ocean circulation caused by a D-O event (and possibly a H-event) could be responsible for the Younger Dryas climate reversal at about 12–11 000 years ago (Figure 8.9). The idea is that a sudden switching off of

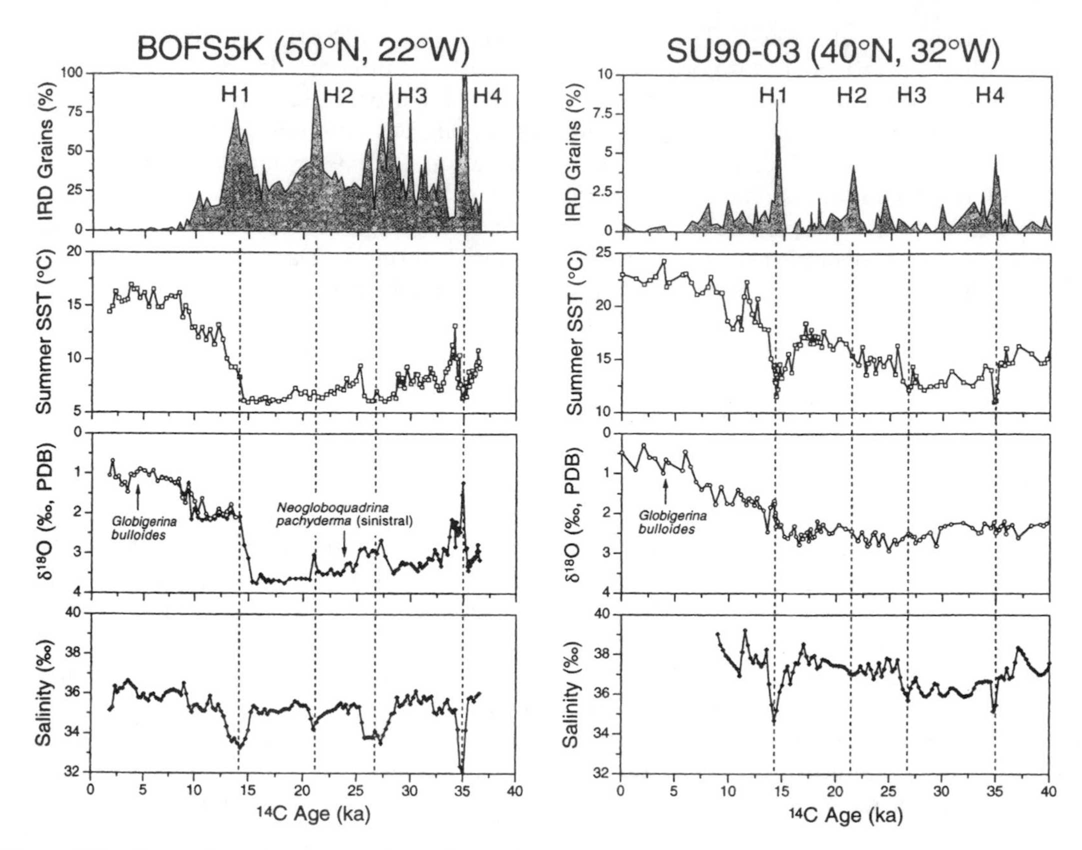

Figure 8.8 Comparison between pulses of North Atlantic IRD (Heinrich events), and sea floor sediment records from North Atlantic (left hand column) and the mid-North Atlantic (right hand column). Taken from Chapman and Maslin (1999), reproduced with permission of Geological Society of America.

the ocean conveyor caused SSTs in the North Atlantic to become much colder, resulting in colder conditions over the adjacent land masses (e.g. Imbrie et al., 1992; Hughen et al., 1998, Clark et al., 1999). This had the effect of reversing the deglaciation that had been going on since the LGM, and glaciers across most of Europe began to re-advance. There is plenty of field evidence to show that this happened in Europe and many other parts of the Northern Hemisphere. The influence of several known Heinrich layers can be commonly observed in palaeoceanographic records in the mid-latitude North Atlantic (Figures 8.8, 8.10). In the Southern Hemisphere there are only occasional records of the Younger Dryas (Alley and Clark, 1999). For example, Newnham and Lowe (2000) reported evidence for the Younger Dryas from a pollen record from New Zealand. However, records of relatively warm conditions at the time of the Younger Dryas are far more common in the Southern Hemisphere. This situation appears compatible with the 'see-saw' ocean model of Alley and Clark (1999), driven by the lack of NADW formation as detailed above. Seidov and Maslin (1999) used a GCM ocean model to show that a release of icebergs from the Laurentide Ice Sheet would lead to a collapse of NADW. They also showed that the North Atlantic responds in a similar manner to the decay of ice from the Barents Shelf, if this ice is transported to the North Atlantic.

An additional process by which fresh water may be input to the North Atlantic is the

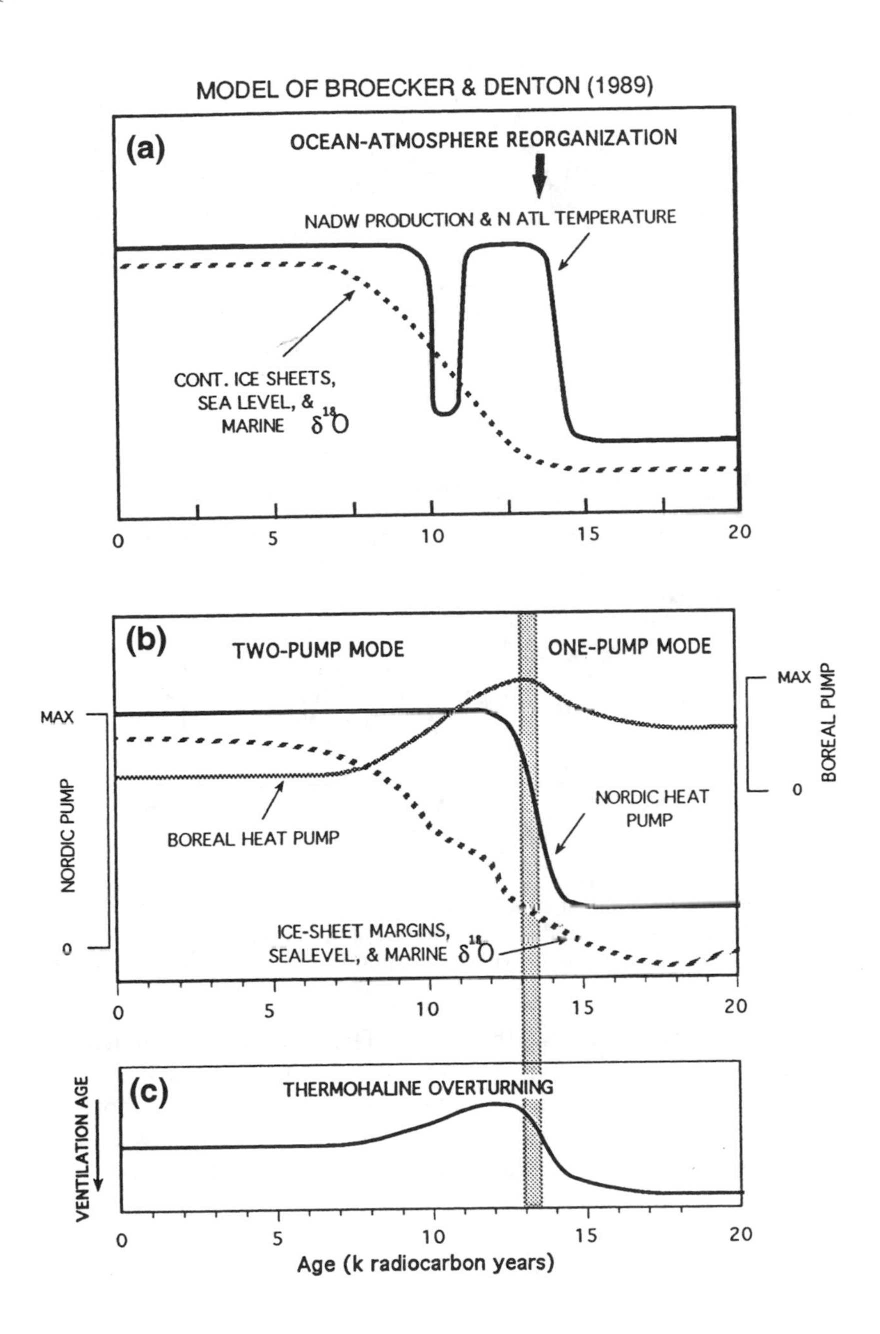

Figure 8.9 Changes in the Nordic and Boreal 'heat pumps' during the last deglaciation, as a potential explanation for the switching-off of the ocean conveyor during the Younger Dryas. Taken from Imbrie et al. (1992), reprinted from *Paleoceanography*, with permission of American Geophysical Union.

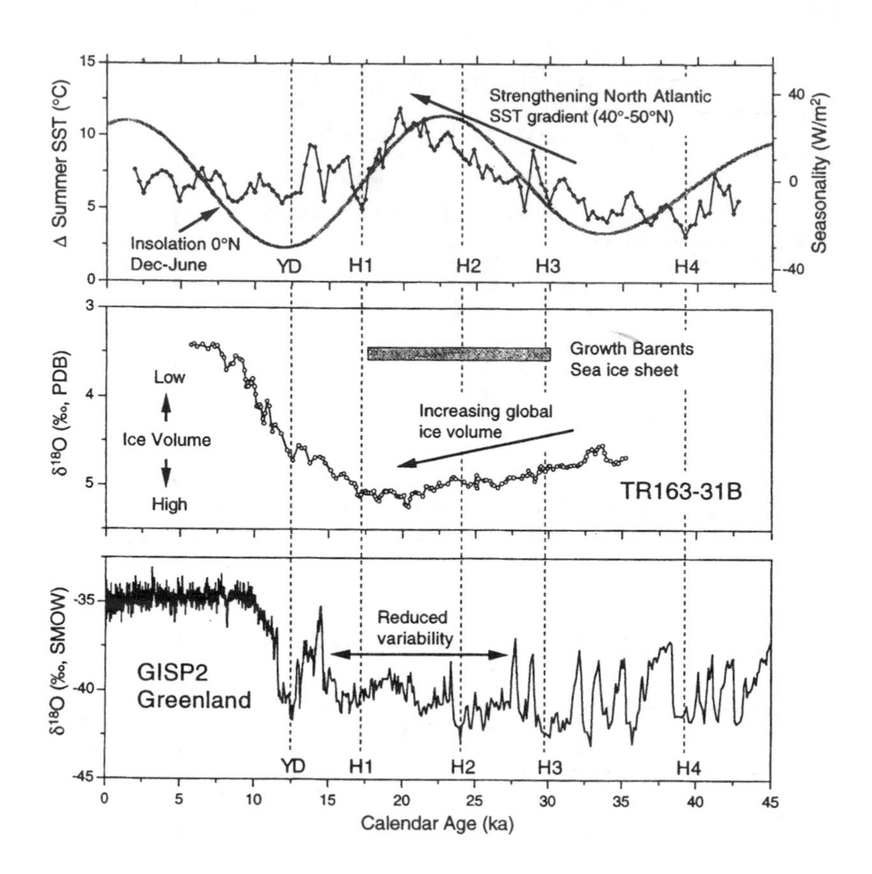

Figure 8.10 Sea-surface temperature gradients (between 40 and 50°N – see Figure 8.8 for details of the full records), sea floor $\delta^{18}O$ record from the mid-North Atlantic and the GISP2 $\delta^{18}O$ record. Taken from Chapman and Maslin (1999), reproduced with permission of Geological Society of America.

discharge of proglacial lakes from North America during the last deglaciation (Chapter 12). One huge proglacial lake (named Lake Agassiz) decanted enough water into the ocean at 8200 years ago, that sea level was increased by around 20 cm in a matter of days. This freshwater input to the ocean correlates with a marked cooling event observed in Greenland ice cores at the same time, suggesting a link between proglacial lake discharge, ocean circulation and climate (Barber et al., 1999).

CHAPTER 9

Ice Sheet Reconstructions I
The Antarctic Ice Sheet

A VERY BRIEF INTRODUCTION TO THE ANTARCTIC ICE SHEET

Antarctica is comprised of two main grounded ice sheets; the West Antarctic and the East Antarctic ice sheets, a relatively smaller ice cap across the Antarctic Peninsula, and dozens of glaciers located in mountainous regions at the margins of the continental land mass (Figure 9.1a). The present volume of ice in Antarctica is around 30 000 000 cubic kilometres (equivalent to around 60–70 metres of global sea level) of which 83 per cent is housed within the East Antarctic Ice Sheet. The maximum surface elevation is just over 4000 metres in East Antarctica. Maximum ice thickness varies between 2800 and 4500 metres in the central regions of Antarctica. The ice thickness and bedrock elevation (at the base of the ice sheet) can vary spatially by hundreds of metres over a few kilometres. The East and West Antarctic ice sheets are separated by the Transantarctic Mountains. If the ice were to be removed from East Antarctica, the bedrock surface would be above sea level. However, if the same were to happen over West Antarctica, even accounting for isostatic rebound, the bedrock would remain below the modern sea level. Thus, the West Antarctic Ice Sheet is often referred to as a marine-based ice sheet. In addition to the two main ice sheets, the other significant region of grounded ice exists over the Antarctic Peninsula where a series of grounded ice caps and glaciers are located.

To recap on glacier dynamics (Chapter 3), grounded ice flows by (1) the deformation of ice, and (2) basal sliding over bedrock and/or (3) deformation of water saturated basal substrate (Figure 3.1). Ice is drained from the domes of the ice sheet interior by fast flowing rivers of ice known as 'ice streams', where the dominant method of flow is by sliding and/or basal sediment deformation. These issue grounded ice into numerous floating 'ice shelves' that surround the grounded ice continent. The largest two ice shelves are the Ross and the Filchner-Ronne ice shelves which have an area of 526 000 and 473 000 square kilometres, respectively (Drewry, 1983). Icebergs, usually of tabular form, are calved out from the marine margins of ice shelves, where ice thickness is usually around 250–300 metres. This process represents the most important mechanism by which ice is lost from the ice sheet system; accounting for 85–75 per cent of ice loss (Jacobs et al., 1993; Paterson, 1994). A recent simple model of ice flow in Antarctica shows the flux of ice from the interior of the ice sheet to the onset of ice streams, into the ice streams and on to the ice shelves (Figure 9.1b) (Bamber et al., 2000a). This model indicates that the transition from slow flowing ice in the interior to fast moving ice in ice streams occurs across the inland region of the ice sheet.

An interesting concept is that it would take 7.6×10^{24} J of energy to melt the ice mass. This is more than twice the annual solar heat absorbed by the entire Earth-atmosphere system. However, this amount of heat will only warm the oceans by 1°C.

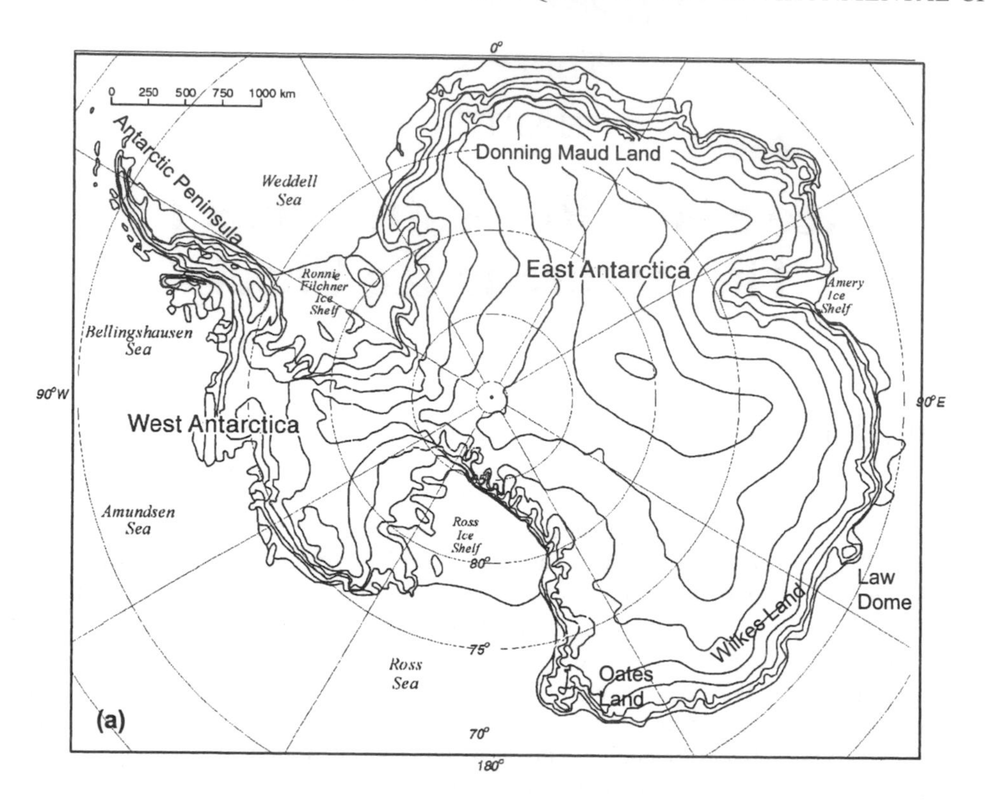

0°
0 250 500 750 1000 km
Antarctic Peninsula
Weddell Sea
Donning Maud Land
East Antarctica
Ronnie Filchner Ice Shelf
Amery Ice Shelf
Bellingshausen Sea
90°W
90°E
West Antarctica
Amundsen Sea
Ross Ice Shelf
80°
Law Dome
Wilkes Land
75°
Ross Sea
Oates Land
(a)
70°
180°

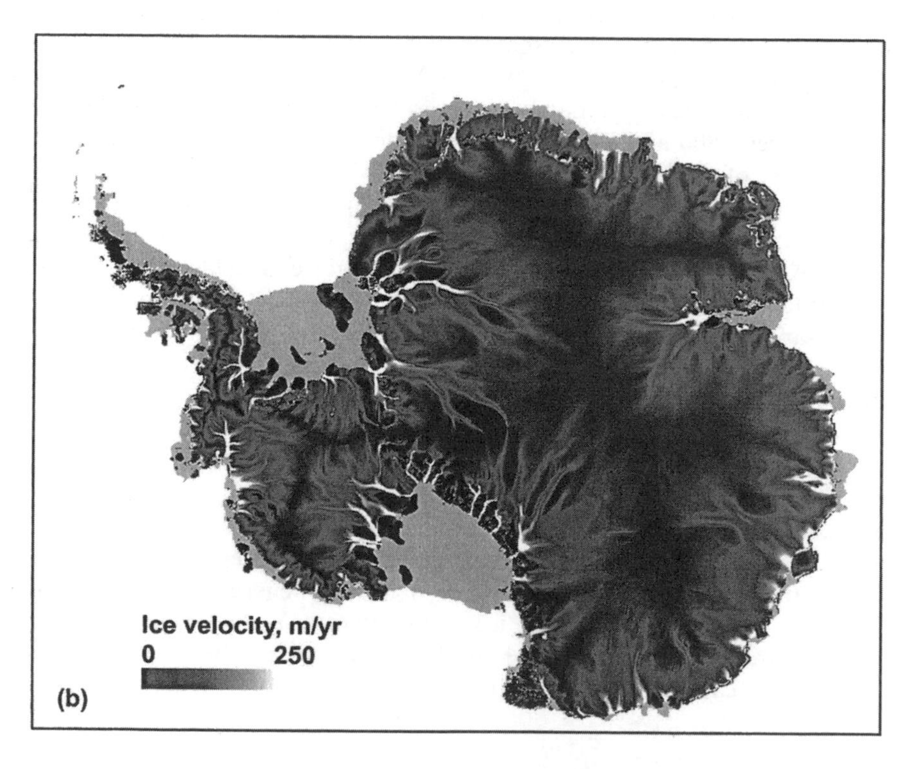

Ice velocity, m/yr
0
250
(b)

A. GEOLOGICAL INFORMATION

Pre Quaternary ice sheet history

It is important to note that, just as there is debate on the size and extent of the LGM Antarctic ice sheet, there is even more contention as to the stability of the Antarctic ice sheet during the past 30 million years. There are two opinions on this matter (Miller and Mabin, 1998). The first is that the ice sheet has remained in a relatively stable, constant configuration for this time. A basic concept in this argument is that the modern-type ice sheet and climate is associated with the Antarctic continent breaking away from South America, allowing the circumpolar currents and winds to isolate the weather systems over the ice sheet from the rest of the planet. Thus the ice sheet and climate system over the Antarctic has been 'mutually sustaining' for the past 30 million years or so. Geological evidence to support this idea comes from uneroded ash beds in the Dry Valleys that are thought to be up to 15 million years old. This is interpreted as being contrary to a dynamic ice sheet since the meltwater and glacier advance that would be associated would erode such features. The second opinion is that the ice sheet has been involved in a number of phases of growth and decay, implying a 'dynamic' ice sheet over the last 30 million years. Evidence for this comes from marine-based diatoms within the Transantarctic Mountains that are interpreted to have been deposited in a fjord- or lacustrine-type environment. The proponents of this argument say that this geological evidence means the ice sheet was 'wet based' and susceptible to unstable behaviour. This issue remains contentious, and may prove to be a major research theme in the next 10 years.

Basic LGM issues

The climate around Antarctica is unlikely to have been significantly different in physical character (if not in absolute numeric terms), to that of today. Therefore, the very 'minimum' reconstruction for the LGM Antarctic Ice Sheet is that which is similar to the present configuration. However, because the growth of Northern Hemisphere ice masses caused sea-level reduction and ocean temperature changes, then ice expansion around Antarctica, to the continental shelf edge, may have occurred as a response. In this situation, the ice shelves bordering the Antarctic Ice Sheet may have thickened, grounded and become part of the ice sheet proper. This scenario is the form of the 'maximum' reconstruction.

Having said this, there is some recent evidence from the ice stratigraphy from the Taylor Dome (East Antarctic, Ross Ice Shelf intersection) which suggests a much lower accumulation rate at the LGM. However, it also shows that the LGM storm paths came from a different direction to the modern day. This suggests a reorganisation of the climate system at least at this part of the ice sheet (Morse et al., 1998).

In order to understand ice sheet behaviour in the Late Quaternary, it is necessary to be aware of the connection between the ice, ocean and atmosphere. For example, if sea ice extent increased due to cooling of the air temperature over the southern oceans (as shown by Armand, 2000), then the moisture supply to the ice sheet may have decreased and so ice sheet growth may have been impeded. This introduces the possibility that independent glaciers on the Antarctic continent may have responded differently to LGM conditions than the areas connected to the ice sheet. In other

Figure 9.1 Surface elevation and ice flow in Antarctica. (a) Location map and surface elevation of Antarctica, with place names mentioned in the text. Contours are in 500 metre intervals. (b) The flux of ice in Antarctica as modelled by Bamber et al. (2000a) (from data supplied by J. Bamber, University of Bristol, England).

words, the ice sheet advance may have been related to sea level fall whilst, at the same time, glacier decay may be related to a significant reduction in rates of precipitation.

Modern accumulation and temperature of the Antarctic ice sheet is provided in Figure 9.2. Palaeoclimate information from the Vostok ice core indicates that the air temperature above the ice sheet at Vostok at the LGM was 8–10°C colder, and probably more arid than at present.

Geological information for Late Quaternary Antarctic ice sheet change

Last interglacial

The oceanic $\delta^{18}O$ record suggests that, between 150 and 120 thousand years ago, there was less ice in the world than at present and global sea level was several metres higher than today (Figure 1.6). Some believe that this indicates the decay of the West Antarctic Ice Sheet, whilst others suggest the decay of the Greenland Ice Sheet is responsible (Chapter 10). Recent evidence to support the collapse of at least part of the West Antarctic Ice Sheet at the last interglacial comes from marine diatom and high concentrations of Beryllium-10 data from the Ross Embayment which have been interpreted as meaning this region was an 'open marine environment'. Beryllium-10 and diatoms will only accumulate when the sea is open. When the sea is ice covered, there is very little deposition of these materials. Moreover, if ice is grounded, any small amounts of Beryllium-10 and diatoms will be eroded. Therefore, these data indicate open ocean conditions in the Ross Embayment. For this to happen, part of the West Antarctic Ice Sheet must have collapsed prior to the advent of open water (Scherer et al., 1998). Evidence in support of this idea comes from marine sediments that have been found beneath the current West Antarctic ice sheet (D. Blankenship pers. comm.). The theory is that these sediments were placed during the last interglacial when marine conditions prevailed. The current dynamics of the ice sheet may subsequently be affected by the location of these weak, now subglacial, sediments.

Last glacial

Regardless of the situation at the last interglacial, there is persuasive evidence from the geological record to suggest ice expansion at 21 000 years ago. Ice core evidence shows that such changes, although probably forced by Northern Hemisphere ice expansion, are in relative phase with the growth of other ice sheets. These records suggest a cold phase at 60 000 years ago, prior to the glacial maximum at 21 000 years ago. However, major phases of ice recession during the Late Quaternary are unlikely in Antarctica, implying that ice growth in Antarctica may have occurred steadily for tens of thousands of years during the last glaciation.

CLIMAP reconstruction for the LGM

Important geological information comes from the Transantarctic Mountain belt, where there is no ice at present. Lateral moraines that are located high above the ice extent at the foot of the glaciers, merge to the surface of the ice sheet near to the head of the glaciers. Hence, the thickening indicated by these moraines increases downstream, and is indicative of a grounded Ross ice sheet, rather than thicker East Antarctic ice (Figure 9.3). Grounded ice from the Ross Sea pushed westwards into the

Figure 9.2 Modern climate of Antarctica (a) mean annual surface air temperature in °C. (b) Surface mass balance (in centimetres water equivalent per year). Adapted from Giovinetto and Bentley (1985).

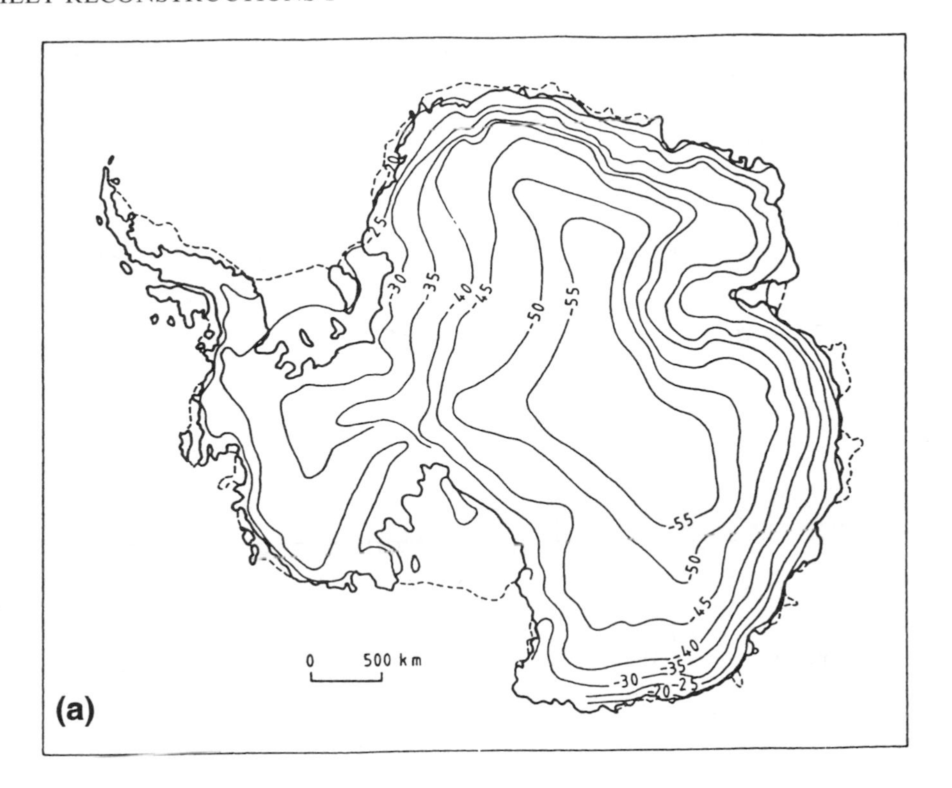
-20
-25
-30
-35
-40
-45
-50
-55
-55
-50
-45
-40
-35
-30
-20
-25
0 500 km
(a)

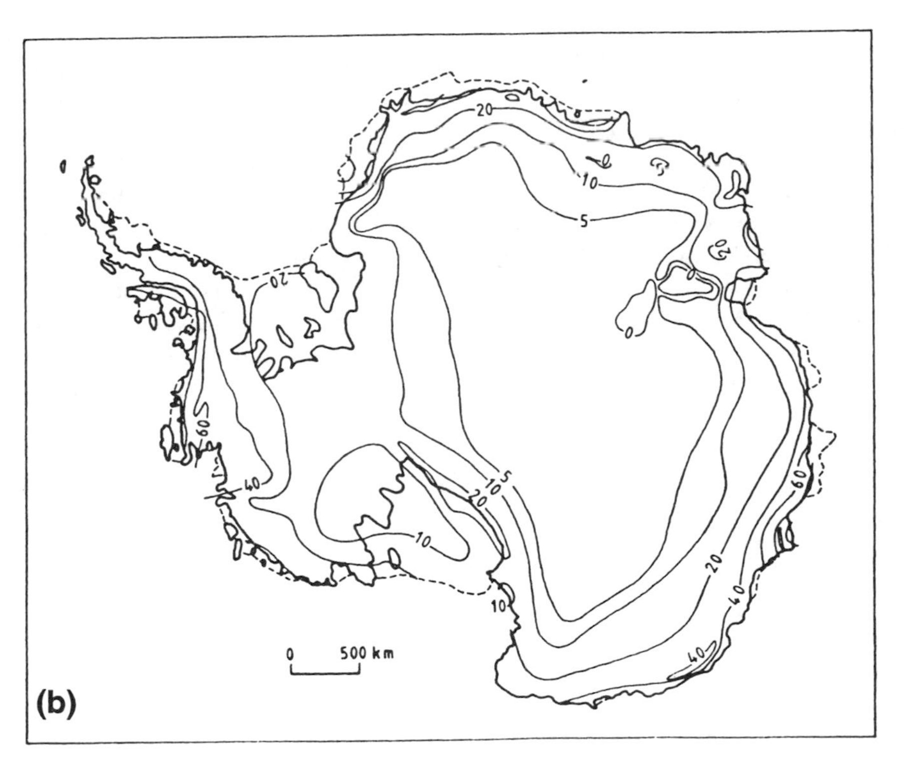
20
10
5
20
60
40
10
5
10
20
10
0
60
20
40
40
0 500 km
(b)

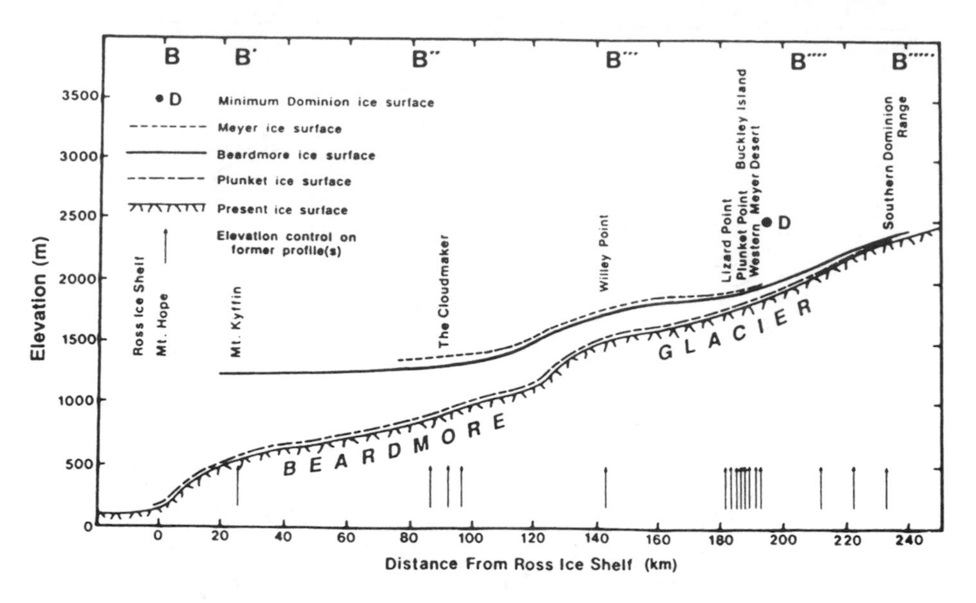

Figure 9.3 Surface of the Beardmore Glacier, and former ice extent mapped from trimlines and lateral moraines. The section is aligned from the Ross Ice Shelf on the left, through the Transantarctic Mountains to the East Antarctic ice sheet on the right. Reprinted from *Quaternary Science Reviews*, 9, W. Broecker and G. Denton, The role of ocean-atmosphere reorganizations in glacial cycles, pp. 305–341, copyright (1990), with permission from Elsevier Science.

valleys, causing lobes of ground moraines (known as the Ross Sea Drift), and evidence of ice dammed lakes in the ice-free Taylor and Ferrar valleys. According to CLIMAP (1976), McMurdo Sound was covered by ice 1325 metres thick, and Byrd Station had ice 1700 metres thicker than at present. In their reconstruction, the Filchner-Ronne Ice Shelf also thickened and grounded, allowing the grounded ice sheet to extend to the continental shelf in most places. To summarise, the CLIMAP group expected the LGM Antarctic ice sheet to be substantially larger than that at present, with grounded ice expansion to the continental margin in many places.

Drewry's reconstruction

Drewry (1979), suggested that, for the Ross Ice Shelf to ground, glaciological considerations demanded that a sea level depression of 120–135 metres existed for more than 10 000 years. As has been noted in the discussion about global sea level (Chapter 1, Figure 1.4), this is hard to justify for the last glacial prior to 20 ka ago. Instead Drewry (1979) proposed limited expansion of the Ross Shelf, which was backed by marine sedimentary studies from below the ice shelf (Figure 9.4b). If the Ross Sea had limited ice expansion, then so must have the West Antarctic Ice Sheet. This idea is supported from information from the Byrd Ice Core suggesting that the ice surface here experienced no significant uplift during the LGM (e.g., Whillans, 1976). The problem of the Ross Sea glaciation has been addressed recently by a series of geophysical and sedimentological investigations. The results of these studies are discussed below.

Post CLIMAP scenarios

In the 20 years since CLIMAP, much more geomorphological information has been gathered. Some data agree with the CLIMAP

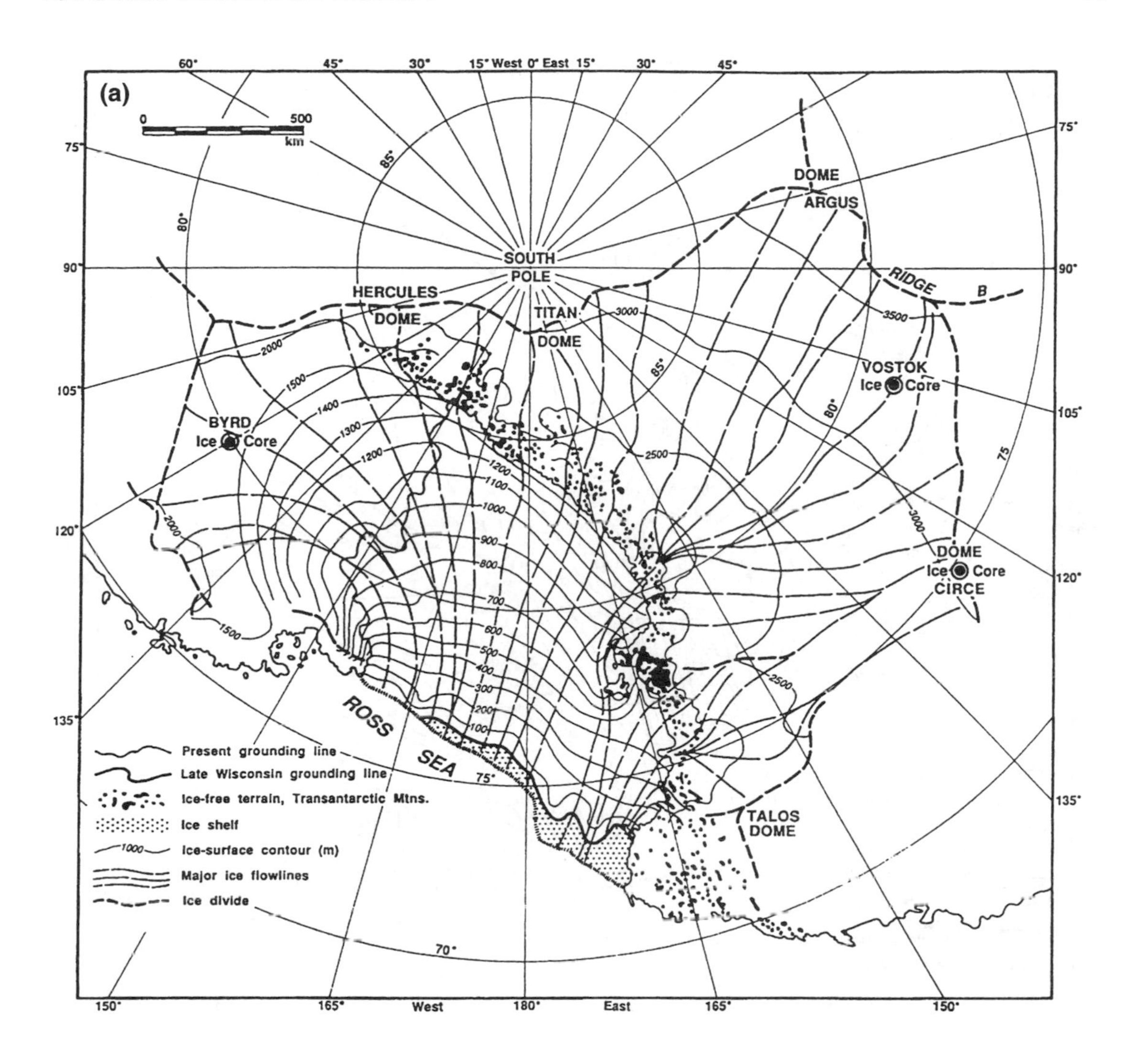

Figure 9.4 LGM reconstruction of the Ross ice shelf region according to (a) CLIMAP and (b) Drewry. Taken from Denton et al. (1989), and reproduced by permission of Academic Press.

reconstruction, others indicate that a revision is necessary.

Ross Sea glaciation

In three field seasons during the 1990s, a comprehensive geophysical and sedimentological dataset was compiled across the Ross Sea (Shipp et al., 1999; Domack et al., 1999). The aim of this investigation was to identify the maximum extent and configuration of the LGM ice sheet over the Ross Sea, to understand the processes operating at the base of this ice sheet and to identify the deglaciation history. Geophysical methods included seismic profiles, sonar profiles and side-scan sonar imagery of the sea floor sediments in front of the present day Ross Ice Shelf. This campaign was able to determine the morphology of the sea floor and, from the landforms recorded, an assessment of

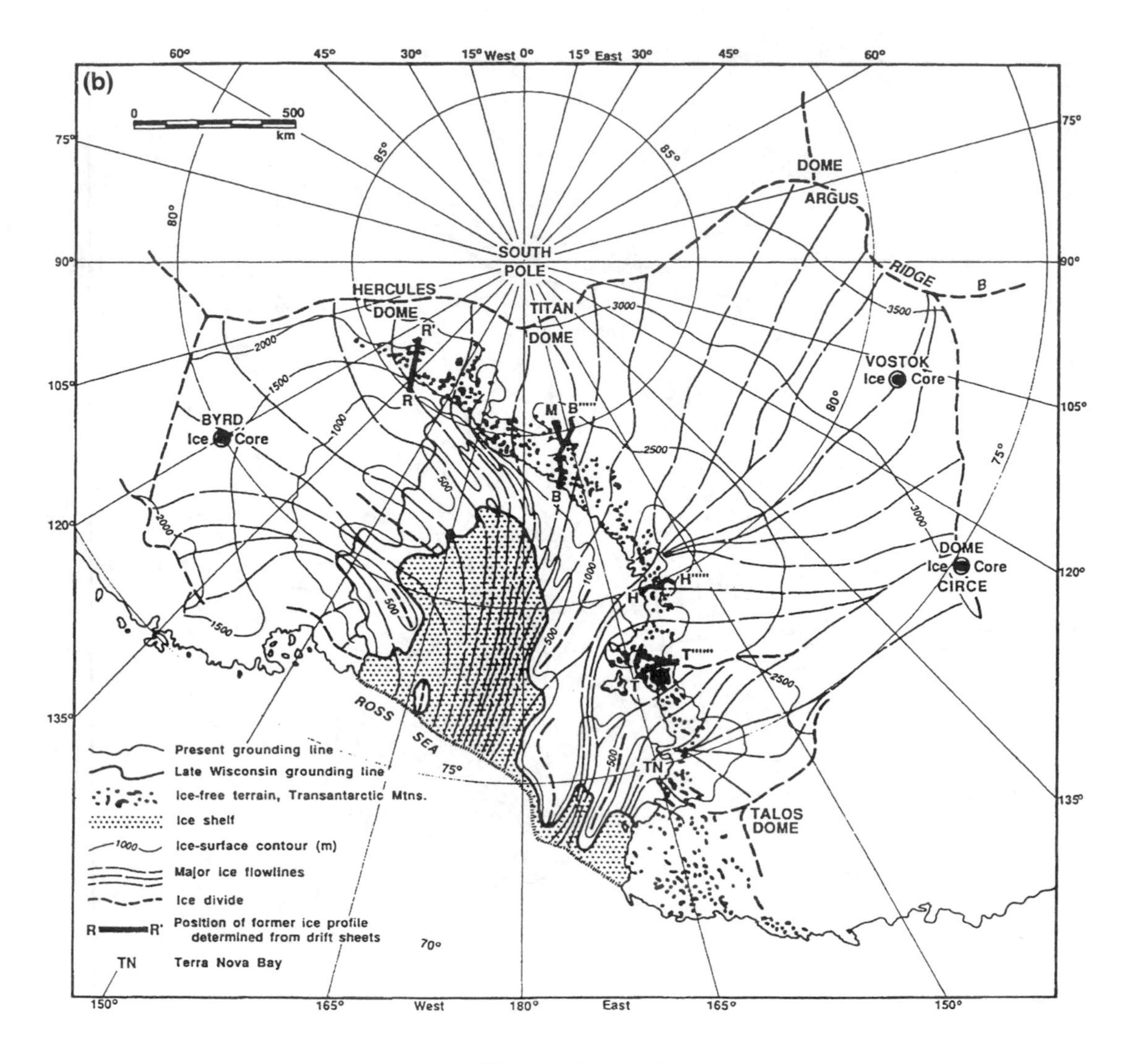

Figure 9.4 (*cont.*)

ice sheet history. Lineations, drumlins and large-scale grooves were identified (Figure 9.5), providing a good basis for the assessment of former glaciation, basal processes and ice flow direction. Importantly, these features are found predominantly in bathymetric troughs, which may have been occupied by fast-flowing ice streams. The ice sheet in which these ice streams were part of would have been marine-based just like the West Antarctic ice sheet. Sedimentary features suggest that a deforming till layer may have been an active component of ice stream flow in this region. The Ross Sea was therefore occupied by fast-flowing ice streams that terminated at the continental shelf edge (Figure 9.6). The majority of ice expansion in the Ross Sea was caused by an enlarged West Antarctic Ice Sheet. These investigations therefore give some support to the CLIMAP reconstruction in this area.

Traditional glacial theory suggests that if the Ross Sea was covered by grounded ice, then

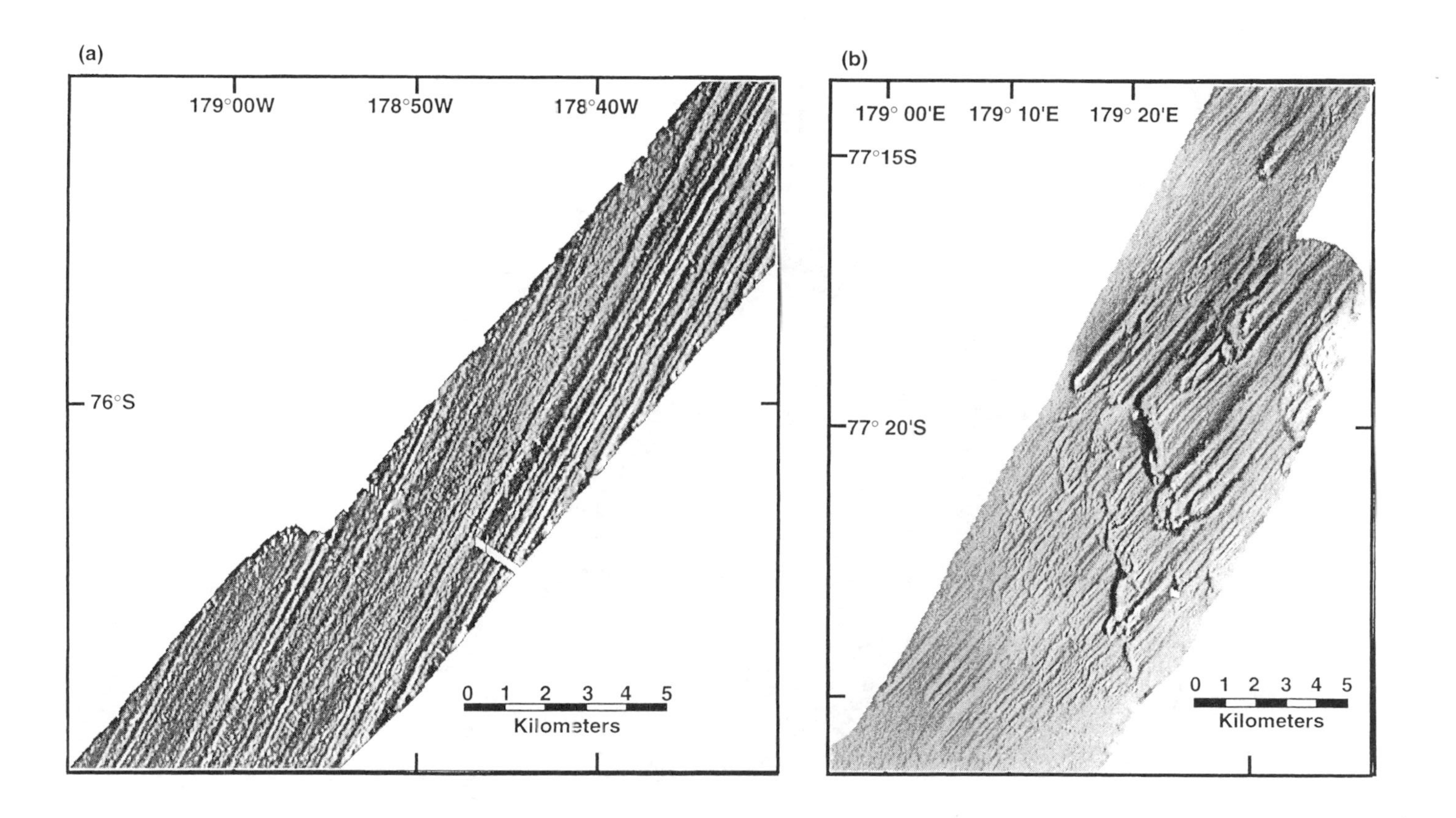

Figure 9.5 Side-scan sonar images from the floor of the central and western Ross Sea showing evidence of former grounded ice. (a) Fluting. (b) Drumlins. Taken from Shipp et al. (1999), reproduced with permission of Geological Society of America.

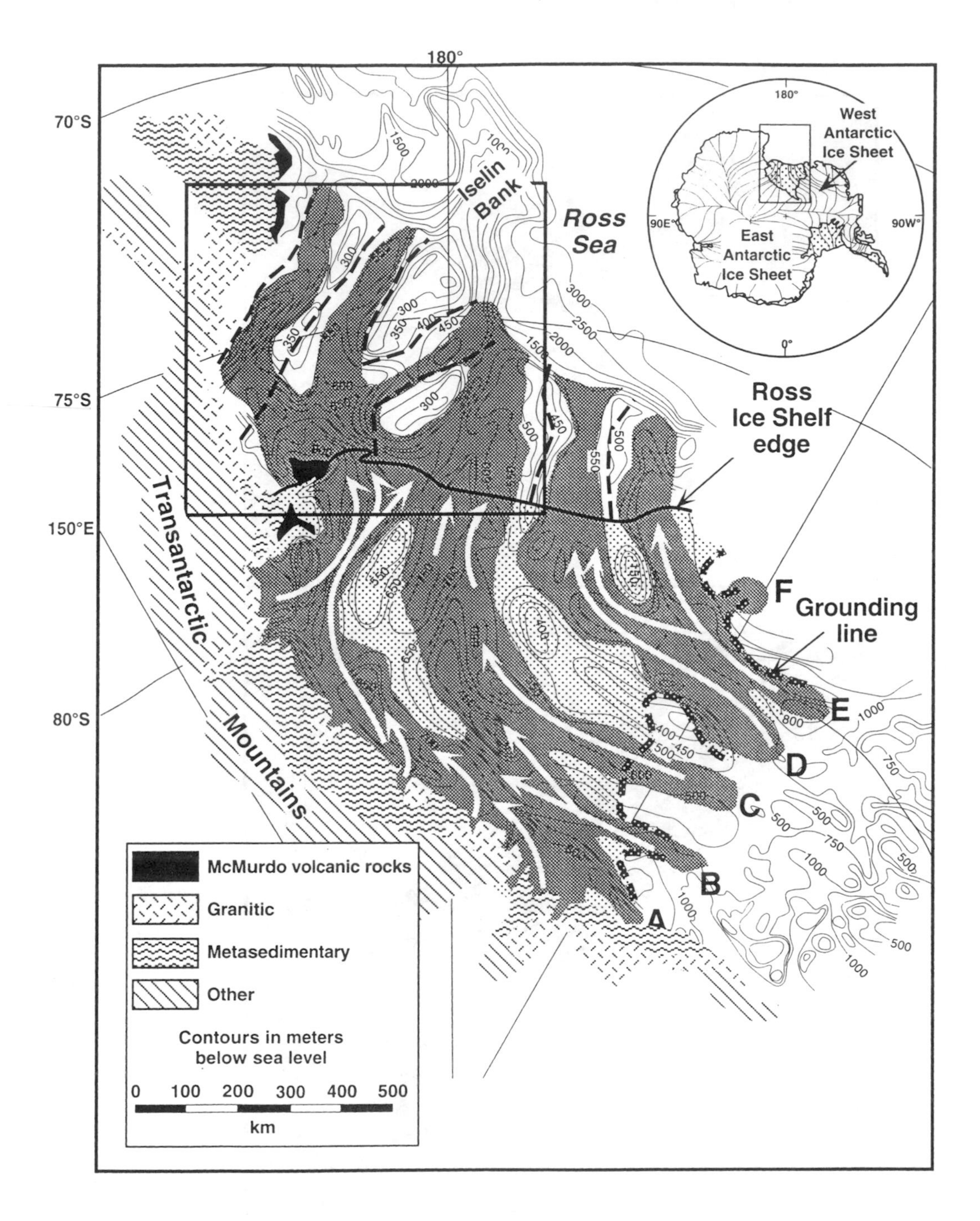

Figure 9.6 Reconstruction of the Ross Ice Shelf region at the LGM. Lightly stippled regions shows the Ross Ice Shelf. The heavy dashed line indicates the present grounding line. The flow of former ice streams is shown by a dark stipple and white arrows (indicating the ice flow direction). The modern bathymetry is superimposed over the map. Letters refer to names of modern Siple Coast ice streams. Taken from Shipp et al. (1999), reproduced with permission of Geological Society of America.

the thickness of the West Antarctic ice sheet is expected to be greater than at present. However, evidence for an increase in ice thickness during the last glacial has been found for 10 000 years ago, much later than the LGM. This evidence comes from Marie Byrd Land, where a lateral moraine exists 45 metres above the present ice surface (Ackert et al., 1999). Therefore, the relationship between West Antarctic ice sheet thickness and the glaciation of the Ross Shelf is not obvious at the moment.

Prydz Bay and the Lambert Amery system

The Lambert Ice Stream drains ice from a basin occupying around 14 per cent of East Antarctica, into the Amery Ice Shelf. The Amery Ice Shelf terminates as a calving wall in Prydz Bay. Recently, there have been a number of marine geophysical and sedimentological investigations of the Prydz Bay area. These studies aimed to determine the LGM history of the ice sheet, as well as former Quaternary and pre-Quaternary glaciations. Domack et al. (1998) used gravity cores to extract a series of sea floor sediment samples across the continental shelf region in front of the Amery Ice Shelf. The bathymetry of the continental shelf is characterised by a number of shallow banks (water depths less than 500 metres) and a 1000 metre deep subglacial trench called the Lambert Deep. This trench is thought by some to reflect an over-deepening erosion feature from an enlarged, grounded Lambert Ice Stream. However, Domack et al. (1998) found glacial marine and biogenic sediments within the Lambert Deep, with no interruption from till at least as far back as 30 ka. This suggests that the Lambert Deep did not have grounded ice within it at the LGM. Thus, the simple CLIMAP extent of grounded ice to the continental margin may be too simple to be applied to the Prydz Bay region, where glaciation from several sources (not just the Lambert Ice Stream) may have been possible, but difficult to reconstruct.

Antarctic Peninsula

The Antarctic Peninsula is an important area in terms of Antarctic ice sheet reconstruction because there is arguably as much exposure of formerly glaciated bedrock here as anywhere in the continent. There have been a number of recent marine geophysical and sedimentological investigations of the continental shelf off the shore of the peninsula.

One region where investigations have focused is the Palmer Deep (an enclosed bathymetric depression in the inner continental shelf). Rebesco et al. (1998) used piston core and seismic analyses to determine the glacial history of this part of the shelf. They found several noticeable changes to the sediment character in this basin, reflecting a number of distinct depositional environments. The first is characterised by biogenic material and slumps/slides of glacigenic sediments, interpreted to be caused by open ocean conditions over the basin, and a grounded ice sheet margin across one end of the basin. Above these sediments, there is a thin layer of fine grained material free of biogenics, thought to be caused by the subglacial melt out of the ice sheet as it overrode the basin. This, it is thought, would leave the basin as a 'subglacial lake'. Ice recession then occurred, resulting in a similar sediment character to the first unit. Above this, open marine conditions are reflected in the modern-type sediments. Thus, the ice sheet may have expanded to the continental margin across the Antarctic Peninsula, but the Palmer Deep was not covered by grounded ice. Instead, it was occupied by a subglacial lake.

Bentley and Anderson (1998) used glacial geological ice flow indicators from exposed surfaces across the Antarctic Peninsula, and glacial marine sedimentary evidence, to reconstruct the Antarctic ice sheet across the Weddell Sea sector. They concluded that the glacial history of the region was not well known. Their geologically-based ice sheet limits are given in Figure 9.8. The ice sheet margin is thought to extend to the shelf edge in the east but, across

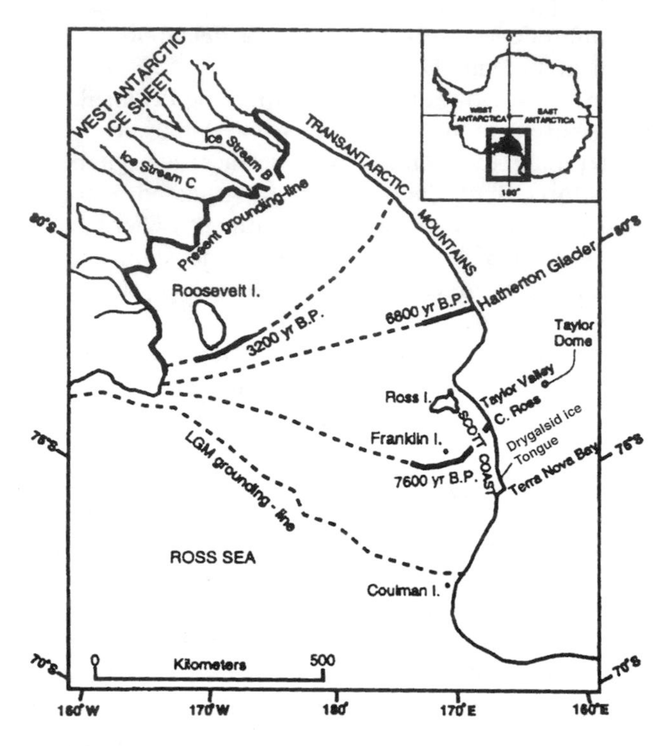

Figure 9.7 The retreat of ice across the Ross Sea embayment from the LGM to mid-Holocene. Taken from Conway et al. (1999).

the west, grounded ice sheet expansion probably only extended to the mid-shelf regions. In the extreme north-east there is some controversy on the glacial history. Opinions are divided between a 'maximum' model where the ice sheet reached the shelf break (Pudsey et al., 1994) based on subglacial morphological evidence, and a more restricted glaciation where the mid-shelf deposition of till was replaced by glacimarine sediments prior to 13 ka (Pope and Anderson, 1992). In conclusion, the LGM ice sheet over the Antarctic Peninsula was probably smaller than the CLIMAP ice sheet, which involved ice extent to the continental shelf break.

Further support for CLIMAP ice extent

There are several other datasets that support the CLIMAP ice sheet reconstruction for Antarctica. For example, grounding of the Filchner-Ronne ice shelf to a water depth of 500 metres is supported by marine sedimentological data (Elverhøi, 1981). Fresh striations and erratics on nunataks up to 500 metres above current ice surfaces in the South-east Antarctic Peninsula, adjacent to the Ronne Ice Shelf and Weddell Sea indicate expansion of the ice sheet across the Peninsula at the LGM. Trimlines on the Whitmore Mountains nunataks suggest the ice thickness in this region was 300–500 metres greater than at present. Further support for CLIMAP reconstruction comes from the Vestfold Hills region of East Antarctica (near the Australian Davis Station). Striations on these rocks indicate recent overriding in a radial fashion suggesting ice sheet development from this region. Also, terminal moraines observed seaward of the Ninnis and Mertz outlet tongues in East Antarctica are in agreement with the CLIMAP reconstruction of major ice expansion, but cannot be used to testify whether ice extended to the continental margin or not.

It can therefore be seen that there is a bulk of recent geological evidence in support of a significant expansion of the Antarctic ice sheet at the LGM, but that the CLIMAP reconstruction of the LGM Antarctic ice sheet is probably too large in a number of areas. Further, as Bentley (1999) calculated, there is a large discrepancy between the ice thickness of CLIMAP and other ice sheet reconstructions, and the contribution to sea-level reduction that the corresponding ice volume is responsible for (see later in this chapter).

Evidence for Heinrich-style IRD events from the Antarctic Ice Sheet

Although the southern ocean floors have not been cored anything like as much as the North Atlantic, there is some IRD evidence from the south-east Atlantic to suggest that the West Antarctic Ice Sheet issued regular pulses of icebergs during the last glacial cycle (Kanfoush et al., 2000). Three sea floor sediment cores reveal that IRD was deposited up to six times between 20 000 and 60 000 years ago, with a periodicity of between 6000 and 10 000 years. There are two main explanations for an IRD pulse in the southern oceans. The first is that the IRD was deposited by an increased number of icebergs. The second is that cooling of sea surface temperatures may have allowed icebergs to survive across the southern ocean thus depositing their material in distal locations. However, one of the cores showing the IRD signal is taken far north of the position of the LGM polar front, which seems to suggest that the former explanation may be relevant. Much work has still to be done to determine the true origin and significance of the southern ocean IRD pulses. However, there is a theory that the IRD reflects the instability of the West Antarctic Ice Sheet during the last 100 000 years (Kanfoush et al., 2000).

Holocene retreat of ice

Antarctic ice sheet recession is thought to have begun relatively late in the last deglaciation, after the majority of Northern Hemisphere-induced sea-level rise had occurred. In fact there is some geological evidence that the West Antarctic Ice Sheet kept growing until about 10 ka (Ackert et al., 1999). The most comprehensive investigation of post LGM ice decay in Antarctica was undertaken across the Ross Sea. Here dated glacial sediments record the margin of the grounded ice sheet as it retreated across the Ross Shelf (Domack et al., 1999; Shipp et al., 1999). Four sedimentological facies have been determined across the Ross Shelf, each related to a different depositional environment. First, subglacial diamicton (deposited beneath a grounded ice mass), 'pelletised lift-off' facies (which reflects when the grounded ice sheet becomes afloat), sub ice shelf silts and open marine muds and ooze. From analysis of these sediments (and the geomorphology determined from geophysical analysis), ice sheet grounding line decay occurred across the Ross Sea by 11 ka. The retreat of the floating ice shelf margin probably lagged behind the recession of the grounding line. Interpretation of sediment facies suggests that the ice shelf margin retreated at a rate of 100 m/yr, and reached its modern limits by 6 ka. However, grounding line retreat is expected to have continued after 6 ka, and this may be related to Late Holocene sea level rise (Domack et al., 1999).

This notion of ice shelf decay late in the last glacial cycle suggests that ice sheet break up was forced by global sea level rise. Sea level rise causes an increase in iceberg calving. This, in turn, causes mass loss and thinning of the ice, retreat of the grounding line and refloating of the ice shelf regions. And so it seems that the marine sections of the ice sheet would have been susceptible to ice decay first. This ice decay was caused by enhanced iceberg calving due to an increase in global sea level at around 12–11 ka.

In other areas of Antarctica, most substantial lowering of the ice thicknesses occurred after 12 000 years, and were complete by 6000 ago. However, the exact timing and speed of recession probably differs across the continent.

One alternative theory about Antarctic ice sheet decay is that the majority of ice loss took place after 6 ka. If this is true, then a mechanism other than global sea level rise is required (since most had taken place by 6 ka), and brings into question the relative stability of the West Antarctic Ice Sheet (Bentley, 1999). The retreat of the grounding line across the Ross Ice Shelf has been dated at three locations, suggesting that migration of the ice sheet back to its present position occurred between the middle to late Holocene (Conway et al., 1999) (Figure 9.7). This is an important theory because it may mean that the West Antarctic Ice Sheet is capable of changing its configuration in the absence of external climate forcing.

The deglaciation story is complicated by consideration of the Hypsithermal (the period of maximum Holocene warmth) in the Southern Hemisphere, which peaked at 9400 years ago, prior to the retreat of grounded ice across the Ross Ice Shelf (Figure 9.7). Because of this warm period, the ice within ice shelves may have become softer and more susceptible to iceberg calving. This warm period could also have affected the surface mass balance of the ice sheet (although this was probably of minor importance). Possibly because of the Hypsithermal, the George VI ice shelf disappeared by 8000 years ago, and reappeared by 6000 years. This region is, consequently, susceptible to climate change and is the current focus of attention for glaciologists who are interested in the mass balance of the ice sheet (cf. Doake and Vaughan, 1991). However, as Domack et al. (1991) explained, this period of warmer temperatures may have lead to an increase in the rate of precipitation and, thus, glacier advance. They suggested that this period could be used as an analogue for future climate change in Antarctica.

B. NUMERICAL MODELLING INFORMATION

Glaciological ice sheet modelling

In 1992, Huybrechts published his doctoral thesis on numerical modelling of the Antarctic Ice Sheet. He used his results to establish whether a CLIMAP-type ice sheet could have existed over Antarctica. Ten years on from this study, Huybrechts' is still regarded as one of the most complete models of the Antarctic Ice Sheet. The ice sheet model used by Huybrechts (1992) is similar to that indicated in Chapter 3, with the improvement of ice sheet thermal regime calculated in three dimensions. It is not the intention to examine further the process or mechanisms of ice sheet modelling. However, it is wise to note a few things concerning the philosophy behind the subject. Computer ice sheet models provide quantitative information about ice sheets. They use information about the rheology of ice to determine how it flows within the ice sheet system. Ice sheet models require inputs of bedrock elevation, initial ice thickness, sea level, climate. The models can be calibrated to modern ice sheets to make sure they are working properly. These calibrated models can then be run with different inputs to see how the ice sheet may respond to climate change. These climate inputs can take two forms. The first is a time-dependent variation in climate conditions, such as that through a glacial cycle. The results from this type of experiment provide information on the time-dependent behaviour of the ice sheet. The second involves controlled change from modern conditions to a steady alternative value. These results allow us to determine how various climatic factors affect the ice sheet. This can be thought of as a sensitivity experiment, where one parameter is changed and all others are held at modern levels. The ice sheet model is then run until a stable ice sheet solution is reached. The results indicate (1) the maximum amount of ice sheet response to that forced change and (2) the response time of the ice sheet to the change. Huybrechts employed both of these modelling procedures in the investigation of the LGM Antarctic Ice Sheet.

Ice sheet sensitivity

Huybrechts ran his ice sheet model under modern-type environmental conditions until stability was achieved with respect to ice sheet extent and mass balance. He then ran the model, with the stable modern ice sheet configuration with different forcing conditions, until stability was again achieved. The subsequent change in ice sheet dimensions indicates the sensitivity of the ice sheet to the type of environmental forcing. There are three fundamental questions that can be answered by this procedure:

1. What is the relative importance of changes in the environmental conditions to ice sheet size?
2. What are the response scales connected with these changes?
3. What is the spatial distribution of the resulting ice sheet fluctuations?

The reference state to which all 'affected ice sheets' are compared with is the Interglacial Run (i.e. the modern-type ice sheet established from contemporary environmental conditions). The stable Interglacial Run ice sheet is very similar to that of the modern ice sheet, but there are *some* differences, notably in the Antarctic Peninsula, where the ice sheet is larger than at present (Figure 9.8). This may be due to the unrealistic subglacial topography that occurs when a 'smoothed' representation of relief is used in the model.

In order to examine how environmental factors combine to produce the LGM ice extent, the ice sheet was subjected to shifts in the imposed accumulation rate, surface temperature and sea level fall, typical of glacial–interglacial magnitude as follows:

- a 10°C uniformly distributed drop of the surface air temperature;
- a reduction in surface snow deposition rates associated with this temperature reduction;
- a eustatic sea level depression of 130 metres from the modern value.

Six experiments are considered, in which the above conditions are applied either singly or in combination. The resulting ice sheet geometries are given in Figure 9.9. The results are interpreted in the following ways:

- accumulation reduction by 50–60 per cent of Holocene values, lowers the ice thickness by several hundred metres;
- the opposite results occur when a temperature reduction of 10°C is imposed;
- therefore, when both these conditions are imposed, their effect on ice sheet size, counter each other to yield relatively unchanged results;
- by far the most important forcing factor is that of the sea level. Lowering the sea level by 130 metres caused grounding of the ice shelves on to the shallow continental shelf areas;
- imposing the full glacial conditions lowers the East Antarctic ice thickness somewhat, but the overall extent of the ice mass is controlled by sea level;
- apart from the arguments concerning the Ross Sea glaciation, this model fits relatively well to the geological interpretation.

The ice sheet response times indicate that ice sheet stability in each of these experiments takes several thousand years. It took, for example, 30–40 000 years for the Ross and Filchner-Ronne ice shelves to become grounded ice sheets. Therefore, it is highly likely that as time-dependent changes to climate takes place, the ice sheet response to that change lags behind. The results of this sensitivity test show a number of interesting features relating to the LGM and the CLIMAP reconstruction as follows:

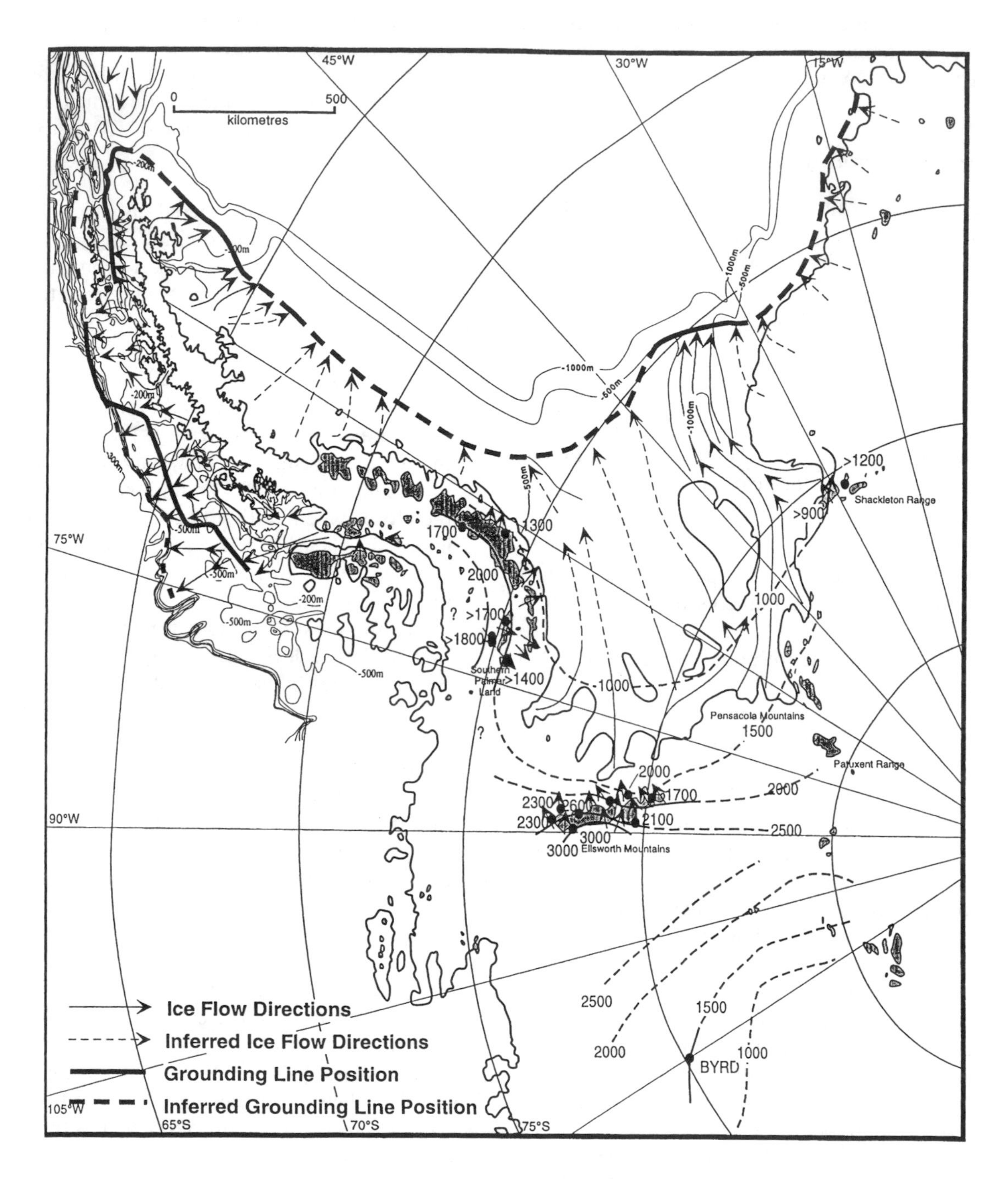

Figure 9.8 Reconstruction of the LGM ice sheet across the Weddell Sea embayment. Taken from Bentley and Anderson (1998) and reprinted from Antarctic Science, vol. 10, pp. 307–323 with permission from Cambridge University Press.

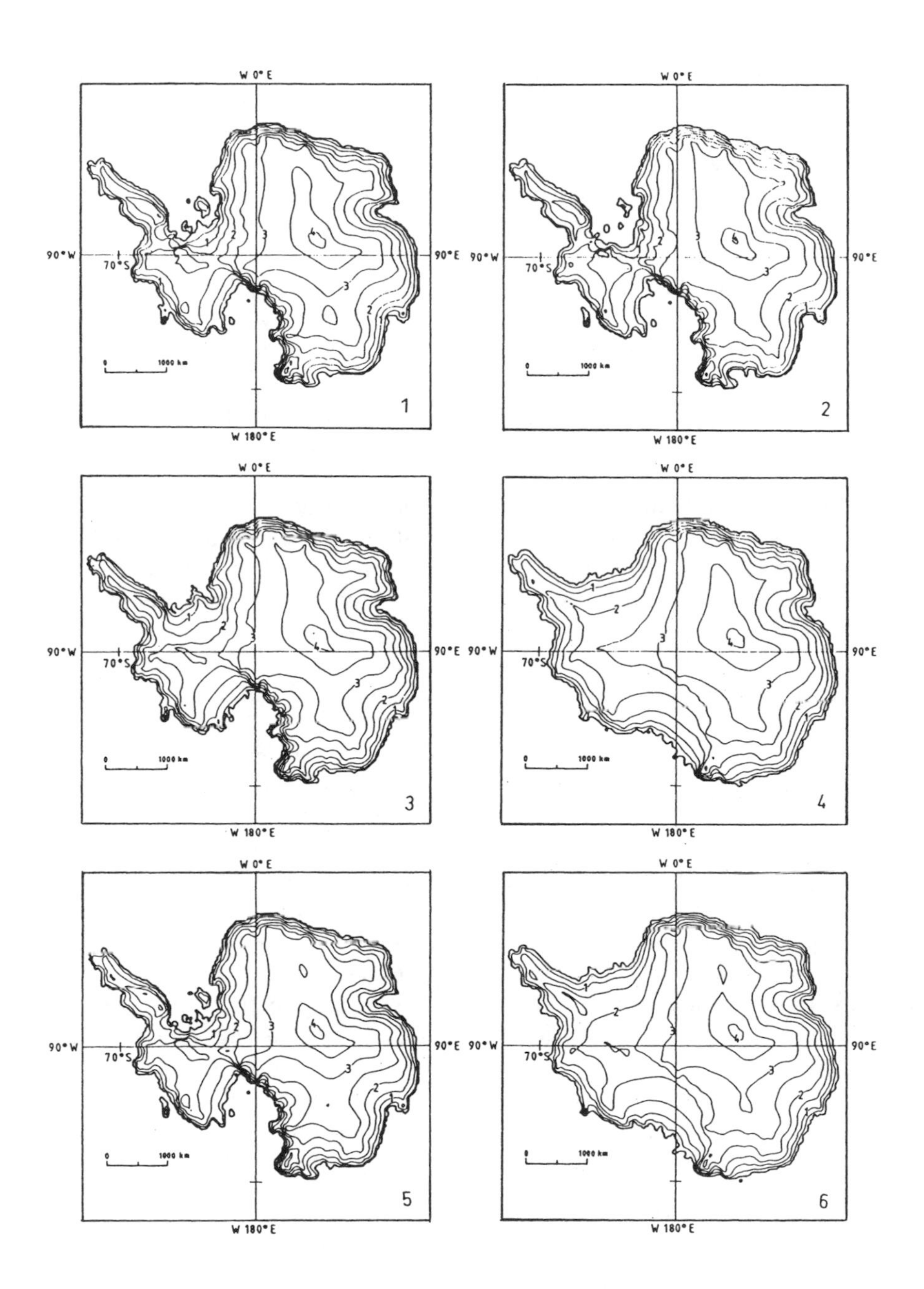

Figure 9.9 Numerical modelling of the response of the Antarctic ice sheet to controlled environmental change. (1) Modern interglacial state. (2) An accumulation rate representative of LGM conditions. (3) An air temperature drop of 10°C. (4) Sea level depression of 130 metres. (5) A combination of conditions in (2) and (3). (6) A combination of all environmental factors (2), (3) and (4). Taken from Huybrechts (1990a), with permission from Springer-Verlag.

- the most pronounced fluctuations to ice sheet size occurred in the West Antarctic Ice Sheet;
- most increase in ice sheet extent can be attributed to sea-level lowering;
- lower ice temperatures lead to an increase in ice extent, but the associated accumulation drop acted to cancel this effect;
- according to Huybrechts steady-state glacial reconstruction, the Antarctic Ice Sheet may have contributed 16 metres to global sea level lowering at the LGM;
- Huybrechts' ice thickness information, thus, does not compare too well with the CLIMAP reconstruction (contribution of 30 metres of sea level fall), although the ice extents match in both experiments;
- since the response times are long, the actual ice sheet may have never reached a 'steady-state'; implying that the 16 metres of sea level fall may be too large. (i.e. further disagreement with CLIMAP).

Antarctic ice sheet reconstruction through the last glacial cycle

The results of Huybrechts' ice sheet sensitivity can be tested further through a time-dependent investigation of the ice sheet growth and decay cycle over the last interglacial–glacial cycle. In order to do this, time-dependent variations in environmental conditions are required as input to the model. Huybrechts used a simplified version of the δ^{18}O sea level record. This comprised a sawtooth function changing linearly with time from modern values at 125 ka to 120 metres of sea level fall by 20 ka, and modern values restored by 10 ka. Surface air temperature was modelled using a temperature record for the last 150 ka derived from analysis of the Vostok ice core record established in the mid-1980s. This also resembles a sawtooth function (with minor oscillations within the signal), with temperature depressions (from modern values) ranging from 0°C at 130 ka and 6 ka, to –8°C at the LGM. Accumulation was linked to the temperature through a simple climate model (Jouzel and Merlivat, 1984; Lorius et al., 1985).

Similar to the ice sheet sensitivity experiments, model results are compared with the 'interglacial' reference run, where the modern-type ice sheet has been constructed. The model results provide a time-dependent evaluation of the ice sheet across the period of the last glacial. Specifically, they show the following brief scenario for ice sheet history (Figure 9.10):

- Ice sheet growth and decay are affected mostly by variations in global sea level.
- Ice sheet volume at 125 ka was less than at present due to a smaller West Antarctic Ice Sheet. This would have left marine sediments beneath the modern subglacial topography, which may be influential in the manner of ice sheet flow today.
- The first stage of ice sheet growth was between 125 and 80 ka, where the West Antarctic Ice Sheet grew to its modern-type size. Subsequently there was up to 50 000 years of much slower, steady ice sheet build-up until 30 ka.
- At 30 ka the ice sheet expanded quickly in response to global sea level fall until the LGM when the contribution of the ice sheet to global sea level fall was around 12 metres.
- LGM ice expansion involved the migration of the grounding line to the continental shelf break in most places.
- The major present day ice sheet domes and divides are maintained throughout the glacial cycle.
- The modelled ice sheet at the LGM is similar in form to that determined by the CLIMAP group (Figure 9.10).

Modelling of the ice sheet in Dronning Maud Land

One area of Antarctica where geological evidence of the LGM ice sheet is hard to find is the western Dronning Maud Land region of

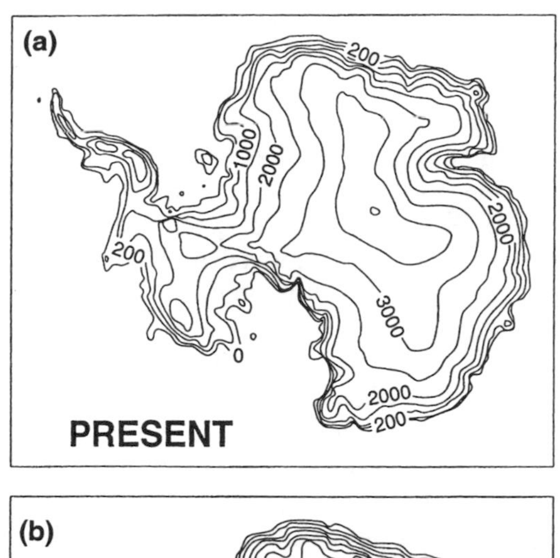

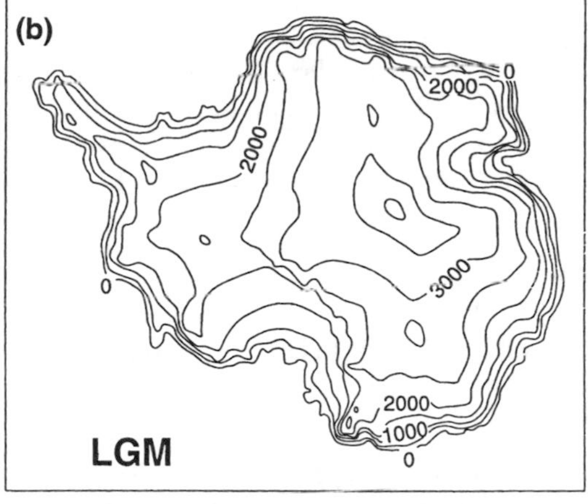

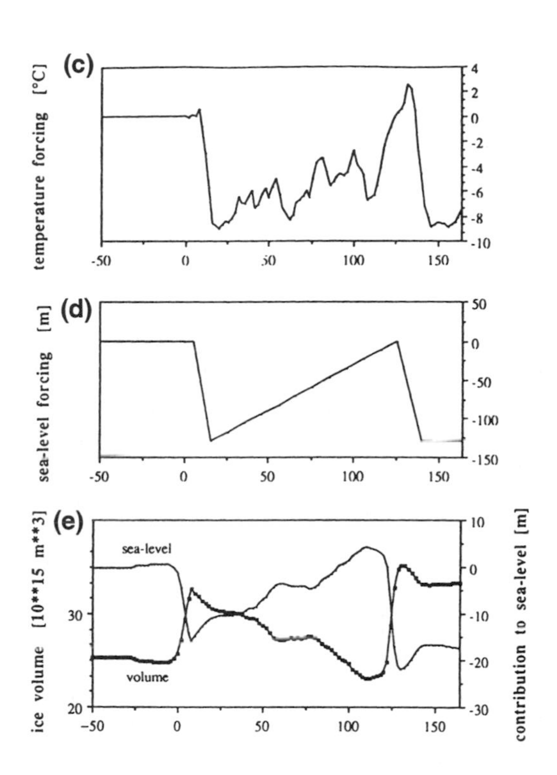

Figure 9.10 Numerical modelling of the Antarctic Ice Sheet through the last glacial cycle. (a) Present-day ice sheet. (b) LGM ice sheet. (c) Temperature change used to force the model. (d) Sea level change used to force the model. (e) Time-dependent change in ice sheet volume and contribution to global sea level. (a–b) Taken from Huybrechts (1992) and reproduced by permission of the Alfred Wegener-Institute and the author. (c–e) Taken from Huybrechts (1990b), and reprinted from the *Journal of Glaciology* with permission of the International Glaciological Society and the author.

East Antarctica (Figure 9.1). Näslund et al. (2000) used a numerical ice sheet model to calculate how this sector of the Antarctic Ice Sheet responds to LGM climate conditions. They found that whilst the interior ice plateau remained relatively unaffected by a 5°C climate cooling, the ice thickness increased substantially between the Heimefrontfjella Mountains and the shelf break, where the ice sheet margin was located. Compared with present day conditions, a larger proportion of the ice sheet base experienced pressure-melting conditions at the LGM. This was especially true in the deep subglacial trough in which the Veststraumen ice stream is located. This may suggest that ice stream dynamics were more active at the LGM that at present. The modelling work of Näslund et al. (2000) is compatible with Huybrechts' large-scale model of Antarctica at the LGM, and the CLIMAP reconstruction.

Results from isostatic modelling of post-glacial uplift and the contribution to global sea level

As discussed in Chapter 3, the Earth's crust is displaced by the loading of ice sheets, which recovers when these loads are taken away. Numerical Earth models can be used to reconstruct former ice loads if information on post-glacial uplift is available (e.g. dated raised beaches). Unfortunately for Antarctica, there is not much reliable uplift information to help in this sort of modelling. The few data that are available come from the Ross Sea, McMurdo Sound, Law Dome (see below) or across the sub-Antarctic islands (where the isostatic signal from the Antarctic Ice Sheet is confused by local glaciation).

Despite the lack of information to run these models, a number of attempts have been made to try and reconstruct late Quaternary ice sheet history using this technique. The results, in order of the contribution to sea level that these models predict, are: Nakada and Lambeck (1988) where 37 metres of sea level was predicted; Tushingham and Peltier (1991) whose ICE-3G model yielded 26 metres worth of ice volume; and Peltier (1994) whose ICE-4G model yielded 22 metres of sea level fall equivalent ice volume. Interestingly, when all the ice sheets modelled by ICE-4G are added together, only 105 metres of sea level fall is modelled (15 metres lower than the eustatic sea level records suggest). These models also yield dissimilar results on the timing of ice sheet decay. Nakada and Lambeck (1998) suggest that deglaciation continued after 6 ka. Tushingham and Peltier (1991) indicated that ice decay began at 9000 years ago, whereas Peltier (1994) calculated this event at 12 000 years ago. There appears to be no clear scenario for the LGM Antarctic Ice Sheet from these models. This is probably due to the paucity of information used to force the model and over-simplifications in the physics of the models themselves (sophisticated though they are).

Goodwin and Zweck (2000) used uplift information from the Windmill Islands and Budd Coast as input to an isostatic model of the Law Dome ice cap in East Antarctica (Figure 9.1). The field data show that since about 7000 years ago there has been around 30 metres of uplift at these sites. A numerical Earth model was used to determine the glacial component of this observed uplift. The model predicts that an ice load between 770 and 1000 metres needs to be removed from the present-day margin of the Law Dome in order for the measured uplift to be calculated. In this situation it is most likely that the former ice sheet margin extended between 40 and 65 kilometres out across the continental shelf towards the shelf break (Goodwin and Zweck, 2000). Since the coast between Wilkes and Oates land (Figure 9.1) is thought to show a similar uplift record to that near the Law Dome, Goodwin and Zweck (2000) postulate that 1000 metres of ice existed across the margin of this part of the East Antarctic Ice Sheet at the LGM.

C. SUMMARY OF GLACIAL HISTORY AND THE CONTRIBUTION TO LGM GLOBAL SEA LEVEL FALL

There is no single agreed value of the Antarctic ice sheet's volume at the LGM and, therefore, its contribution to sea level fall (Bentley, 1999). Earth models yield values ranging from 37 to 22 metres worth of sea level fall. The CLIMAP project calculated a contribution to the eustatic level of 25 metres, whilst Huybrechts' ice sheet model expected about 12 metres of sea level fall from the Antarctic ice sheet. Bentley (1999) summarised these results and undertook a new investigation of the ice volume of LGM Antarctica, based on geological data and GIS procedures. He concluded that the ice volume of the LGM Antarctic ice sheet was only large enough to reduce global sea level by between 6–13 metres (see Chapter 13). This figure ties in well with the maximum value predicted by the numerical ice sheet model of Huybrechts (1992). With the agreement in mind, it is possible to list a few important aspects relating to the late Quaternary Antarctic ice sheet history as follows:

- There is significant evidence to suggest the Antarctic ice sheet experienced a major expansion at the LGM.
- Ice growth was controlled largely by sea level fall, induced by the growth of Northern Hemisphere ice sheets.
- However, the grounded ice margin did not extend to the continental margin in all places. In fact there is geological evidence to suggest a mid-shelf position for the grounded ice sheet edge at several key areas.
- Both the Ross and Filchner-Ronne ice shelves thickened and became part of the grounded ice mass at the LGM. They left behind a series of glaciological landforms that have been observed recently through marine geophysical experiments.
- A grounded Ross Ice Shelf has implications for an enlarged West Antarctic Ice Sheet since the majority of ice within the Ross Ice Shelf originates from this region.
- Numerical ice sheet modelling indicates that the present configuration of major ice sheet domes and divides were active at the LGM.
- The contribution of the Antarctic Ice Sheet to sea level fall prior to deglaciation was between 6 and 13 metres.
- Deglaciation probably occurred at around 12 000 years ago in response to sea level rise induced by the decay of Northern Hemisphere ice sheets. Ice decay was probably completed by 6000 years ago, although a final phase of ice decay after this time has been suggested, which, if correct, has implications for the present stability of the West Antarctic Ice Sheet.

CHAPTER 10

Ice Sheet Reconstructions II
The Greenland Ice Sheet

A VERY BRIEF INTRODUCTION TO THE GREENLAND ICE SHEET

Present volume of the Greenland Ice Sheet is 2 825 000 cubic kilometres and represents 10 per cent of the world's glacierised surface area (at 1 755 640 square kilometres), making it the second largest ice sheet (Figure 10.1). It is constrained by its surrounding topography, which effectively acts as a 'cage' holding the ice sheet within it. The thickness at the centre of the ice sheet is over 3200 metres. Understanding of the climate during the Late Quaternary has increased in recent years due to the high resolution deep ice cores that have been extracted from the central regions of the Greenland Ice Sheet.

The ice sheet loses mass via surface ablation in the southern-most region, and by iceberg calving of its outlet glaciers in the east and west. The modern climate setting of Greenland is characterised by very cold and dry conditions in the north of the ice sheet, and warmer moist conditions in the south-east due to its proximity with the North Atlantic (Figure 10.2). The climate around Greenland is likely to have been significantly different during the last glacial compared with the present situation, due to its location in the Northern Hemisphere (taking into account what we have already acknowledged about the LGM climate in this region).

The source of moisture for the snow that accumulates on the ice sheet is the North Atlantic and the neighbouring seas that border the ice sheet. This moisture source affects the present day distribution of precipitation (snowfall) over Greenland. There is significantly more accumulation in the south (where a maritime climate dominates with snow accumulation rates up to 80 centimetres per year) than in the north (which is influenced by a more polar environment allowing only 3–40 centimetres of snow each year). Over glacial cycles, changes to the conditions of these marine regions result in an alteration in the accumulation and temperature regimes of the ice sheet, which can be measured from ice cores. Repeat airborne laser altimetry surveys in 1994 and 1999 have shown that under modern climate warming, it is the lower elevations of Greenland that experience the greatest ice sheet response, whilst the higher elevations show little change (Krabill et al., 1999, 2000). This is especially true in the south of Greenland where changes to the ELA, and variability in the accumulation of snow, are most likely to influence ice sheet mass balance (Davis et al., 1998; McConnell et al., 2000).

Ice flow in the central regions of the Greenland Ice Sheet is similar to the Antarctic Ice Sheet (Figure 9.2). The slow-moving ice sheet interior feeds fast-flowing outlet glaciers and ice streams. Recent numerical models and interferometric synthetic aperture radar (InSAR) data have been used to identify that these drainage features extend several hundred kilometres into the parent ice mass (Bamber et al., 2000b). The position of fast flowing outlet glaciers is constrained by topography. The Quaternary glaciations of Greenland have resulted in a well-established series of mega-scale erosional landforms that form a topographic path through the

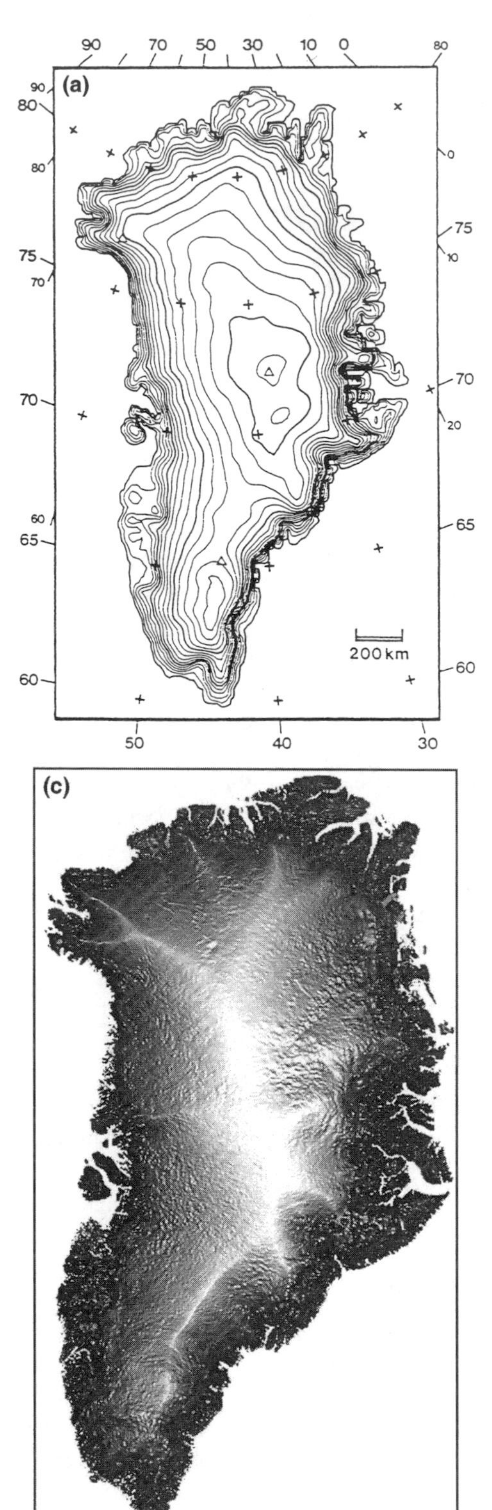

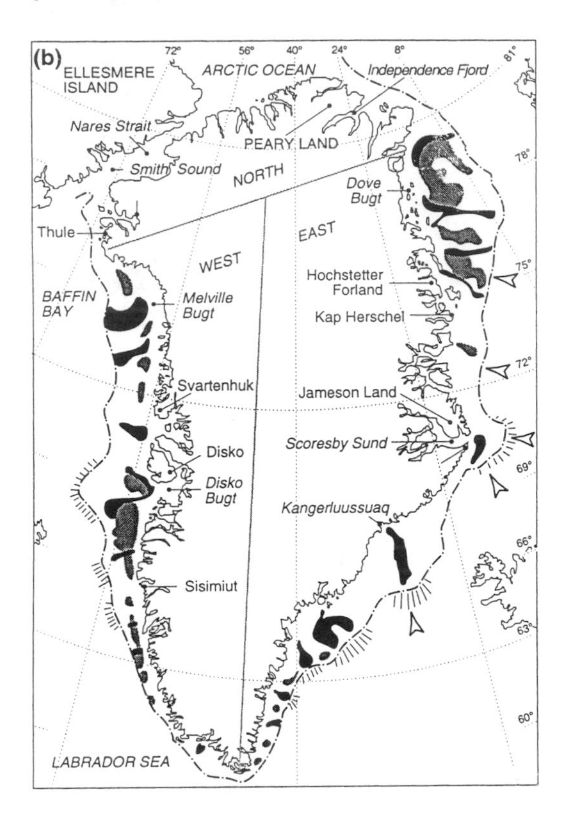

Figure 10.1 (a) Modern surface elevation of Greenland. Contours are provided in 200 metre intervals. Reprinted from *Global and Planetary Change*, 90, A. Letréguilly et al., The Greenland ice sheet through the last glacial-interglacial cycle, pp. 385–394, Copyright (1991) with permission from Elsevier Science. (b) Greenland continental margin bathymetric features. Dark shaded regions are troughs. Light shaded regions are shallow banks. Reproduced by permission of the Geological Society of Denmark. (c) Surface morphology of the Greenland Ice Sheet (from data supplied by J. Bamber, University of Bristol, England).

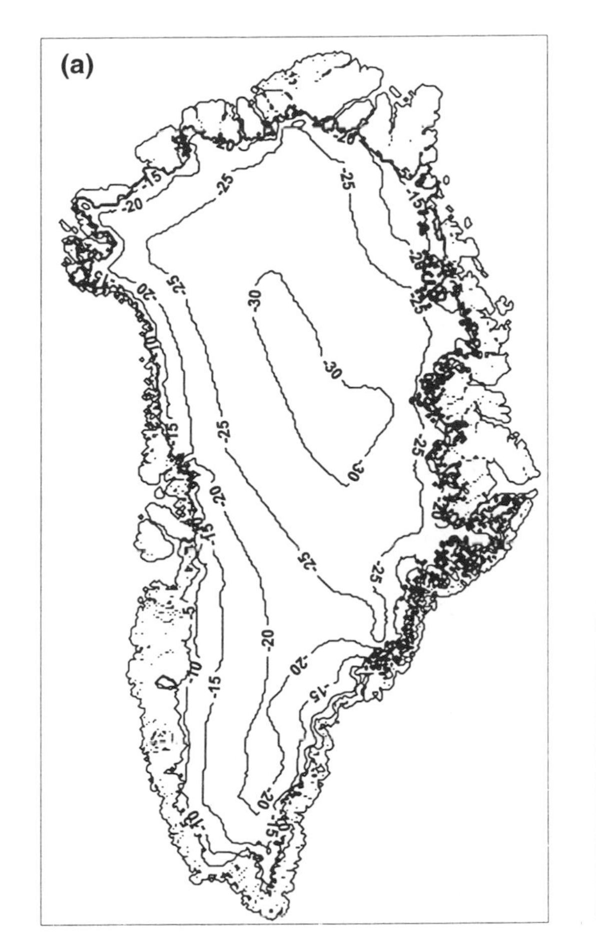

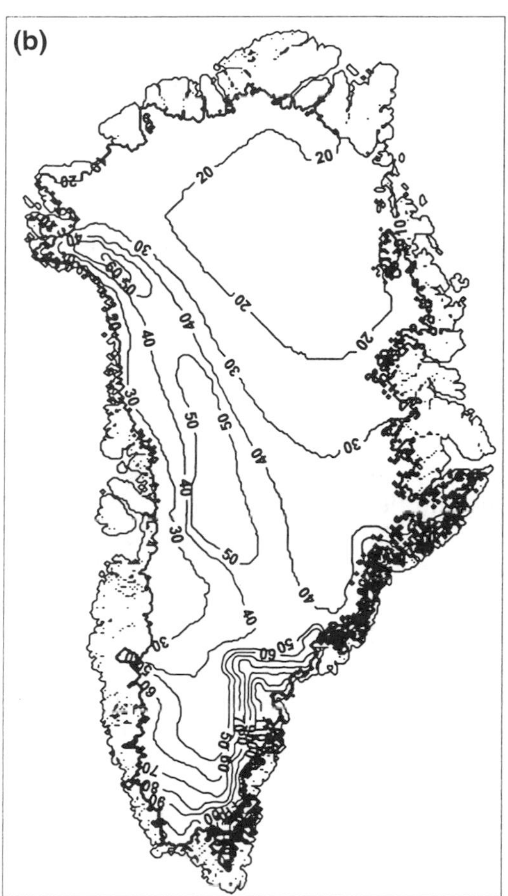

Figure 10.2 Present day climate conditions over Greenland. (a) Mean annual surface temperature (in °C), (b) mean annual accumulation of ice (in cm yr^{-1} water equivalent). Taken from Funder et al. (1998) from data presented in Ohmura and Reeh (1991). Reprinted from *Quaternary Science Reviews*, 17, S. Funder et al., History of a stable ice margin – Greenland during the upper and middle Pleistocene, pp. 77–123, copyright (1998), with permission from Elsevier Science.

mountains, to the open ocean to the north, west and east. The coastal regions of Greenland are characterised by a suite of glacially derived fjords. At present, many outlet glaciers terminate as calving ice walls within these fjords. However, their presence testifies to the expansion of the ice sheet that must have occurred at some time in the Quaternary. For example, the Scoresby Sund fjord complex, at 13 400 square kilometres, is the largest in the world and has several outlet glaciers leading into it, supplying icebergs to the Norwegian–Greenland Sea. The continental shelf break lies only a few tens of kilometres from the mouth of the fjords. Bathymetric troughs often occur between the fjords and the continental margin (Figure 10.1).

Although the flow of ice within the central regions of the Greenland Ice Sheet is measured at a few metres per year, the velocities of ice in the outlet glaciers that drain huge interior basins within the ice sheet are hundreds to thousands of metres per year. Indeed the fastest known outlet glacier, Jakobshaven Isbræ, located in West Greenland, drains 6.5 per cent of the Greenland Ice Sheet (Echelmeyer et al., 1991, 1992) and has a surface velocity of between 1 and 7 km yr^{-1} (Iken et al., 1993). Approximately 400 000 square kilometres of the ice sheet drains through the outlet glaciers of East Greenland, 600 000 square kilometres of the ice sheet drains via the west coast and 700 000 square kilometres drains to the north (Weidick, 1985). As well as the Greenland Ice Sheet, there is a further 65 000 square kilometres of ice located in a series of ice caps and glaciers within this landmass (Weidick, 1985). Around 20 per cent of Greenland is currently free of ice.

The sea around Greenland may exhibit a control on the maximum extent of ice that can be grown. The GIN seas are to the east, with the continental shelf break only a few kilometres from the fjord mouths. In the north-west, a complex suite of islands connects Greenland with the Canadian High Arctic. A narrow strip of water separates Arctic Canada from Greenland (known as the Nares Strait). Mid-west off Greenland there is a very deep water region between mainland Greenland and Canada.

Oxygen isotope analysis indicates that the sea level was several metres higher at 130 000 years ago than at present (Figure 1.6). Although marine sediments beneath West Antarctic indicate that the ice sheet may have been smaller during the last interglacial, a smaller Greenland ice sheet at this time cannot be ruled out because the ELA in the south of the ice sheet is most likely to have increased in response to the sea level rise. Indeed Cuffey and Marshall (2000) demonstrate this point by comparing GRIP ice core records and numerical ice sheet modelling results of the Eemian glaciation of Greenland. They suggest that the Greenland Ice Sheet was responsible for 4.5 metres of the known last interglacial sea level rise; a conclusion in agreement with the earlier work of Letréguilly et al. (1991a) (Figures 10.3, 10.4). It should be noted that, at present, there is no consensus as to which of the Greenland and West Antarctic ice sheets were responsible for the sea-level rise during the last interglacial. It could well be that they were both involved in some way to raising the levels of the global oceans at this time.

As in the last chapter on Antarctica, the discussion on the reconstruction of the Greenland Ice Sheet is split in two parts (a) geological information detailing the extent of ice during the LGM and (b) numerical modelling providing information on central ice thickness at this time.

A. GEOLOGICAL INFORMATION

Introduction

Between 1990 and 1995, a scientific programme called the *Polar North Atlantic Margins: Late Cenozoic Evolution* (PONAM) ran in order to identify the glacial history of East Greenland and Western Barents Sea. The programme enabled several field expeditions to the East Greenland fjords. These were aimed at examining terrestrial glacial geology (including Holocene uplift patterns), fjord sedimentology and the continental shelf sedimentology. Much of the discussion of the Late Quaternary geological history of the eastern Greenland margin has been established from this programme. Information from other regions of the ice sheet is available from a variety of literature sources, as summarised in Funder and Hansen (1996).

The last glacial maximum

During the LGM, only southern Greenland experienced a significant growth in ice extent and thickness, with expansion on to the continental shelf. This is intuitively reasonable given that the position of the southern margin of the ice sheet is controlled by the ELA. Reduction of the ELA in response to falling air temperature would cause expansion of the ice sheet in this region. In the north of the ice sheet, outlet

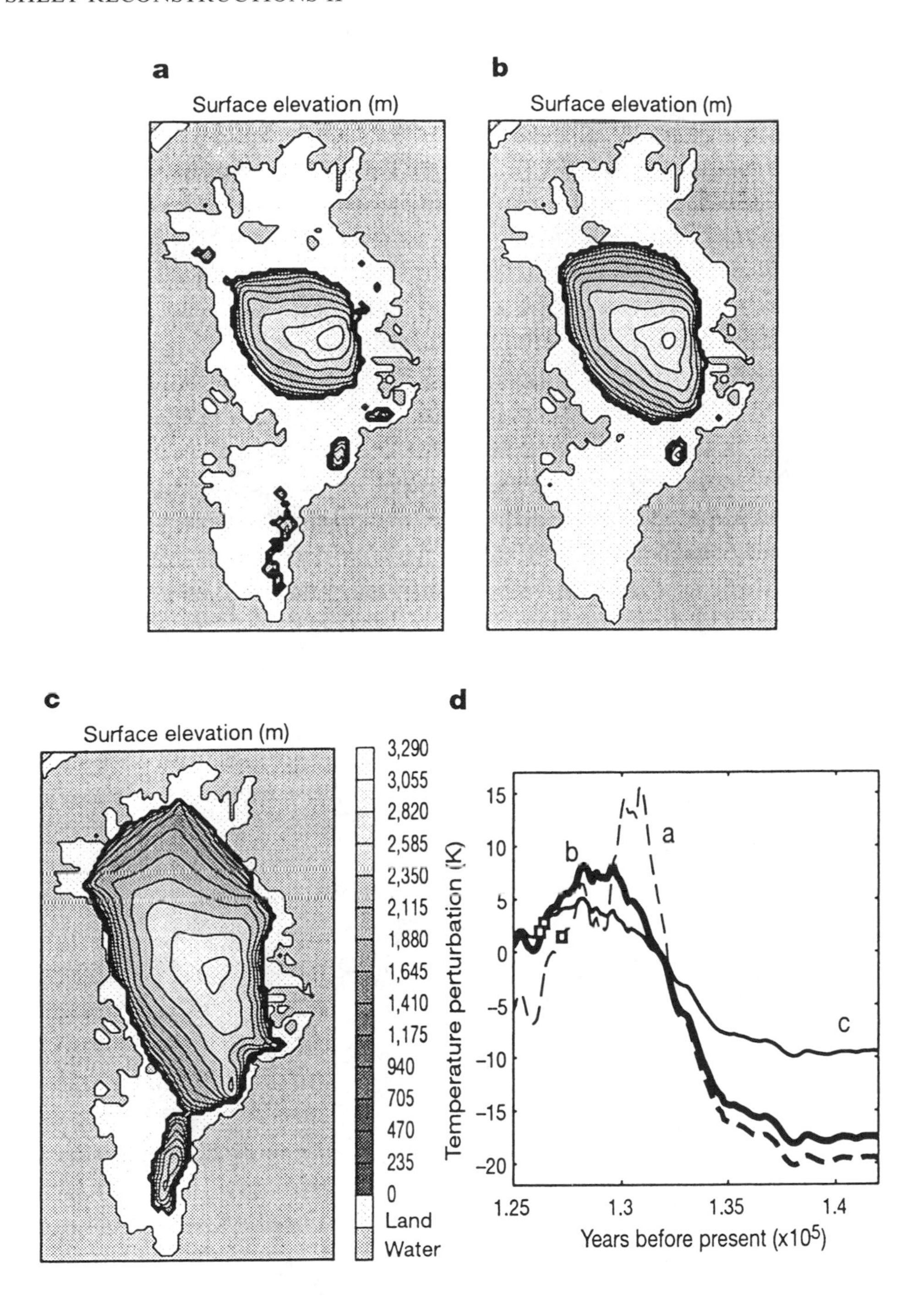

Figure 10.3 Modelled response of the Greenland ice sheet to Eemian climate conditions as indicated in ice core records. a–c represent ice sheet at the Eemian minimum under three different interglacial temperature reconstructions (Cuffey and Marshall, 2000). Reprinted with permission from *Nature* vol 404, pp. 591–594, copyright (2000) Macmillan Magazines Limited.

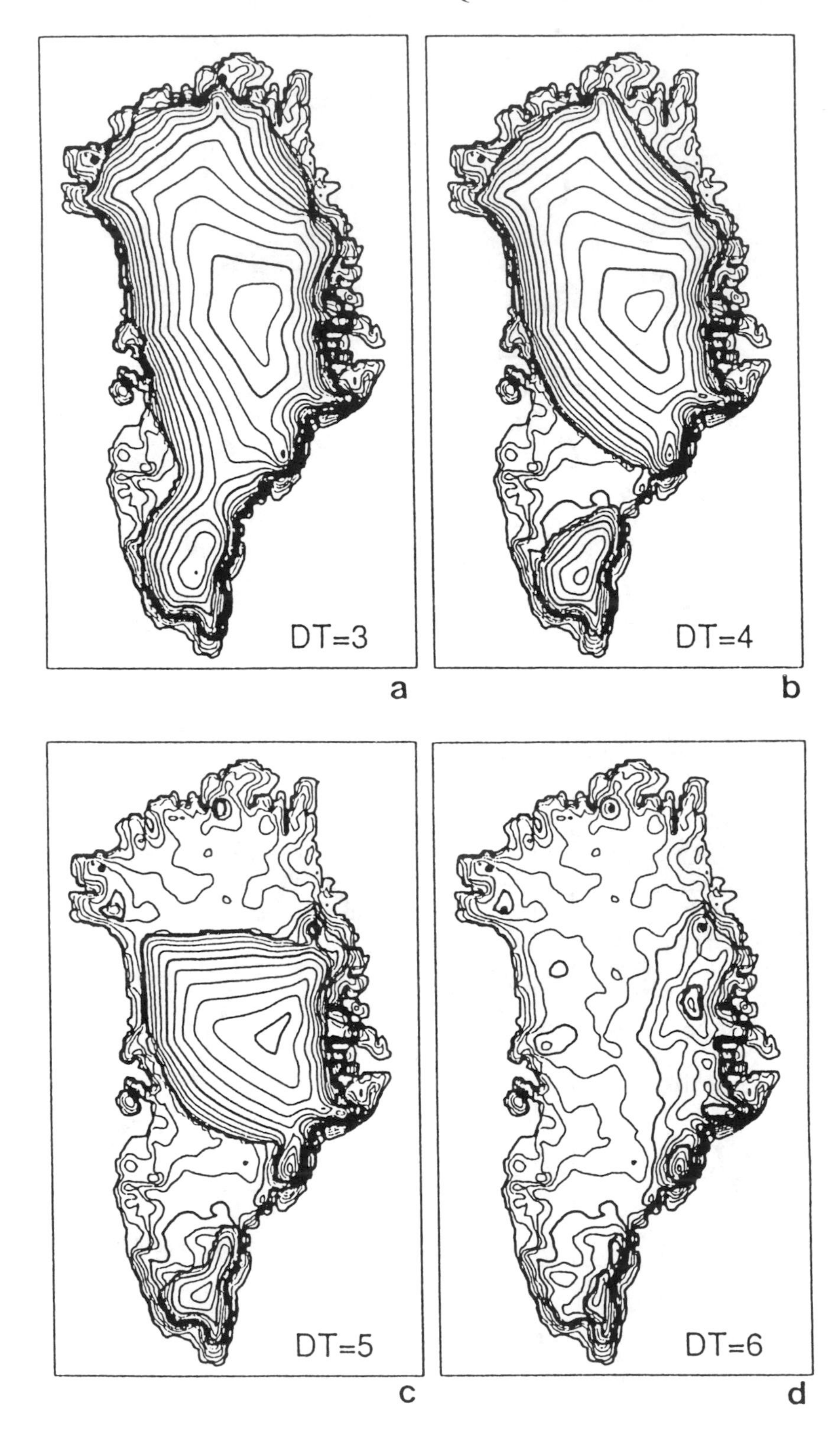
DT=3
a
DT=4
b
DT=5
c
DT=6
d

glaciers filled fjord basins, including the Nares Strait between Canada and Greenland, and piedmont glaciers descended from coastal mountains on to the coastline. However, glaciers did not expand to the continental shelf margin in the north of Greenland. In the west, outlet glaciers expanded to the shelf break. However, the ice in this region was relatively thin, and decoupled (effectively) from the interior ice mass by the rim of mountains. The glacial geology of Greenland provides information on the LGM and last deglaciation. However, due to the scarcity of field data, the discussion here is rather limited to the work of only a few important recent field campaigns over the periphery of the continent.

West Greenland

The most obvious glacigeological features across the mountains of West Greenland are the weathering limits or trimlines (where highly weathered rocks border regions of glacial striations). These indicate altitudes and locations of the ice sheet surface at the LGM. The weathering limit is found between present day altitudes of 300 and 1100 metres above sea level, rising steadily from north to south. This implies the LGM ice cover in the west was relatively thin (between 300–600 metres of ice thickness). Marine sediments indicate that fjords were filled with ice, but only to the inner shelf, where a number of moraine-type features have been observed. The 10 000 year ice margin in this region is based, rather loosely, on interpolation between dated moraine segments within the fjords.

One recent investigation has examined sediments within isolation basins across a small part of central West Greenland (Long et al., 1999). Isolation basins are formed when surface lakes depressed below sea level due to the isostatic response to ice sheet growth. This causes a change in the character of sediments from lacustrine to marine. As the ice sheet decays, isostatic recovery leads to the emergence of marine sections, and the re-establishment of lakes, and lacustrine sediments. Dating the transitions between lacustrine and marine sediments within a number of these lakes from a variety of locations and elevations leads to an understanding of the pattern and rates of isostatic behaviour during the Late Quaternary. Long et al. (1999) examined nine isolation basins around Disko Bugt. They found that the north–south isostatic signal seen in trimlines was observable even within their relatively small study area. The information also showed that the isostatic signature was probably more complex than the trimlines indicated. However, despite the significance of their findings, a lack of data from a wide area of West Greenland currently precludes the extension of the results to other regions.

North Greenland

Fresh glacial landforms and striations, combined with a large Holocene isostatic rebound, show that a major glacier occupied the Nares Strait between Greenland and Ellesmere Island during the LGM. Dates from shell fragments found in till suggest that Smith Sound and Nares Strait were free of ice at 29 000 years ago. The glacier was derived from the Innuitian Ice Sheet over Arctic Canada and the Greenland Ice Sheet; reaching a maximum elevation of 550 metres on the present adjacent mountains. This glacier flowed south toward Baffin Bay, where an iceberg calving front existed, and blocked a present-day connection between the Arctic and Atlantic oceans. The ice stream may have been 1500 metres thick. It is termed the Smith Sound Ice Stream, and drained ice from the northern

Figure 10.4 Modelled response of the Greenland ice sheet to increases in the modern mean annual air temperature. Contours are provided in 200 metre intervals. Taken from Letréguilly et al. (1991b) and reprinted from the *Journal of Glaciology* with permission of the International Glaciological Society and the authors.

regions of the inland ice sheet (Figure 10.1; also see Chapter 12). The Nares Strait deglaciated about 10 000 years ago (Zreda et al., 1999). Along the other end of Nares Strait, at the Arctic Ocean edge, the ice extent was limited to fjord glaciers advancing toward the fjord mouths. Piedmont glaciers within the mountains covered the coastline between the glacier filled fjords. However, as mentioned in the discussion on Antarctic glaciation, some of the glaciers may have actually been smaller at the LGM than at present.

East Greenland

Marine, fluvial and glacigenic sediments exposed in coastal cliffs and stream-cut sections in East Greenland (between 69° and 78°N) display a record of Quaternary climate and environmental change dating back to pre-Saalian time (>240 ka), with the most information dating from the last glacial. PONAM activities were based largely within this region. Onshore studies at the mouth of Scoresby Sund (the largest fjord complex in the world) indicate a large grounded outlet glacier occupied the fjord at the LGM. The thickness of the glacier is likely to be less than 400 metres. The edge of the Scoresby Sund glacier was probably at the 'Kapp Brewster sedimentary ridge'; a 20 kilometre wide, 175 metre high and more than 30 kilometre long ridge of Quaternary sediments located from airgun and bathymetric data at the fjord mouth (Figure 10.5) (Dowdeswell et al., 1994). A similar but smaller moraine ridge occurs north, at the mouth of Kong Oscar Fjord, but ridges are lacking in other fjords. Coring of sediments over the continental shelf off East Greenland indicates that significant iceberg fluxes and meltwaters were derived from East Greenland between 21 000 and 16 000 years ago. Coring also has provided evidence that suggests deglaciation began before 13 600 years, leaving the 250 kilometre broad shelf free of ice by 10 000 years ago. From Scoresby Sund and northwards, glaciation was limited to outlet glaciers that filled fjords, but left adjacent uplands free of ice. To the South, inland ice reached the shelf, and expanded to the shelf break.

South Greenland

To reiterate, southern Greenland experienced the most increase in ice volume, and ice extending to the continental shelf was connected directly to the interior parent ice mass.

Evidence from ice cores

In the past few years, two deep ice cores from the centre of the ice sheet have been drilled through over 3 kilometres of ice to the subglacial rock. They are the GRIP and GISP2 ice cores (Chapter 2). Palaeo-environment evidence from these cores allow a detailed understanding of the climate during the last glaciation. Information from the last glaciation is not limited to these new ice cores. There are at least nine other ice core sites from Greenland and the Canadian Arctic detailing the climate conditions for the last 50 000 years (Paterson, 1994). Statistical correlation from information within the GRIP ice core has resulted in a surface air temperature history for the centre and edge of the ice sheet, over 50 000 years. The GRIP ice core is located on the summit of the ice sheet, where the ice is over 3 kilometres thick. These data indicate that LGM temperatures were around 23°C colder than at present.

Summary of deglaciation

The marine sections of the ice sheet, many of which were effectively decoupled from the interior ice mass by the topographic control

Figure 10.5 (a) The location of Scoresby Sund and the Kapp Brewster moraine. (b) Seismic section of the Kapp Brewster moraine and (c) interpretation of the seismic data. Taken from Dowdeswell et al. (1994) with permission from Taylor & Francis AS.

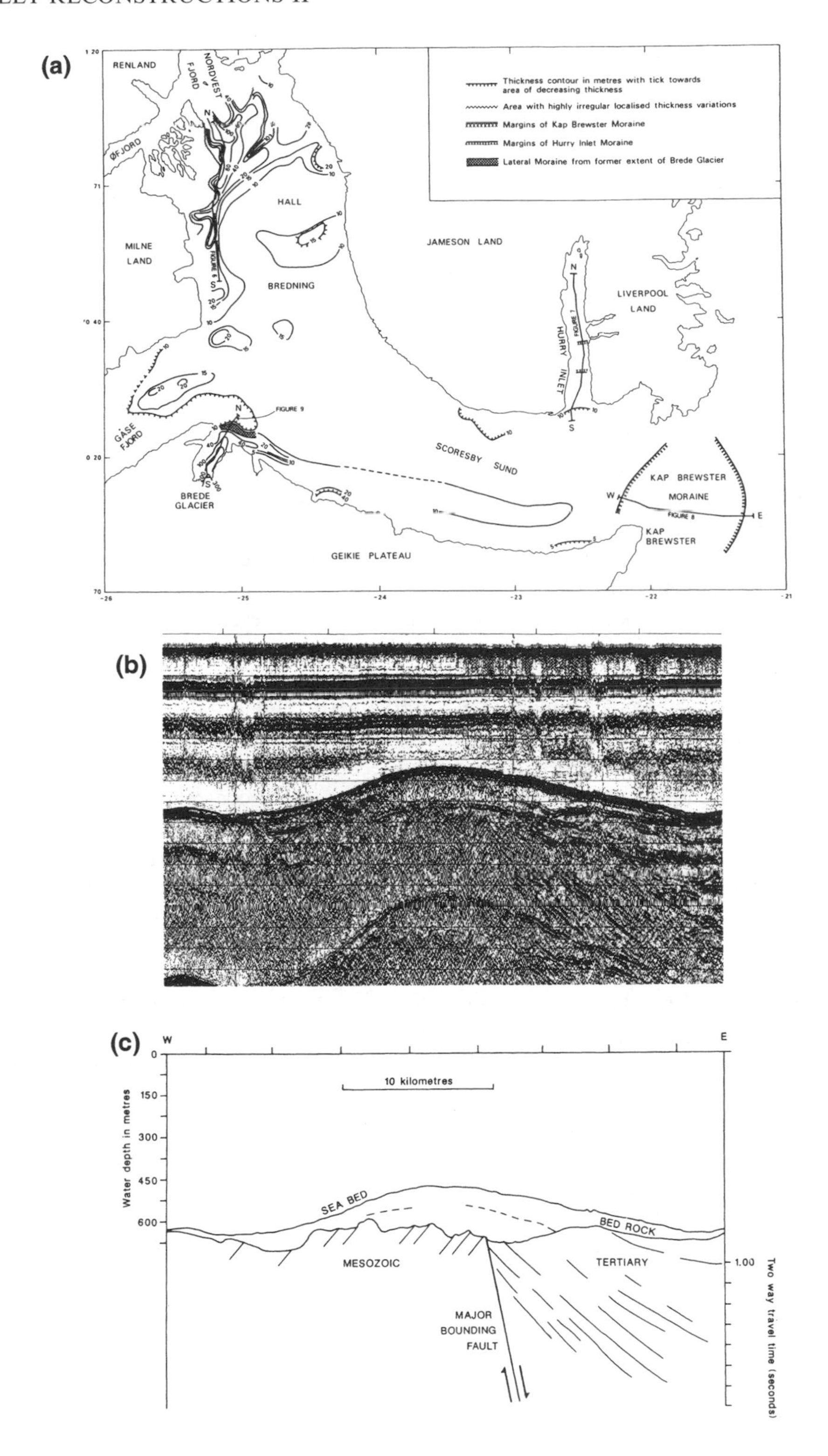
(a)
RENLAND
NORDVEST FJORD
ØFJORD
MILNE LAND
HALL
BREDNING
JAMESON LAND
LIVERPOOL LAND
HURRY INLET
GÅSE FJORD
BREDE GLACIER
SCORESBY SUND
KAP BREWSTER MORAINE
KAP BREWSTER
GEIKIE PLATEAU
Thickness contour in metres with tick towards area of decreasing thickness
Area with highly irregular localised thickness variations
Margins of Kap Brewster Moraine
Margins of Hurry Inlet Moraine
Lateral Moraine from former extent of Brede Glacier
(b)
(c)
W
E
10 kilometres
Water depth in metres
0
150
300
450
600
SEA BED
BED ROCK
MESOZOIC
TERTIARY
MAJOR BOUNDING FAULT
1.00
Two way travel time (seconds)

exerted by the rim of mountains, decayed first due to sea level rise and increase in iceberg production.

Ice core evidence for rapid climate changes over Greenland

The GRIP and GISP2 ice cores, separated by about 30 kilometres, indicate significant climate variability for the last deglaciation. This variability has been linked to oscillations in the North Atlantic which, as will be discussed later, may be due to ice sheet conditions across North America (e.g. Alley and MacAyeal, 1994; Adkins et al., 1997; Appenzeller et al., 1998). For example, Alley et al. (1993) indicate decadal temperature changes of 10–15°C during deglaciation. At present, the North Atlantic acts as a source of heat for the Greenland Ice Sheet, due to the present set up of the ocean circulation system (Chapter 8). However, if this system were altered such that the northward transport of warm mid-Atlantic water was curtailed (e.g. Jones, 1991a; Lehman and Keigwin, 1992), this heat source may be halted. This idea has been used to explain Dansgaard-Oeschger oscillations, referred to in Chapter 8 (e.g. Alley and Clark, 1999).

At the end of the last deglaciation, the return to warm conditions was reversed for about 1500 years (the Younger Dryas). This appears to be different to a D–O event because it is not cyclic. However, the brief return to glacial conditions may too have been caused by ice-ocean-atmosphere interaction (see Chapters 8 and 12 on Heinrich events). Evidence for the Younger Dryas can be found throughout the Northern Hemisphere. It was a period of ice sheet growth at the very end of the last deglaciation. Many regions that were covered by ice at the LGM and subsequently deglaciated, witnessed glacier readvance. Unfortunately, in many areas, the Younger Dryas caused some reworking of the LGM glacial geology. This is especially true in the case of Greenland, where the interference of the Younger Dryas to the deglacial isostatic signal has still to be resolved.

It should be noted at this point that there is some evidence for synchronicity of ice core records for rapid climate changes between Greenland and some parts of Antarctica (Steig et al., 1998), suggesting that the climate reversals originating from the Northern Hemisphere may have a global consequence. For example, an ice core from the Taylor Dome shows a remarkable similarity to the GISP2 ice core in a step-wise increase in δD across the Bølling-Allerød period (between 15 and 13 000 years ago) and the Younger Dryas shortly afterwards (Steig et al., 1998).

Uplift history

The isostatic recovery of the Greenland continent during and after deglaciation is complicated by the presence of the ice load that exists today. During the LGM, when the ice sheet had expanded across the mountains to the west, east and north, and on to the continental shelf in many places, the periphery of the continent was subject to isostatic depression. However, ice thickness in the centre changed by only a small amount. Subsequent ice sheet decay led to the recovery of the crust at the margins of the ice sheet, but generally not beneath the ice sheet. Thus, the isostatic depression and recovery occurred as a series of 'domes', centred over the position of maximum change in ice thickness between the LGM and Holocene (Figure 10.6). Because of this, the rates of isostatic uplift recorded in raised beaches and isolation basins may vary significantly across relatively short distances. The marine uplift around Greenland takes the form of three main domes; Scoresby Sund region of East Greenland (maximum limit of 135 metres), Sisimut area of West Greenland (maximum limit of 140 metres) and Hall Land in North Greenland (maximum limit 130 metres). The domes coincide with dateable moraines indicating that (1) large-scale decay of ice had occurred before 12 ka, (2) a re-advance occurred around 12 ka, and (3) after 11 ka, 2000 years of further rapid retreat to modern limits. These

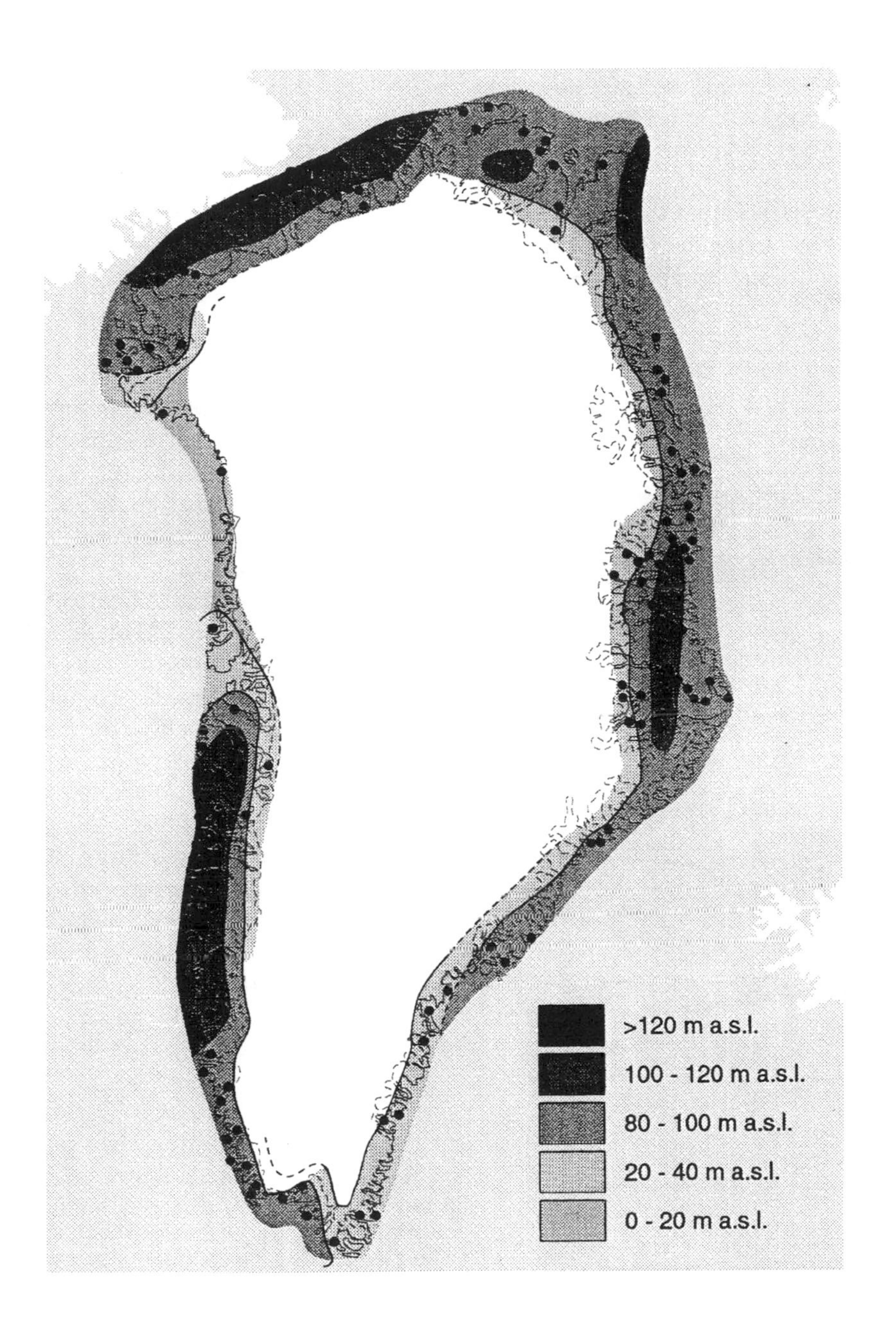

Figure 10.6 Post-glacial uplift of Greenland. Reproduced by permission of the Geological Society of Denmark.

data show the main ice loading zones across Greenland, and demonstrate the likely areas of ice sheet advance. A simple model of how uplift is organised in 'domes' as a consequence of ice retreat in Greenland is shown in Figure 10.7.

Geological information has allowed a relatively good understanding of the glacial history during the LGM of the ice sheet edge, but relatively little information about the ice sheet centre. For discussion on this topic, we use numerical model information.

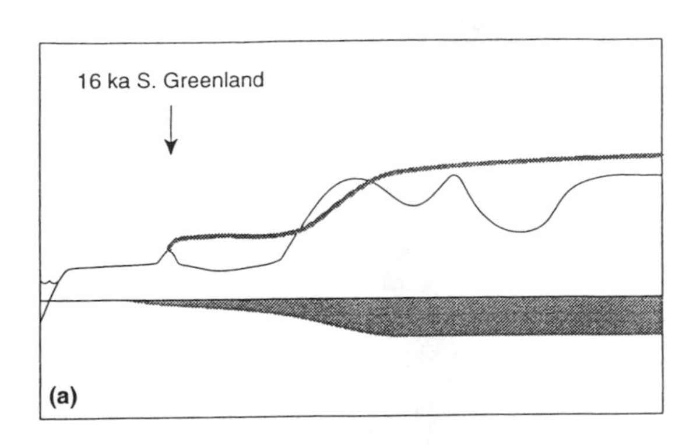

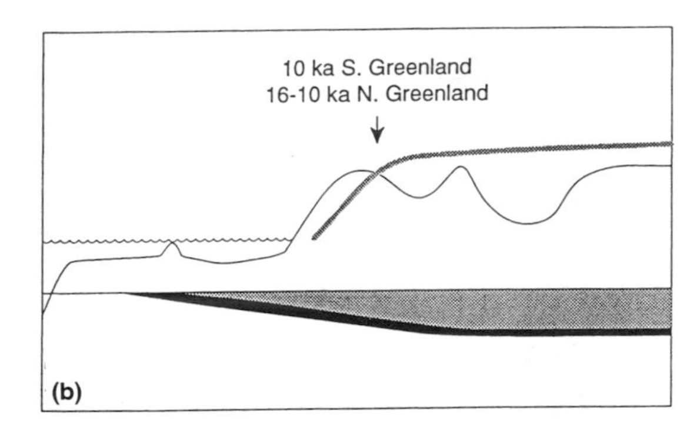

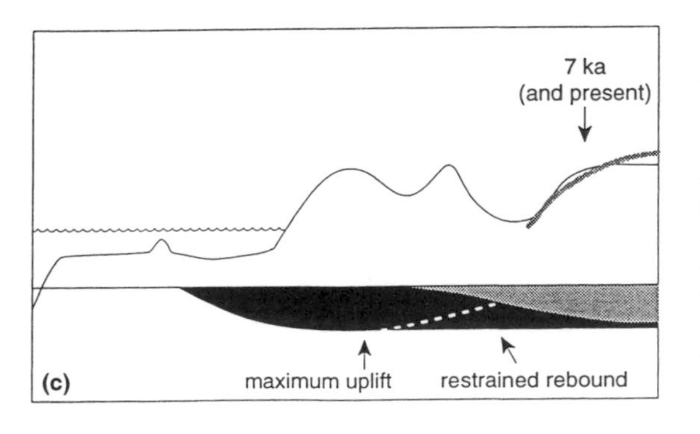

Figure 10.7 Conceptual model of postglacial uplift in Greenland. At the LGM, the ice sheet extended across the continental shelf. The relatively thick ice cover across the continental shelf caused a small amount of depression relative to that beneath the main body of the ice sheet further inland. As the sea level rose, the ice sheet retreated to the fjord mouths and the coast of Greenland. This resulted in a small amount of uplift across the continental shelf but most uplift had yet to occur. Holocene warming caused ice sheet retreat, significant in the south of Greenland and across the coastal regions, causing uplift between the continental shelf and the new position of the ice sheet. Hence, the uplift was limited to the edges of Greenland. Reproduced by permission of the Geological Society of Denmark.

B. NUMERICAL MODELLING INFORMATION

Glaciological ice sheet modelling

Introduction

Numerical ice sheet models are able to describe the flow and extent of contemporary ice masses. By adjusting the input conditions (climate) to the model to account for changes in the environment at, say, the LGM, we can use the same models to tell us information about former ice sheets.

The most recent modelling of the Greenland Ice Sheet through the LGM was performed by Letréguilly et al. (1991a,b). As in numerical models of the Antarctic Ice Sheet, the main problem in reconstructing the ice sheet is that the subglacial bedrock topography is known only from airborne radio-echo sounding data (e.g. Legarsky et al., 1998). This means that the subglacial morphology used in the model represents a heavily smoothed topography which may be unrealistic in places. However, the advantage of modelling this ice sheet (as with Antarctica) is that the modern ice sheet surface elevation is well known. This provides a dataset with which to compare and calibrate model results.

Glacial–interglacial results

It must be noted that the model used by Letréguilly et al. (1991a,b) did not account well for the dynamic change that occurs at the ice margin. Consequently, the model is capable of determining changes to the central ice thickness, but not necessarily the ice extent. The main points of the modelling are as follows:

- the model begins at 150 000 years ago, prior to the last interglacial, where environmental conditions were similar to that at present (Figure 10.8);
- the ice sheet was forced by a climate function derived from an oxygen-18 record measured on surface ice samples collected at Pakitsoq, Central West Greenland (Figure 10.8);
- this forcing function guides the modern precipitation regime through the glacial cycle;
- the model shows two periods of ice growth and decay prior to 100 000 years ago. Minimum ice volumes were modelled at 130 000 and 100 000 years ago (the Eemian and last interstadial, respectively). The volume of ice involved was enough to raise global sea level by about 2 metres in both cases. This ice sheet variation was observed mainly in the south of Greenland (Figures 10.8 and 10.9);
- it should be noted that Cuffey and Marshall (2000) have also used numerical modelling, combined with recent ice-core data, to suggest that the Greenland ice sheet was small enough at the last interglacial such that sea level would have been as much as 4–5 metres higher than at present (Figure 10.3);
- ice surface elevations across the middle and north of the ice sheet remained largely unaffected during the last glaciation;
- ice volume increased steadily between 100 000 and 20 000 years ago. However, this ice volume contributed less than 1 metre to global sea level fall (Figure 10.8; Figure 10.9);
- by 10 000 years ago, the ice sheet had returned to its present configuration.

These model results indicate that ice sheet expansion across fjords and the continental margin was not caused by significant changes to the interior ice sheet. Instead, sea-level fall is the likely driver for ice sheet growth across the margins of Greenland. Since the changes in volume of the interior ice sheet caused a reduction in global sea level of only 1 metre, it is highly likely that the expansion of the Greenland Ice Sheet occurred as a result of large-scale glaciation elsewhere (i.e. North America – Chapter 12).

Ice sheet sensitivity experiments

A second study by Letréguilly et al. (1991b) determined the present ice sheet response to steady-state conditions of future climate warming. This can be regarded as a sensitivity experiment similar to Huybrechts' experiments with the Antarctic Ice Sheet. The main results from this exercise are as follows:

- the model indicates that the ice sheet extent is sensitive to climate warming;
- however, the model is also relatively stable, since it requires a temperature increase of at least 6°C to effect collapse of the ice sheet;
- very interestingly, and controversially, the model indicates that the ice sheet is not a relic from the last ice age, but would reform on bare bedrock under the modern, or even slightly warmer, climate conditions.

Future work

The EISMINT group of ice sheet modellers have recently used the present day Greenland

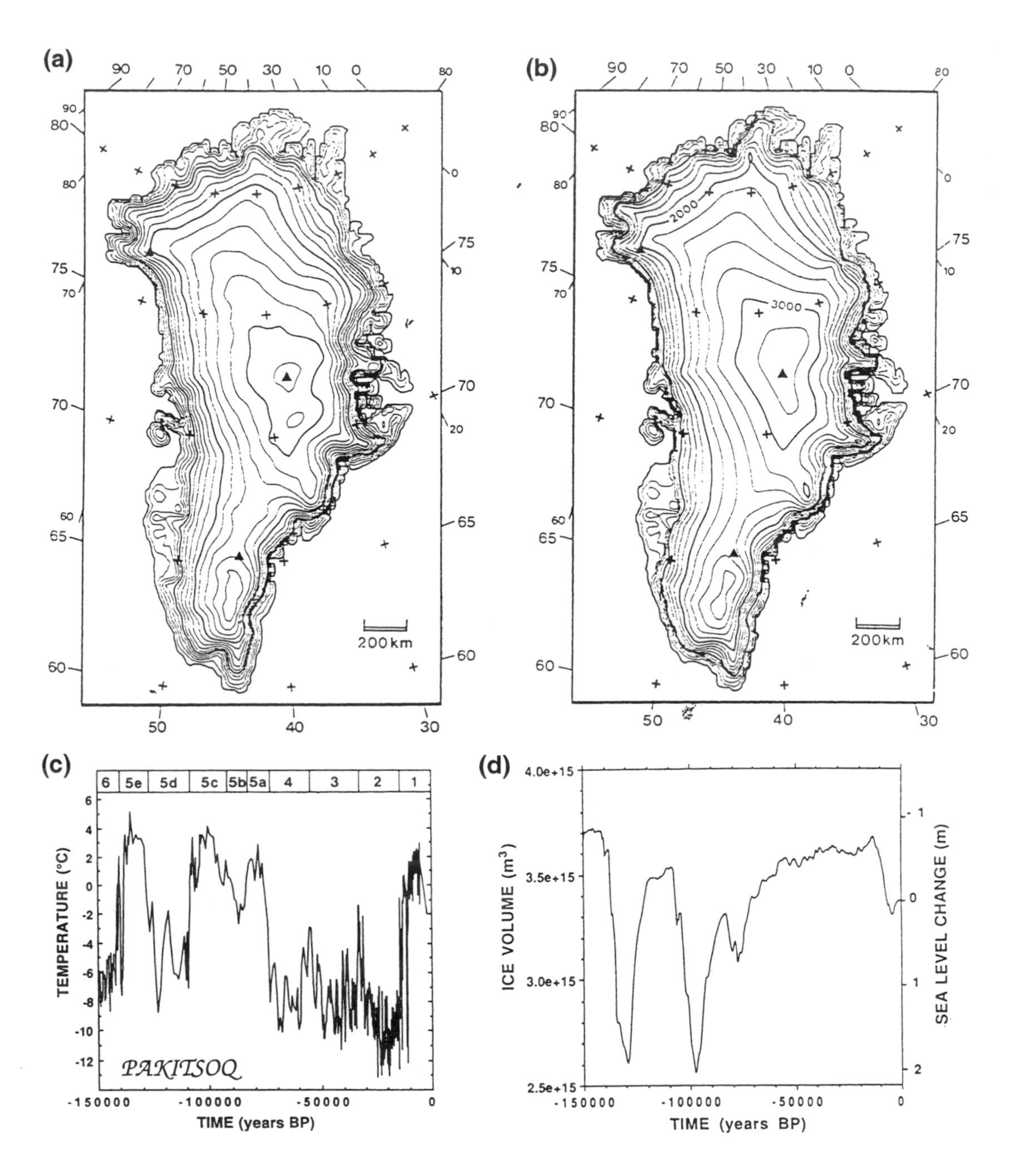

Figure 10.8 (a) Present surface elevation of Greenland. (b) Modelled surface elevation of Greenland under present-day environmental conditions. Contours are provided in 200 metre intervals. (c) Temperature record from the Pakitsoq blue ice region. (d) Modelled time-dependent change in Greenland ice sheet volume, with the sea level equivalent. Reprinted from *Global and Planetary Change*, 90, A. Letréguilly et al., The Greenland ice sheet through the last glacial-interglacial cycle, pp. 385–394, copyright (1991) with permission from Elsevier Science.

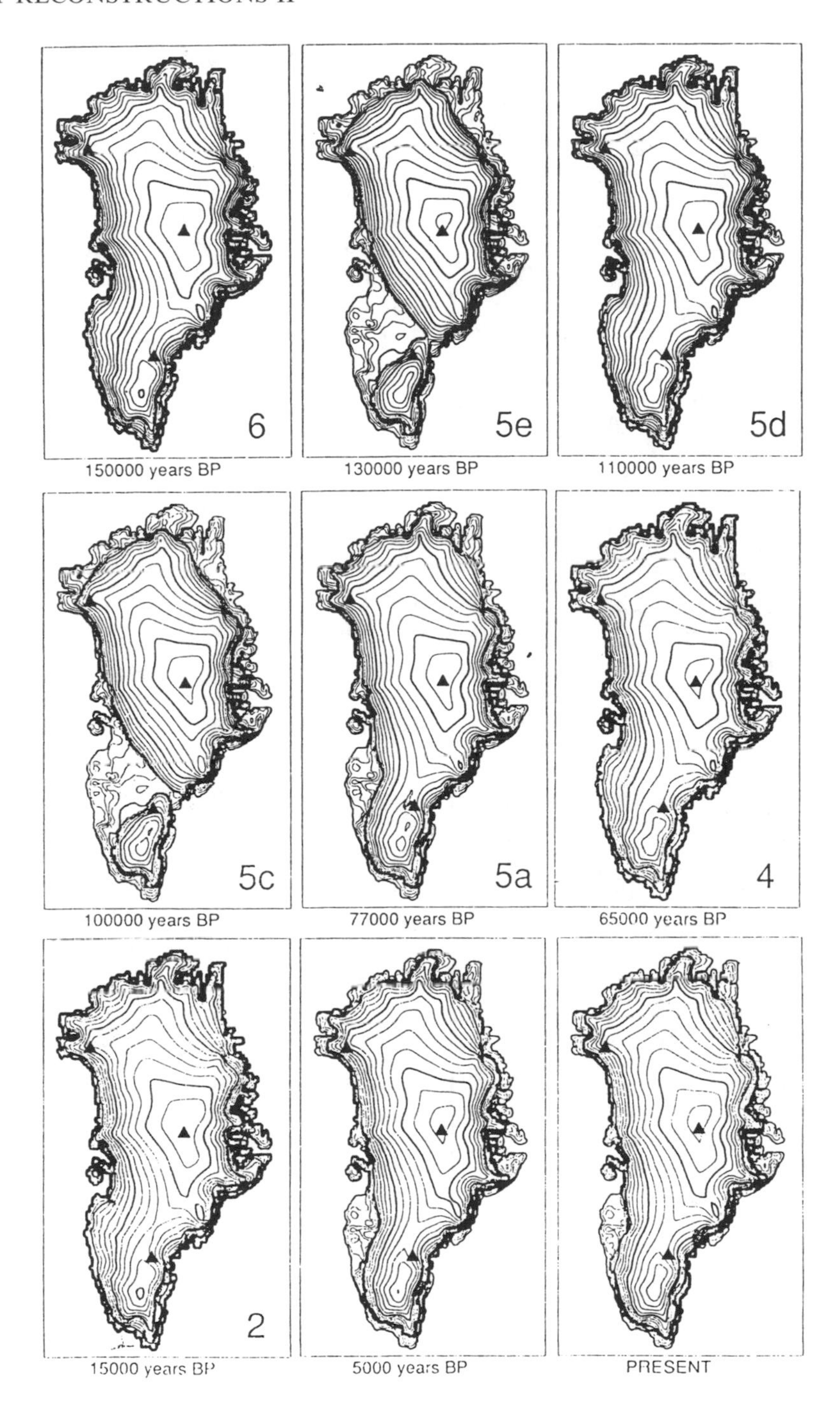

Figure 10.9 Maps of the Greenland ice sheet surface at several oxygen isotope stages during the last 150 000 years. The date of the modelled ice sheet, and the oxygen isotope stage, are indicated for each panel. Triangles refer to the locations of the Camp century, Dye 3 and Summit ice core sites. Contours are provided in 200 metre intervals. Reprinted from *Global and Planetary Change*, 90, A. Letréguilly et al., The Greenland ice sheet through the last glacial-interglacial cycle, pp. 385–394, copyright (1991) with permission from Elsevier Science.

Ice Sheet to inter-compare ice sheet models. Further ongoing modelling work involves matching ice sheet results to the ice sheet balance flux (e.g. Bamber et al., 2000b); a calculation of the ice sheet flow required to maintain the present ice sheet in 'steady state' given the modern mass balance regime. Future research will also link InSAR ice velocities with numerical model results.

The geological evidence collected recently by PONAM members, and a number of other researchers, has yet to be used to guide ice sheet modelling. As it stands at the moment, these geological datasets provide information on the ice sheet margins, whilst the ice sheet modelling yields a reconstruction of the ice sheet interior. Future work will doubtless unite these areas within a single glaciological ice sheet model.

Isostatic modelling

Isostatic modelling provides a way of reconstructing former ice loading across the whole of the Greenland continent. The most recent of these models is the ICE-4G model of Peltier (1994). In this model, ice sheet expansion to the continental margin is predicted. The ICE-4G model is several hundred metres thinner over Greenland than the CLIMAP reconstruction (Figure 10.10). However, the complex nature of the isostatic signature across the margin of Greenland provides the context in which the reliability of these results should be viewed. Future research by A.J. Long and W.R. Peltier aims to solve this problem by coupling the isostatic signal from isolation basins with numerical isostatic model results.

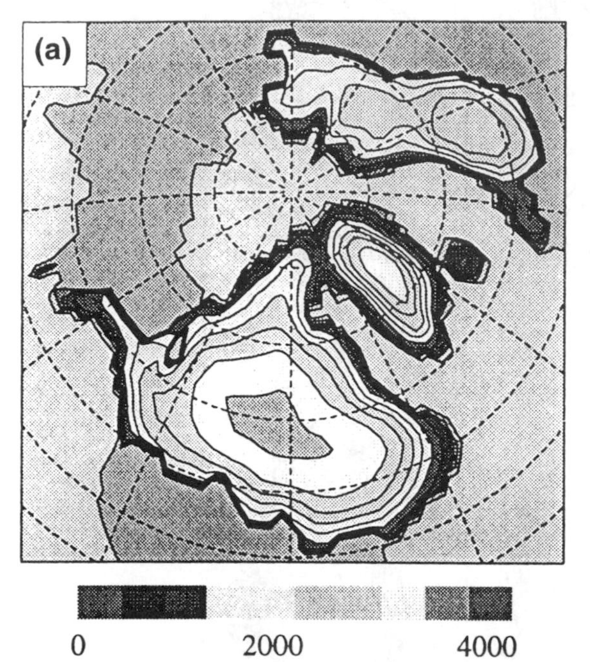

Figure 10.10 (a) Northern Hemisphere ice-sheet reconstruction according to CLIMAP. Contours are provided in 400 metre intervals. (b) Difference between the CLIMAP and ICE-4G ice sheet reconstructions. Contours are provided in 300 metre intervals. Taken from Clark et al. (1999).

C. SUMMARY OF GLACIAL HISTORY AND THE CONTRIBUTION OF GREENLAND ICE SHEET TO LGM GLOBAL SEA LEVEL FALL

Evidence for an enlarged Greenland Ice Sheet at the LGM comes from geological data at the margins of the continent and glaciological modelling studies of the main body of the ice sheet. Uplifted domes across the western, northern and eastern edges of the continent, observed from marine terraces and trimlines, demonstrate ice sheet advance in these areas. Thus, the Greenland ice sheet flowed across the rim of mountains at its current edge, into fjords and across the continental shelf in several places. Numerical modelling, however, shows that interior ice sheet growth was limited to the south of the ice sheet. The position of the southern ice sheet margin is controlled by the ELA. Air temperature depression of as much as 23°C would have caused the ELA to lower, and the ice sheet to grow. Ice thickness in the centre of the ice sheet probably only fluctuated by a few tens of metres. During the last glacial cycle, the largest fluctuation in the size of the ice sheet was prior to 100 000 years ago, when two warm periods (the Eemian and the last interstadial; at 130 000 and 100 000 years ago) caused the ice sheet to decay across its southern section. The water melted during this ice sheet decay caused 2 metres of global sea level rise. This compares to 1 metre of sea level fall as a result of ice sheet expansion at the LGM.

The Greenland Ice Sheet was connected to the Innuitian Ice Sheet (Chapter 12) in the north, by an ice stream within the Nares Strait, called the Smith Sound Ice Stream. During ice sheet decay, ice cores records show that climate over Greenland was highly variable. This may be due to the ice-ocean-atmosphere interactions between large decaying ice sheets located over North America and Eurasia, and the seas that surround Greenland.

CHAPTER 11

Ice Sheet Reconstructions III – British Isles Ice Sheet, Scandinavian Ice Sheet, Eurasian High Arctic Ice Sheets (Barents Ice Sheet)

INTRODUCTION

The majority of ice within the Eurasian Arctic is located in the high Arctic archipelagos of Norwegian Svalbard, and the Russian islands of Franz Josef Land, Novaya Zemlya and Severnaya Zemlya (Figure 11.1). Ice in these areas totals 92 000 square kilometres. Very little ice is present east of Severnaya Zemlya because here the annual precipitation is minimal. To the south, on the Arctic mainland, small ice caps and glaciers are found in the mountains of Norway, Sweden and the Russian Urals.

Examples of large ice caps within the Eurasian High Arctic are found in Nordaustlandet, Svalbard, where Austfonna (8105 square kilometres) and Vestfonna (2510 square kilometres) are located (Dowdeswell et al., 1986). The maximum thickness of these ice caps is around 580 metres. Vestfonna is drained by several outlet glaciers at its southern margin (Dowdeswell and Collin, 1990), where ice velocities of around 200 m/yr^{-1} have been recorded. The southern margin of the larger Austfonna ice cap forms a 130 kilometre-long marine-based iceberg calving wall. The percentage cover of ice over Svalbard increases towards the east, as one moves progressively away from the relatively warm climatic conditions that prevail over the eastern Norwegian–Greenland Sea as a consequence of the northward drift of relatively warm North Atlantic waters. For example 56 per cent of Spitsbergen is covered by ice, whereas in Nordaustlandet it is 75 per cent, compared with 99 per cent coverage over Kvitøya, the most easterly island in Svalbard.

Other examples of ice caps within the high Arctic can be found over the eastern Eurasian sector. For example, the islands comprising Franz Josef Land (Figure 11.1) contain 13 700 square kilometres of glacierised area. Recent airborne geophysical investigations have provided the first detailed information about the ice thickness in this little known archipelago (Dowdeswell et al., 1996). There is a further 23 600 square kilometres of glacierised area over Novaya Zemlya, whilst on Severnaya Zemlya there is 18 300 square kilometres of ice (Dowdeswell, 1995). Information about the surface morphology of such ice caps has been obtained recently from satellite observations (Dowdeswell et al., 1999).

The absence of an ice sheet across the Eurasian Arctic means that numerical models of the former ice sheet cannot be calibrated to a modern analogy (like Greenland or Antarctica). However, this lack of ice cover has one major advantage for ice sheet reconstructions, and that is the former subglacial topography (a model boundary condition) can be measured extremely accurately across this region. In recent years, PONAM and QUEEN programmes have resulted in a vast increase in glacial–geologic knowledge about the ice sheets across Eurasia and, from this, numerical ice sheet modelling studies have developed a reliable scenario for the

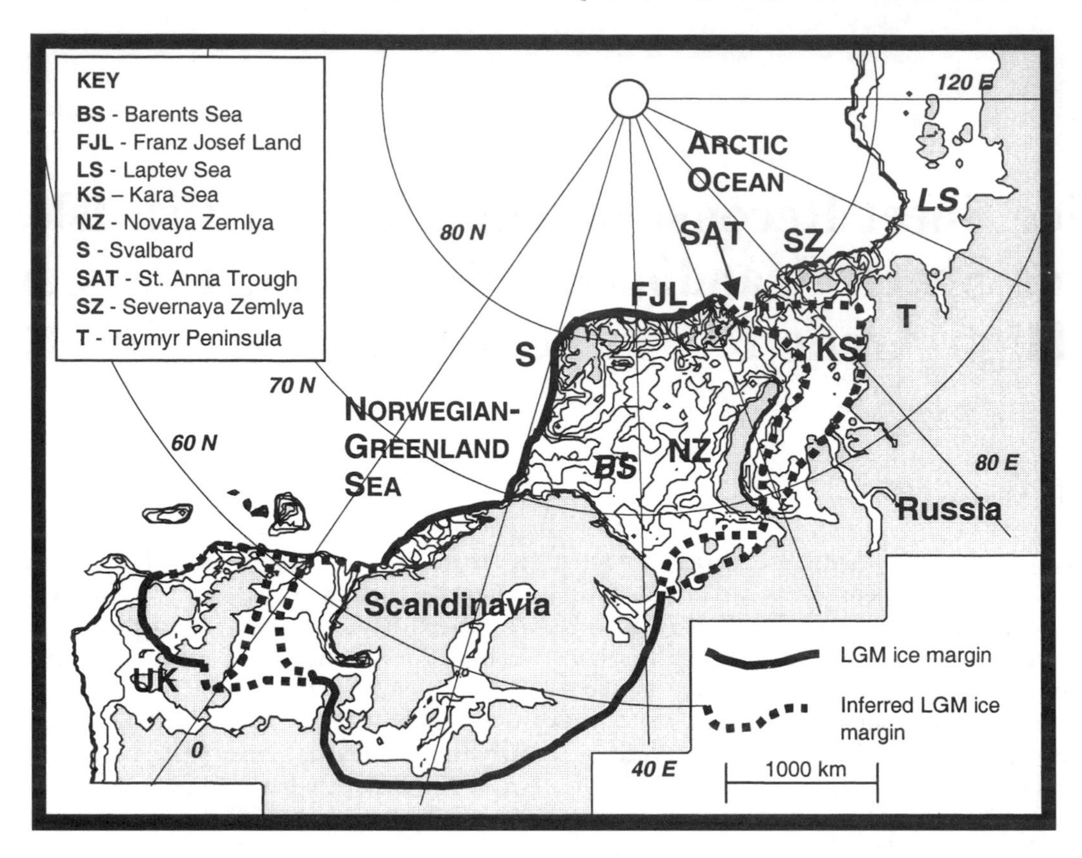

Figure 11.1 Location of the Eurasian continent and the extent of ice at the LGM determined from analysis of the geological record according to Svendsen et al. (1999). Taken from Siegert et al. (1999), and reproduced by permission of Academic Press.

last glaciation. However, prior to PONAM, the glaciation of Eurasia was the subject of much debate and speculation.

Modern and ancient climate controls for the European Arctic

Modern climate

Numerical models rely on a series of model inputs (including variations of ice accumulation rate, iceberg calving and air temperature through the glacial cycle, and a representation of the bedrock topography on which ice accumulates and flows) in order to determine information concerning ice thickness, extent and surface elevation through time. It is therefore essential that such input data are as accurate as possible to ensure that deficiencies in model output do not result from those associated with the input. For the case of the Barents Shelf, one can determine the ice sheet basal topography with certainty (since little ice exists in this region today). However, of particular relevance to the problem of uncertain model input is the reliability of climate information for the Eurasian High Arctic during the Late Weichselian glaciation.

Present day Svalbard has, in relation to its high latitude, an anomalously mild climate when compared with areas of equivalent latitude (such as Greenland). This is due to relatively warm south-westerly prevailing winds which receive heat from the North Atlantic Ocean and the Norwegian–Greenland Sea. The temperature of the Norwegian–Greenland Sea is influenced heavily by the Norwegian Current (a northerly branch of the North Atlantic Drift). Summer sea surface temperatures in the region of the Norwegian Current are typically around 8–10°C, whilst outside the current the temperature is several degrees lower. Large fluctuations in the Svalbard air temperature, of between 20–25°C within 2–3 days (Hisdal, 1985), which are not uncommon, are due to the presence or absence of the prevailing winds. The mean annual air temperature of Svalbard is –5°C, with a summer maximum and winter minimum of +6°C and –17°C, respectively. The modern precipitation on Svalbard is generally less than 400 mm/yr^{-1} (Hisdal, 1985). Thus, the climate of Svalbard and the Barents Sea is usually regarded as 'maritime Arctic'.

The influence of the Norwegian Current on the climate of Svalbard and other high latitude areas is thought to be related to the thermohaline-propelled Atlantic circulation (e.g. Broecker, 1992). Within the ocean circulation pattern, the Norwegian Current is a component of the Nordic Heat Pump (the ocean circulation system that feeds relatively warm water from the North Atlantic into the cooler Norwegian–Greenland Sea (Imbrie et al., 1992)). Broecker (1992) proposed that relatively constant air temperatures on Greenland during the Holocene are evidence that the Nordic Heat Pump has remained stable during the last 9000 years. However, Broecker (1992) also suggests that previous rapid cooler episodes (such as the Younger Dryas event) may have been due to the switching off of this ocean conveyor.

Ancient climate (geological and oceanographic information)

Deep sea sediments, because they are not subject to the glacial erosion that affects shallow sea and terrestrial sediments, often provide a more continuous record of climatic and palaeoceanographic history. Hebbeln et al. (1994) analysed eight sediment cores within two Fram Strait (Norwegian–Greenland Sea) transects with respect to organic carbon accumulation rate, carbon-nitrogen ratio, the change in organic ^{13}C ($\delta^{13}C_{org}$) and grain size fraction. There is a very good correlation in the core signals both between cores, and between the information derived from different methods of analysis within a single core. The locations of all the cores were west of the Knipovitch and Lena troughs (which run parallel to the Svalbard continental slope). Thus, terrestrial material that is carried down slope will not be present within the Fram Strait cores (Hebbeln, 1992), which implies that any terrestrial sediments, observed within the core sections, must have been transported by processes other than near-bottom effects. Hebbeln (1992) identified such terrestrial material as IRD. Typically, carbon-nitrogen ratios greater than 8 are indicative of terrestrial origin, as are values of $\delta^{13}C_{org}$ in excess of 24 per cent (PDB) within the cores. Peaks in C/N ratio and $\delta^{13}C_{org}$ content above these values correlate well with coarser fractions within the cores, and correspond to the input of land-based sediments through the process of iceberg rafting. Three such peaks occur in the Fram Strait at around 60 000 24–22 000 and 17–15 000 ^{14}C years ago. Hebbeln et al. (1994) has found that chalk fragments occur within the cores at the times of high IRD input, which have an origin from below 59°N. Benthic foraminiferal studies of the same cores also suggest the advection of warm Atlantic water, thus providing the northerly oceanic transport required for chalk deposits within Norwegian–Greenland Sea IRD. An ocean condition that would permit the northerly transport of IRD is the presence of (at least seasonally) sea ice free waters within the eastern margin of the Norwegian–Greenland Sea. This has two consequences: first, relatively warm surface temperature conditions within the Norwegian–Greenland Sea at around 24 000 ^{14}C years ago; and second, a moisture source available for precipitation, and therefore fast glacier ice growth, over the Barents Sea at between 24–22 000 ^{14}C years ago. Open ocean conditions within the eastern Norwegian–Greenland Sea at around the LGM have also been suggested from palaeoceanographical studies of foraminifera by Hald et al. (1994); and from dropstones within sea floor sediments (Bischof, 1990). The Norwegian–Greenland Sea has also been proposed as a moisture source of precipitation for the Fennoscandian Ice Sheet prior to the last glacial (for example Duplessy and Labeyrie, 1992).

The recent discovery of Late Weichselian open ocean conditions along the eastern margin of the Norwegian–Greenland Sea is in contrast to previous proposals that the sea was cold and covered by an ice shelf, fed by the Barents Sea Ice Sheet, during the last ice age (e.g. Ruddiman and McIntyre, 1977; Grosswald, 1980; Lindstrom and MacAyeal, 1986; Lindstrom and MacAyeal, 1989).

There are several lines of evidence indicating that precipitation conditions similar to today existed over the Barents Sea during three episodes within the last glaciation. In Andøya, northern Norway, dated pollen assemblages within a core from Lake Æråsvatnet suggest that relatively warm, moist climate conditions were present over Andøya between 24 000 and 18 000 ^{14}C years ago (Vorren et al., 1988). This is supported by sedimentological and foraminiferal assemblage data from the eastern Norwegian–Greenland Sea, indicating sea ice free conditions at this time, and also between 17 000 and 15 000 ^{14}C years ago (Hebbeln et al., 1994). Additionally, after 14 000 ^{14}C years ago, diatom records from sea floor sediment cores which indicate that ice free conditions remained on the eastern side of the Norwegian–Greenland Sea into the Holocene (Koç et al., 1993).

As part of the CLIMAP reconstruction of the Earth's climate during the last glacial maximum, the size and volumes of the ice sheets at 18 000 years ago were calculated in conjunction with a determination of the LGM climate (CLIMAP project members, 1976). However, over the last 20 years (and especially during the last five years), a number of new datasets have been established which indicate that the climate and ice sheet scenario predicted by CLIMAP may be prone to errors for the case of the Barents Sea region.

BRITISH ISLES ICE SHEET

The UK and Ireland are covered by a wealth of well-understood glacial geological landforms. The southern limit of the ice sheet at the LGM has been established from moraine deposits (Bowen, 1978). The direction of ice motion is understood from indicators of ice flow (such as striations, drumlins, roche moutonnées, crag and tails). The centre of ice loading can be understood through examination of the large number of uplift curves from raised beaches (Lambeck, 1993a,b). Ice thickness can be constrained in mountain regions through trimlines and erosion features on mountain tops (e.g. Ballantyne and McCarroll, 1995; Sugden, 1968). Finally, subglacial conditions can be established from analysis of the subglacial landforms such as drumlins and meltwater erosion features (e.g. Glasser and Sambrook-Smith, 1999). There is an abundance of literature available on the last glaciation of the UK. One widely cited reconstruction of this ice sheet is by Boulton et al. (1977), who used a very simple model, guided by flowlines derived from inspection of the glacial geology, to establish the ice surface topography (Figure 4.8). This ice sheet is likely to have been initiated over the mountain regions of Scotland, Wales and Ireland, where there are many features indicative of the subglacial environment at the centre of an ice sheet. By the LGM, most of northern Britain was covered by an ice sheet with a maximum thickness of around 1.5 kilometres. This model is generally accepted as a 'maximum' reconstruction. Although the southern margin of the ice sheet is well constrained by moraines, there remains a great deal of uncertainty about the nature of the ice sheet over the northern edge of Scotland and the North Sea. In contrast to the maximum reconstruction, the 'minimum' model involves limited ice cover within the North Sea and ice free conditions across Caithness in Scotland (Boulton et al., 1985). The minimum model accounts for distinct ice caps over Scotland, Wales and Ireland, connected by a relatively thin ice cover. In this model, the Irish Sea is covered by an ice shelf, and the grounded eastern margin terminates no more than 100 kilometres off the present coast (Figure 4.8). This problem has yet to be resolved fully. For example, isostatic modelling by Lambeck (1993 a,b) has given support to the 'minimum' model. On the other hand, Sejrup et al. (1994) found evidence for grounded ice coverage in sea floor sediment cores in the northern North Sea between 28 000 and 22 000 years ago. Although the debate on the Late Weichselian ice cover in the North Sea still continues, the relatively early decay of ice predicted by Sejrup et al. (1994) could actually be compatible with the more minimum ice cover later as predicted by Lambeck (Figure 11.2).

The isostatic model utilises uplift rates from raised beaches around the coast of Britain. These data are used to force the isostatic model to reconstruct former ice loading responsible for the measured isostatic behaviour. The model indicates that it is unlikely that the 'maximum' model is responsible for the isostatic patterns (indicating that ice thickness was no more than 1.5 kilometres, and that no significant ice load occurred over the North Sea).

Lambeck's isostatic model also calculates the ice loading during deglaciation (Figure 11.2). The model is guided by the geologically-inferred decay of the ice sheet (Anderson, 1981). The reconstruction indicates that the southern section of the ice sheet decayed first due, presumably, to ELA increase in response to the rise in air temperature. By 18 000 years ago, the Irish Sea was free of grounded ice as was much of central England. By 16 000 years, the small Welsh ice cap (over the Cambrian Mountains) was dislocated from the ice sheet over Scotland and north Ireland. By 13 000 years ago, the only remaining ice cap was in Scotland, centred over Rannoch Moor. Shortly after this time, there was very little left of this ice.

One problem with this type of modelling is that the influence of the Scandinavian ice load on the raised beaches across eastern Britain needs to be accounted for. Because the Scandinavian ice sheet may have decayed after the British ice, it cannot be ruled out that the raised beaches in the UK may be influenced by an isostatic forebulge, resulting in an underprediction of the isostatic recovery due to British ice. However, isostatic modelling by Fjeldskaar (1994) suggests that the forebuldge from the Fennoscandian ice sheet, with a maximum uplift of 60 metres, is limited to the Norwegian Sea and to mainland northern Europe.

Recent isostatic modelling work accounts for the effect of Scandinavian ice loads on the UK. These model results can be compared with a series of new uplift data from north-west Scotland. However, Shennan et al. (2000) found that model results did not match well with the uplift curves consistently around the UK coast, even when a variety of combinations of earth and ice models were used as input. The problem was probably confounded by the *ad hoc* manner in which the ice model (the input to isostatic models) is usually adjusted in order to obtain an isostatic result that can be compared with real data. However, Johnston and Lambeck (2000) have recently developed an automatic means by which the ice model input, in accordance with the uplift data, is derived. When they applied this model to the case of the British Isles Ice Sheet the model showed some accord with the 'minimum' reconstruction, by Boulton et al. (1985), and the earlier model of Lambeck (1993a,b).

Because of the nearby North Atlantic acting as a supply of moisture for precipitation over the ice sheet, the British Isles Ice Sheet can be regarded as a maritime ice sheet. A return to cooler conditions during the Younger Dryas, in conjunction with the continuing precipitation from this moisture source, meant a re-initiation

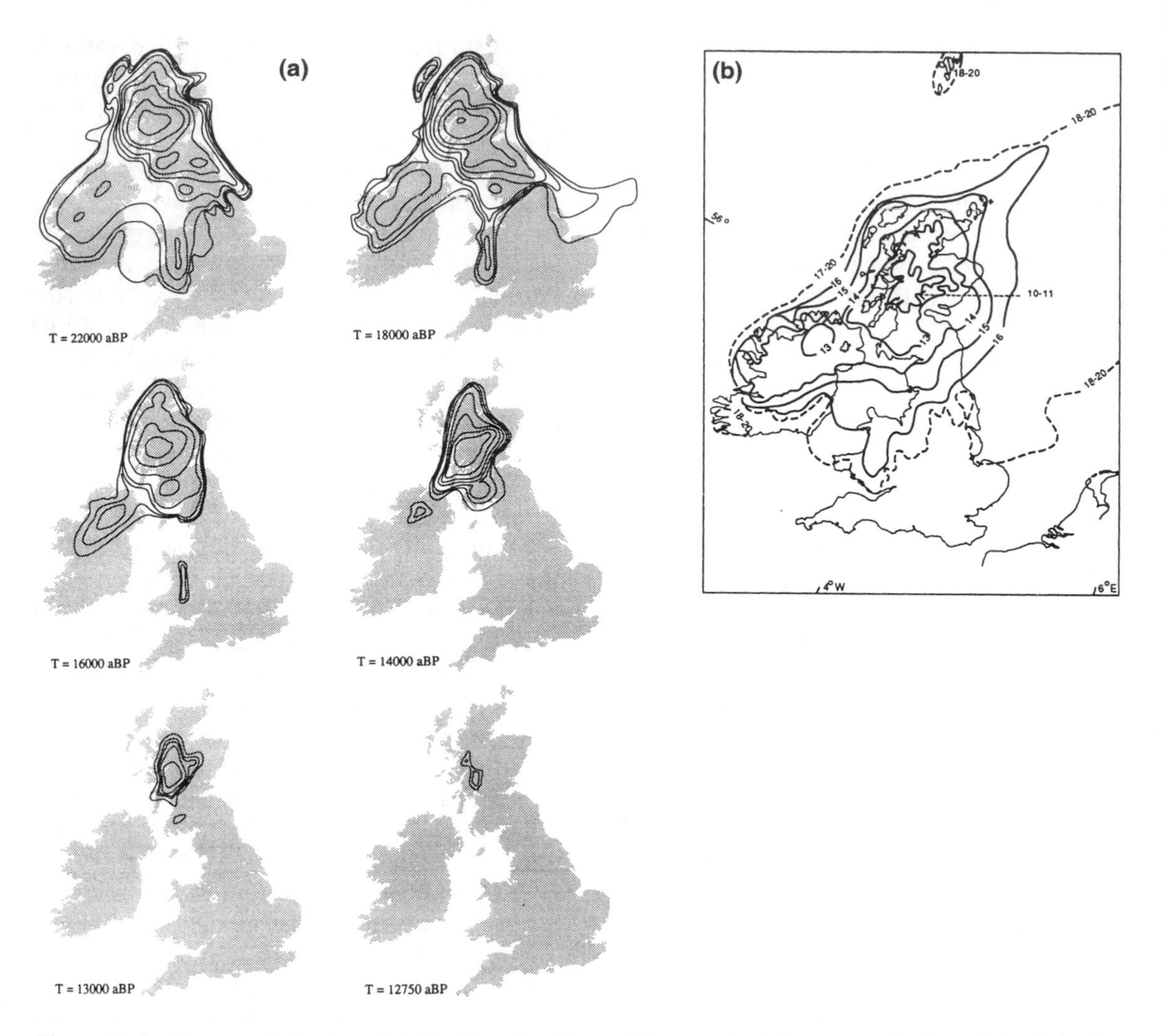

Figure 11.2 Decay of the last British Isles Ice Sheet. (a) isostatic-driven numerical model of the last glaciation. Contours are in 200 metre intervals (b) Geologically-derived ice-sheet limits. Taken from Lambeck (1993a,b).

and advance of glaciers across Rannoch Moor, the Lake District and Snowdonia. The Younger Dryas advance of ice over Scotland has been reconstructed recently by an ice sheet model (Hubbard, 1999). This model suggests that it would take only a 2.5°C temperature decrease to allow glaciers to exist over Ben Nevis today.

It should be noted that no recent numerical ice sheet model has been developed for the British Isles at the LGM. Siegert et al. (1999) included the British Isles in their model of the Eurasian ice sheet, but the ice sheet over Britain was not examined in detail.

SCANDINAVIAN ICE SHEET

Geological records

There is an excellent geological record available for the reconstruction of the Scandinavian Ice Sheet during the Weichselian. Terminal moraines mark the position of the ice sheet in the south and east, and sediments across the continental margin detail the ice sheet in the west and north. Isostatic uplift patterns allow the former centre of ice loading to be established confidently, whilst glacial geological indicators of ice flow enable the ice sheet morphology to be identified. These data constrain numerical modelling of the ice sheet.

Evidence for an extensive interstadial prior to the LGM within Norway is less obvious than in Svalbard to the north (Mangerud, 1991b), possibly because of the lower latitude, and the consequent sensitivity to both 23 and 41 ka periods of astronomically forced glaciation, of which only the 41 ka period is observed on Svalbard (Mangerud and Svendsen, 1992). Sedimentary sections at Fjøsanger (Mangerud, 1991b), Karmøy (Andersen et al., 1983) and Sunnmøre (Larsen et al., 1987) have enabled a composite stratigraphy to be produced for northern Norway. TL dating (and infinite ^{14}C values) of the sections form the conclusion that in Norway, within the period from 50–40 ka BP, a glacial advance occurred. During this period Svalbard experienced an interstadial. Thus, there is a difference between the timing of the Middle Weichselian interstadials on Svalbard and Norway. It is also probable that a small ice cap would have existed over Scandinavia prior to the last glacial.

LGM

Good geological information in the form of striations and uplift curves have lead to a relatively sound understanding about the last ice sheet over Scandinavia. Kleman et al. (1997) used till fabrics and striated bedrock to determine ice sheet dynamics throughout the Weichselian. Absolute dating of some of these ice flow features was possible because some occur in dated sequences elsewhere, whilst relative ages of others was possible where they cross cut one another. Thus, a relatively accurate depiction of dated glaciological sequences could be mapped. This method of 'glacial geological inversion' also allows the nature of the subglacial environment to be established from the identification of cold-based and warm-based landforms. The subsequent ice sheet reconstruction developed by Kleman et al. (1997) shows the following characteristics (Figure 11.3):

- Scandinavia was probably covered by at least a small ice cap throughout the Weichselian;
- western Scandinavia was covered by a thin maritime ice sheet at 110 000 years;
- this ice sheet reduced in size by 100 000 years ago. At 65 000 years, a larger ice sheet is thought to have existed over western Scandinavia, centred over the southern region;
- at the LGM, the maximum phase of glaciation occurred, where the ice sheet was centred over the Gulf of Bothnia;
- identification of cold-based morphology shows that most of the mountain regions of Norway experienced cold-based glaciation (Figure 11.3).

The isostatic uplift pattern indicates that a large ice dome was in existence over Scandinavia at around the LGM. However, at the western and northern limit of Norway, the Late Weichselian ice sheet thickness has recently been estimated to be small enough as to preserve nunataks, and the timing of maximum glacial extent has been suggested to be late in the Late Weichselian glacial cycle.

Late Weichselian ice expansion over Scandinavia is estimated to have started at around 30 ka BP, reaching a maximum at around 22 000 years ago. Vorren et al. (1988b), after studying the biostratigraphy of lake floor cores, claimed that Andøya (northern Norway) was ice free prior to 24 000 years ago, and that it became deglaciated by the LGM. However, Møller (1992) indicated, through geomorphological and lithostratigraphic investigations, that the period of glaciation on Andøya was from 25 000 to 20 000 years ago, which represents at

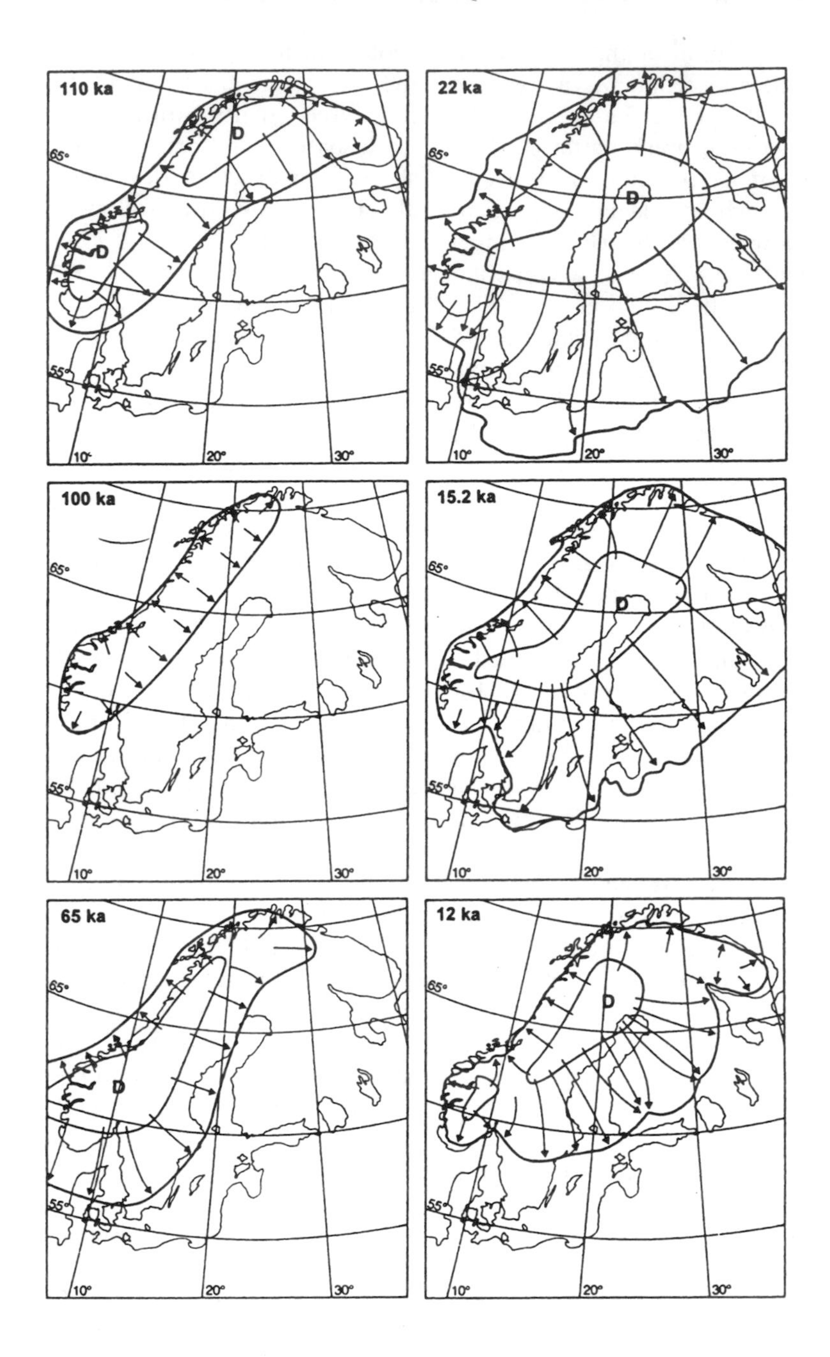

Figure 11.3 Glacial-geologically derived maps of the flow of former ice sheets across Scandinavia. Taken from Kleman et al. (1997) and reprinted from the *Journal of Glaciology* with permission of the International Glaciological Society and the authors.

least three thousand years longer than Vorren et al. (1988b) had previously suggested. Whether Vorren and other's (1988) or Møller's (1992) period of glaciation on Andøya is the more accurate is not important here, since both authors suggest that North Norway experienced a considerably shorter period of glaciation than that of central Fennoscandia.

The ice sheet at 20 000 years ago occupied the whole of Scandinavia, the Baltic and part of Denmark and the northern European mainland. The northern margin would have extended into the Barents Sea, thus connecting to the Eurasian High Arctic Ice Sheet (see later in this chapter).

The position of southern and eastern margins of the LGM ice sheet would have been controlled by the ELA. Thus it was an ablating margin. However, ice extended to the continental shelf margin. The ice sheet here flowed through bathymetric troughs as a series of outlet glaciers and ice streams. These ice streams terminated as calving ice walls, issuing icebergs into the Norwegian Sea. There is also evidence for warm based fast flowing ice streams across the east of the ice sheet, especially during deglaciation (e.g. Punkari, 1997). The model of Kleman et al. (1997) does not really provide useful information on ice thickness. This requires numerical ice sheet or isostatic modelling (discussed below).

At the LGM, a large ice stream, located within a bathymetric trough (the Norwegian Channel) occurred along the southern margin of Scandinavia. This ice stream had many tributaries running into it from the southern margin, through fjords terminating in the Norwegian Channel. The ice stream flowed rapidly due to the action of a deforming bed (King et al. 1999). This process caused the transport of glacigenic material to the continental shelf break, where it was deposited over a large sedimentary fan (the North Sea Fan). Side-scan sonar surveying of this fan indictes that gravity-driven slumps and slides transported the glacial sediments from the shelf break, across the fan. Provenance studies of these sediments suggest that they came from southern Scandinavia (Sejrup et al., 2000) (Figure 11.4). This means that the ice stream would have only become active after the ice sheet had reached quite close to its maximum size (i.e. late in the glacial cycle). Another complex of fast ice flow units were active across the mid-Norwegian continental shelf (e.g. Henriksen and Vorren, 1996). These outlet glaciers fed a large continental slope fan with sediment from a deforming subglacial bed and from IRD. Thus there are two main sedimentary fans associated with the LGM Scandinavian ice sheet, one at the mouth of the Norwegian Channel, the other located off the mid-Norwegian Shelf (these ice streams have been reconstructed by simple numerical modelling, and are referred to later in the chapter).

Deglaciation

The pattern of ice sheet decay after the LGM is well known from the extensive proglacial geology deposited across the mainland as the ice margin retreated (Anderson, 1981). The first point to note about ice sheet decay, is that the Norwegian Channel ice stream is likely to have become deactivated prior to 15 000 years, because the ice sheet margin had retreated into the Norwegian Channel around this time. Between 20 000 and 11 000 years ago, southern ice sheet margin retreated northwards to a position half way across the Baltic (Figure 11.5). During this retreat, the southern ablating margin released large volumes of meltwater, dammed by the ice sheet. Thus, a proglacial lake existed across the Baltic during ice decay. As mentioned in Chapter 5, lacustrine sedimentation is distinct and easy to identify. During the final phase of deglaciation, this proglacial lake was spread across most of the 1500 kilometre long south-eastern ablating margin of the ice sheet. Delta moraine complexes at the glacier margin are also very easy to interpret. Three such delta sequences associated with the Baltic proglacial lake have been

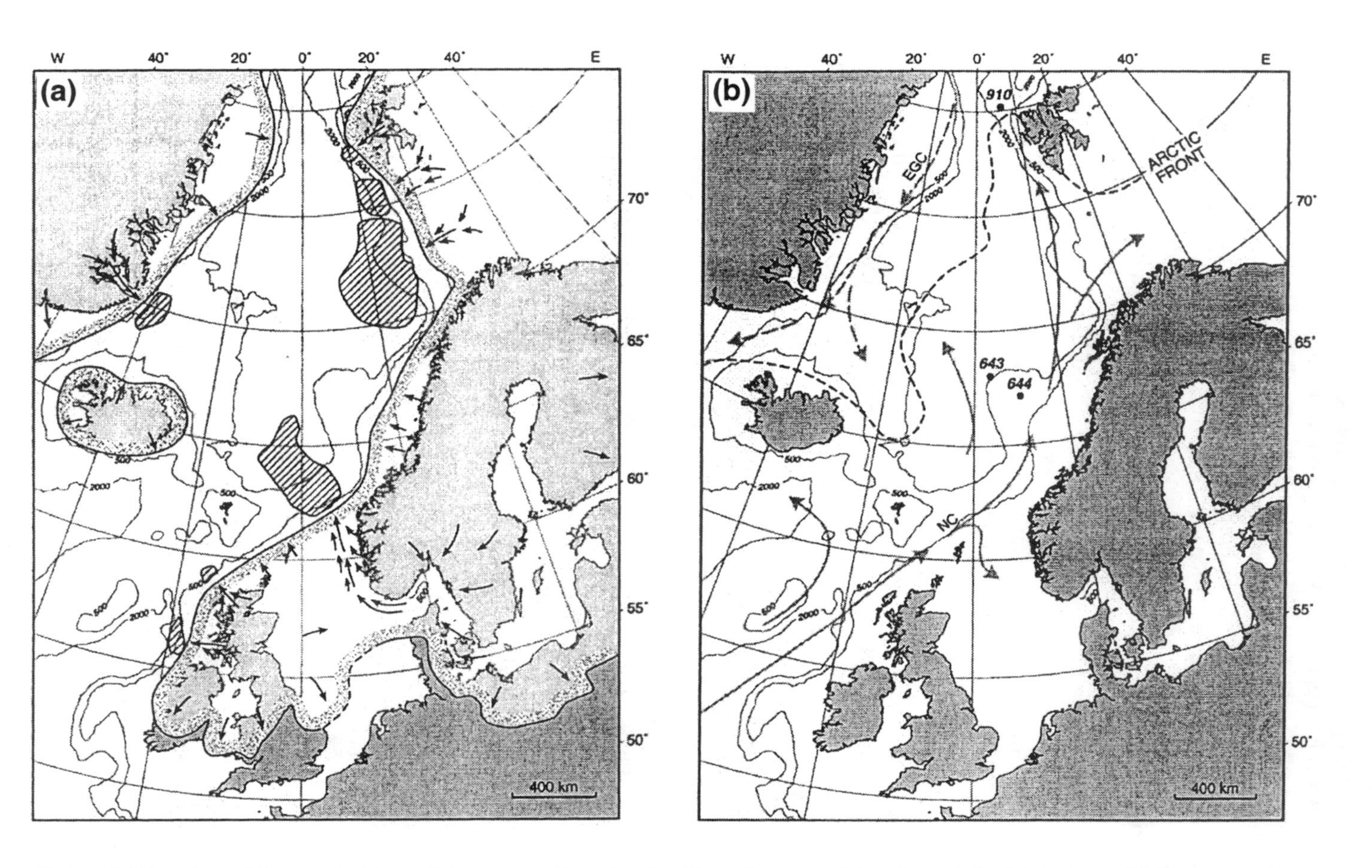

Figure 11.4 (a) Direction of former ice flow of the eastern Eurasian Ice Sheet from consideration of glacial geology. (b) Present surface ocean currents in the North Atlantic and the Norwegian–Greenland seas. Reprinted from *Quaternary Science Reviews*, 19, H. Sejrup et al., Quaternary glaciations in southern Fennoscandia: evidence from southwestern Norway and the northern North Sea region, pp. 667–685, Copyright (2000), with permission from Elsevier Science.

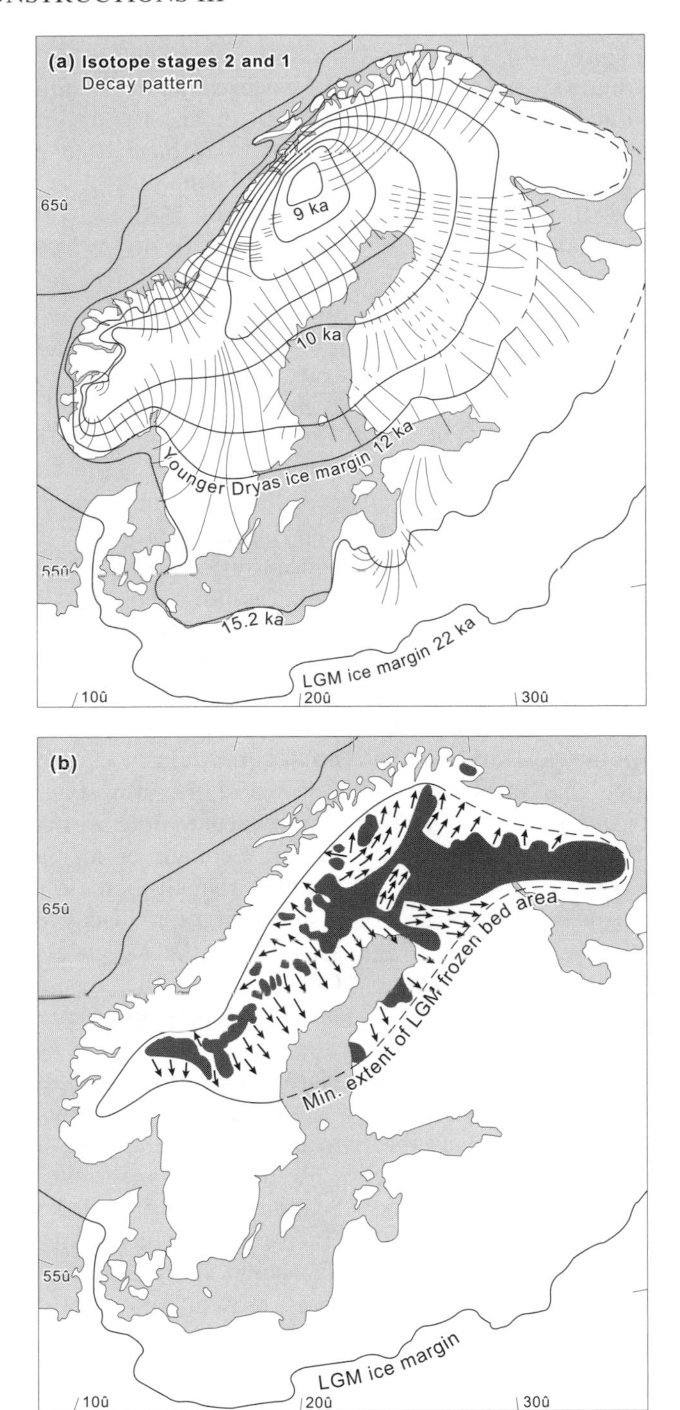

Figure 11.5 (a) Glacial-geologically-derived flow of ice during the last deglaciation of Scandinavia. (b) Zones of subglacial melting and freezing at the LGM in Scandinavia. After Kleman et al. (1997).

identified (Fyfe, 1990) (Figure 11.6). These geological data mark the position of the retreating ice front between 10 500 and 10 200 ^{14}C years BP (just after the Younger Dryas).

As the southern ice sheet margin retreated across Scandinavia, and the north-western margin remained relatively stable, the centre of ice sheet loading would have migrated northwest during deglaciation. This is important because the isostatic signal within raised beaches around the Scandinavian coast, and the former Baltic pro-glacial lake, will be influenced by this shift in ice load position. At the LGM, the maximum ice thickness, and ice sheet divide, were located over the northern Baltic. However, by the final phase of deglaciation, the ice load were positioned over central northern Sweden and Norway. This means that postglacial uplift is likely to have occurred across the European mainland and the Baltic several thousand years before the final uplift over northern Scandinavia. However, the general isolines at 10 000 years ago appear relatively symmetrical around the northern Baltic, suggesting that the LGM ice sheet loading is the major factor for post glacial uplift in this region. This pseudo-symmetrical isostatic pattern is even observed today, where the northern Baltic experiences 8 mm of uplift each year (Figure 5.3).

The decay of ice and development of a large proglacial lake across the south-eastern side of the ice sheet has implications for stable ice sheet flow. Fast flow of ice is possible if water is available to lubricate sediments so that they are geotechnically weak. It is highly likely that water from the proglacial lake may have reduced the strength of the lake floor sediments, and subglacial sediments beneath the margin of the ice sheet. This could have resulted in the sudden rapid flow of portions of the ice sheet (or ice sheet lobes: see Chapter 12). Geological evidence for such lobes is found in divergent patterns in indicators of ice sheet flow across the eastern side of Scandinavia (Punkari, 1993; Kleman et al., 1997) (Figure 11.3). These ice streams may have been partly responsible for the shift of the central ice load westwards from the Gulf of Bothnia, since they would have drained ice from the eastern side of the ice sheet. Once the ice feeding these ice streams was drained, they would have become inactive. Thus, they may be regarded as 'surge-type' features.

Between 20 000 and 11 000 years ago, the north-western margin of the ice sheet retreated across the continental shelf to the present coastline. The massive discharges of icebergs that this decay produced can be seen in the IRD concentrations and δ^{18}O depletions measured from sea floor sediments in the Norwegian Sea (e.g. Bauch and Weinelt, 1997). It is highly likely that the release of fresh water caused by the melting of these icebergs also resulted in a stratification of the water column, allowing a non-equilibrium to exist between the sea water below and the atmosphere. The onset of this activity began quite soon after the LGM (Jones, 1991b; Lehman et al., 1991). This could be because a small increase in sea level may have resulted in rapid decay of the deeper marine sections of the ice sheet (e.g. the Norwegian Channel and the western continental shelf break). Iceberg production is known to have stopped at 8500 years ago, because subpolar planktonic foraminifera are present in sea floor sediments from this time. Thus, the ice sheet would have largely decayed by 8500 years.

A summary for the Weichselian glaciation of Scandinavia is provided by Mangerud (1991) (Figure 11.7). The glaciation curve developed in this paper illustrates the major periods of ice sheet growth and decay throughout the last 100 000 years based on glacial geological evidence.

Figure 11.6 (a) Location of deltaic sequences and moraines left by the receding Baltic proglacial lake. Taken from Fyfe (1990) with permission of Taylor & Francis AS. (b) Isobases from the Baltic lake at 10 300 years BP, just before it drained (in metres a.s.l.). Taken from Svensson (1989).

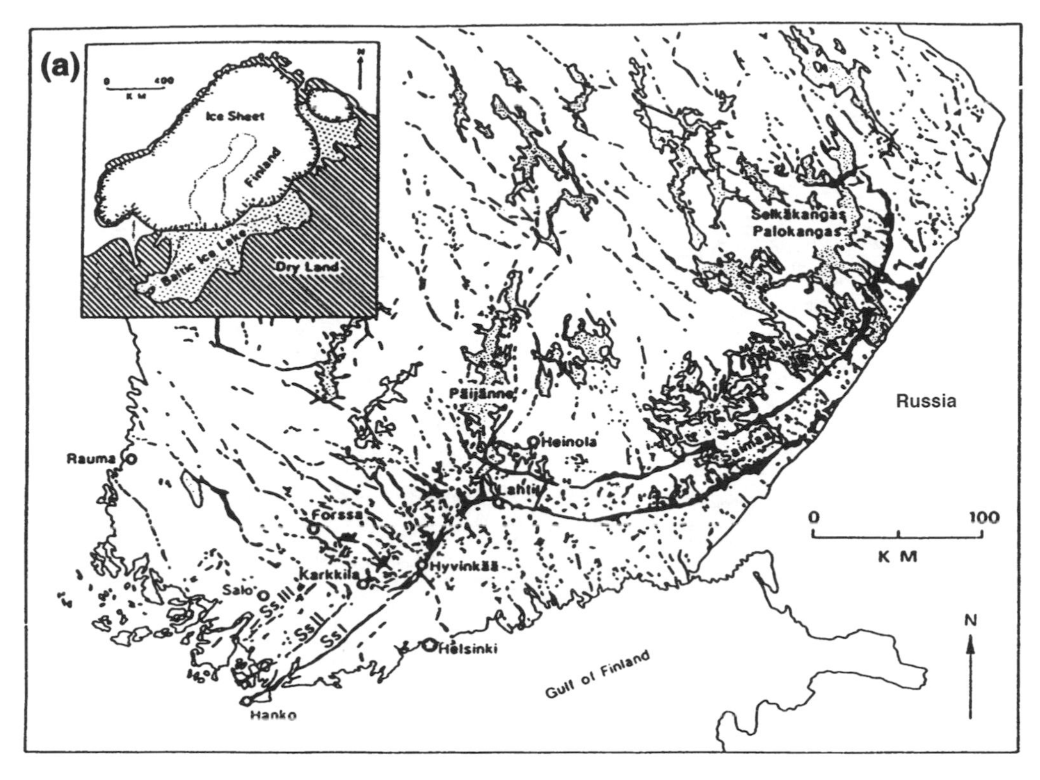
(a)
Ice Sheet
Finland
Baltic Ice Lake
Dry Land
Selkäkangas
Palokangas
Russia
Päijänne
Heinola
Saimaa
Lahti
Rauma
Forssa
Karkkila
Hyvinkää
Salo
Ss III
Ss II
Ss I
Helsinki
Gulf of Finland
Hanko
0 100
K M
N

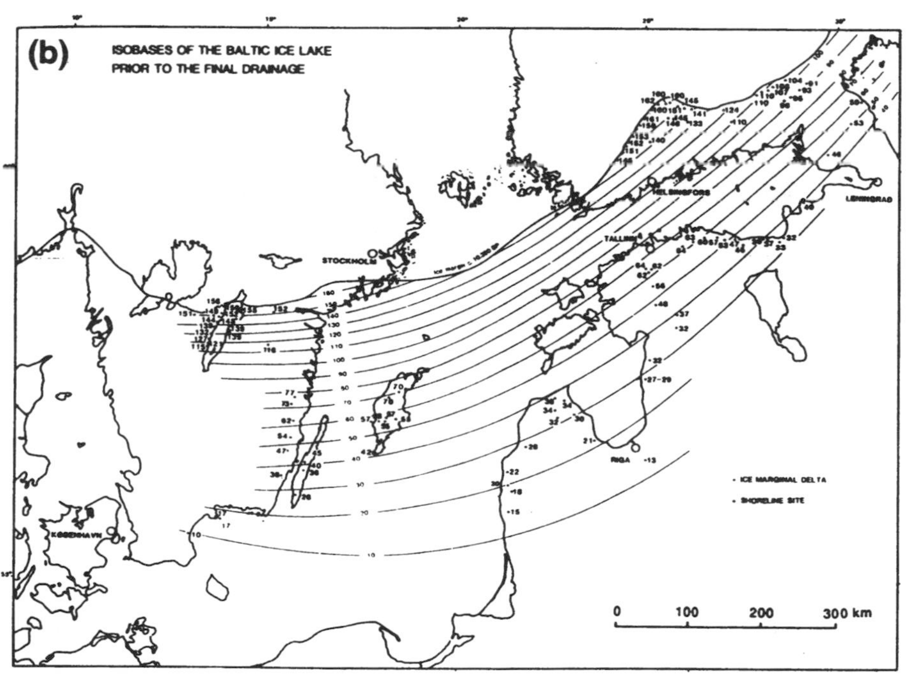
(b)
ISOBASES OF THE BALTIC ICE LAKE
PRIOR TO THE FINAL DRAINAGE
STOCKHOLM
HELSINGFORS
LENINGRAD
TALLINN
RIGA
KØBENHAVN
0 100 200 300 km

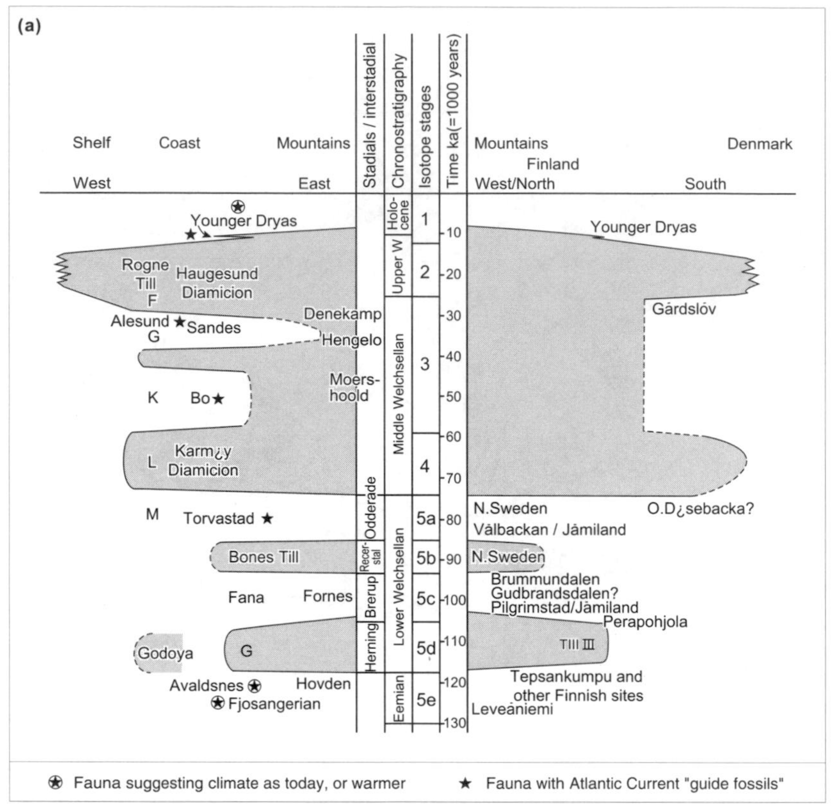

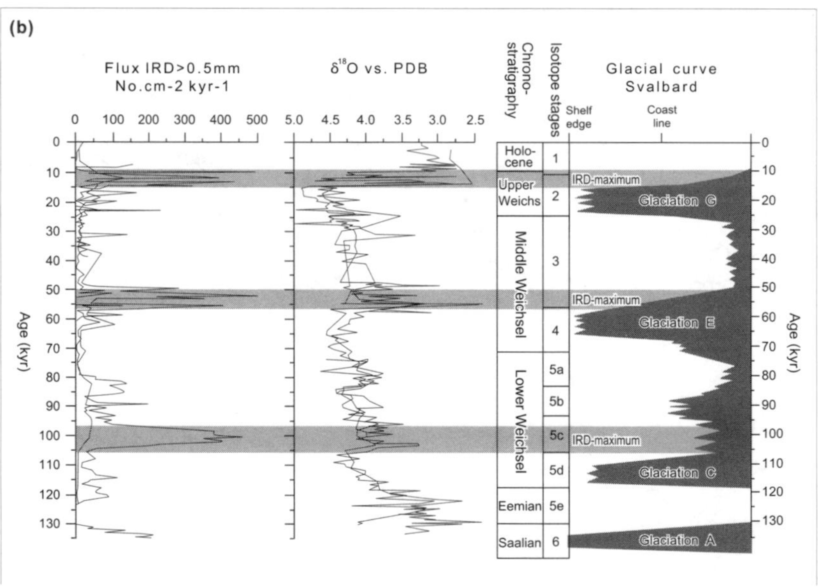

Figure 11.7 Glaciation curves for (a) Scandinavia and (b) Svalbard. Also shown are marine isotope stages (^{18}O), the flux of IRD and the δ^{18}O values west of Svalbard. Reprinted from *Quaternary Science Reviews*, 17, J. Mangerud et al., Fluctuations of the Svalbard-Barents Sea Ice Sheet during the last 150 000 years, pp. 11–42, copyright (1998), with permission from Elsevier Science.

Numerical modelling studies

Both glaciological and isostatic numerical modelling experiments have been undertaken for the Late Weichselian Scandinavian Ice Sheet. The results from these investigations compare relatively well with one another. These investigations make it possible to reconstruct the former ice thickness at the LGM and the changes to ice thickness during deglaciation. Glaciological ice sheet modelling also allows us to calculate time-dependent information about the past climate and ice sheet products (e.g. icebergs and meltwater).

Elverhøi et al. (1993) used the isostatic model of Fjeldskaar and Cathles (1991) to calculate the ice sheet history of the Eurasian Arctic. The Scandinavian component of this ice sheet was modelled with some confidence because of the well-known present day uplift patterns. These uplift patterns are matched by the model when an ice sheet, with extent controlled by the geological limits, has a maximum ice thickness of 3 kilometres over the northern Baltic. The decay of this ice sheet, according to the isostatic model, is illustrated in Figure 11.9.

A recent numerical ice sheet modelling investigation of the Scandinavian ice sheet shows a similar maximum ice thickness to Feldskaar's model (Siegert et al., 1999). This ice sheet model shows large ice streams within the Norwegian Channel and the mid-shelf west of Norway. These ice streams fed sediments to the continental margins where two large sedimentary fans exist (Figure 11.10). Ice sheet modelling has also been used to detail the decay of the Scandinavian Ice Sheet. Geological evidence for the retreat of the ice sheet margin is used to validate model results. The Norwegian Channel ice stream becomes inactive shortly after the LGM. The ice sheet model shows that this pattern of ice sheet decay is possible when an increase in surface air temperature between the LGM and 8000 years ago causes an increase to the ELA. This induces an increase in the area of the ablation zone and, so, the ice sheet retreats. The water developed from the melting of the south-eastern margin collected in front of the ice sheet to form the Baltic proglacial lake (Siegert and Dowdeswell, submitted). The model also shows that iceberg calving is the dominant ice sheet ablation mechanism across the western marine-based region of the ice sheet.

Other theories regarding the dynamics of the ice sheet, developed from interpretation of geological evidence, have also been examined using numerical modelling and interpretation of glacial landforms. Donglemans (1996) identified a number of former ice streams across south-western Scandinavia at the LGM and found that the pattern of ice stream flow had few geological controls such as such as deep unlithified sediment or glacially eroded channels. Payne and Baldwin (1999) then showed how unstable fast ice sheet flow could be initiated across areas such as south-western Scandinavia. In such regions, when the ice sheet is thickest, the ice base becomes 'warm' and fast flow occurs, draining the ice sheet interior until the ice thickness thins enough for a change in the basal conditions to a cold-based regime. This effectively switches off the fast-flowing lobes of the ice sheet. The connection between ice stream activity and subglacial thermal regime is also applicable to the unstable flow of ice along the southern margin of the Laurentide Ice Sheet (see Chapter 12).

European ice sheets and Heinrich layers?

A final comment regarding the terrestrial European ice sheets is that they may have been instrumental in the unstable flow of ice in North America that is thought to have been responsible for Heinrich layers. Dowdeswell et al. (1999) analysed sea floor sediment core records across the North Atlantic and Polar North Atlantic Margin and established that there was no stratigraphic link between European and Laurentide IRD. This suggests that there is no common climate cause of rapid ice sheet surging between North America and Europe. However, Grousset et al. (2000) reported that several

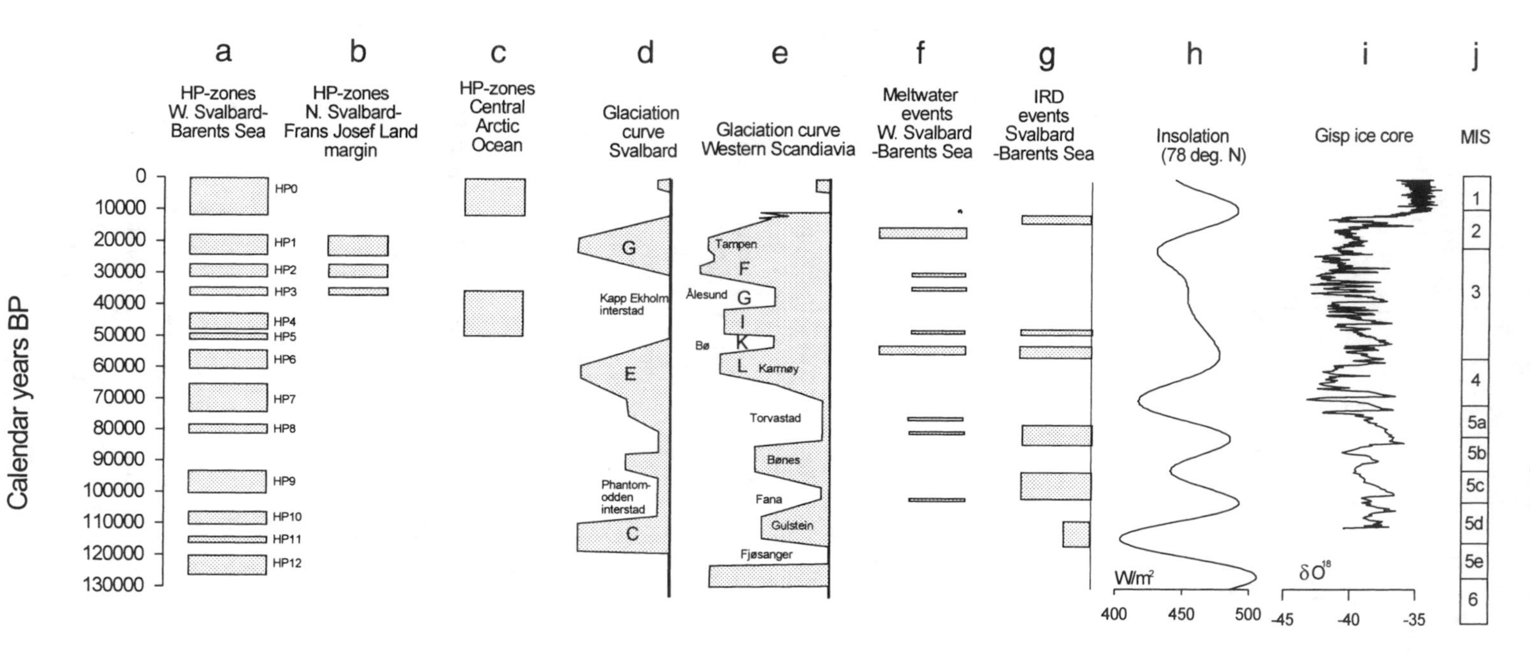

Figure 11.8 Correlation between main periods of (a) advection of Atlantic Water represented by High productivity (HP zones 1–12), and (b) HP zones north of Svalbard and Franz Josef Land, (c) HP zones in the Arctic Ocean, (d) the glacial history of Svalbard and the Barents Sea, (e) glacial history of Western Scandinavia, (f–g) main meltwater and IRD events off western Svalbard and the western Barents Sea margin, (h) July insolation at 78°N, (i) the δ^{18}O record of the GISP2 ice core on summit Greenland and (j) marine isotope stages (MIS). The time scale is given in calibrated years by converting the radiocarbon years <30 000 years. Taken from Siegert et al. (in press).

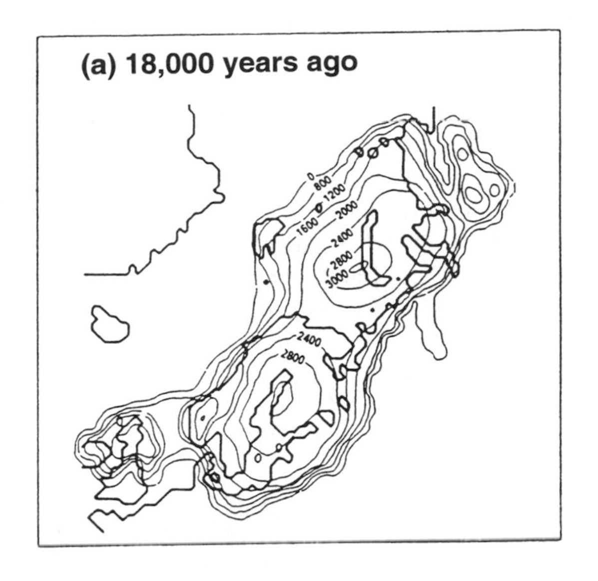

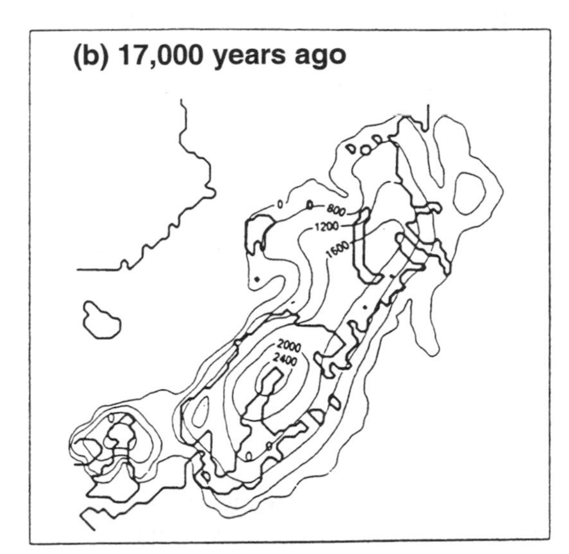

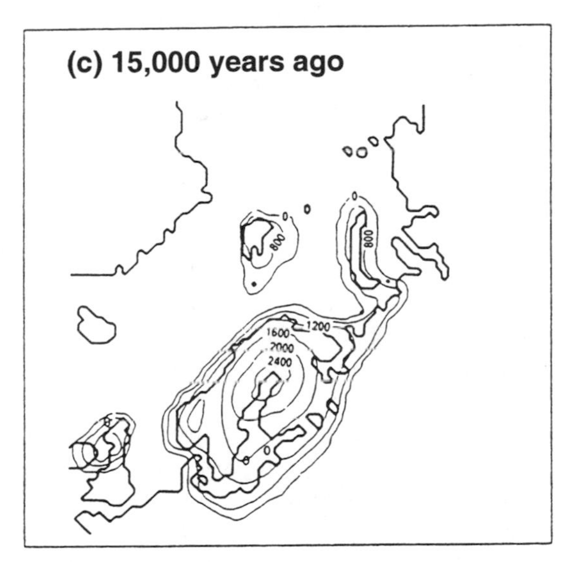

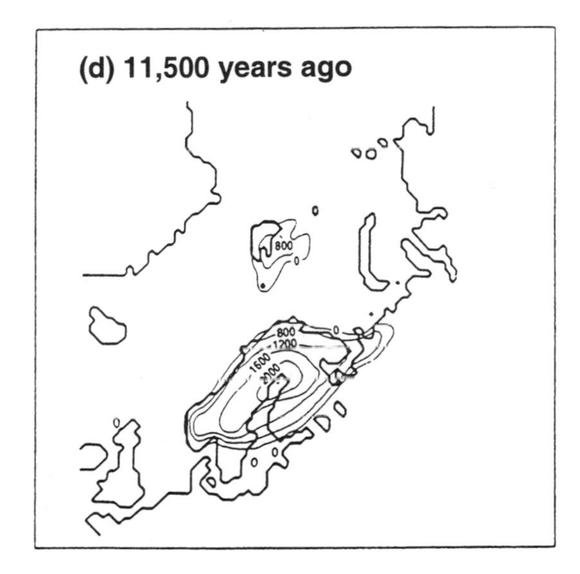

Figure 11.9 Isostatic-driven numerical model of the last British Isles/Scandinavian/Eurasian Ice Sheet. The dates of the modelled ice sheet are indicated for each panel. Reprinted from *Quaternary Science Reviews*, 12, A. Elverhøi et al., The Barents Sea Ice Sheet – a model of its growth and decay during the last ice maximum, pp. 863–873, copyright (1993), with permission from Elsevier Science.

Heinrich events actually have European-derived sediments at their beginning. This has led to a new hypothesis (alternative to the binge-purge theory), whereby the European ice sheets respond to climate change by surging, producing the first part of the Heinrich layer. This caused sea level rise, initiating instability in the Laurentide Ice Sheet and, from this, the remainder of the IRD to the Heinrich layer in the North Atlantic.

EURASIAN HIGH ARCTIC ICE SHEET

Between 1995 and 2000, a follow-up programme to PONAM, called Quaternary Environments of the Eurasian North (QUEEN), was organised to establish the glacial history of the Eurasian High Arctic. This programme aimed to establish the southern and eastern margins of the last ice sheet. Up until QUEEN there has been considerable debate as to the maximum size of the Eurasian Ice Sheet. This debate resulted, in the 1980s, in two extreme ice sheet reconstructions: a maximum scenario involving the complete glaciation of the Eurasian Arctic by a huge 3 kilometre thick ice sheet, and a minimum ice sheet where the growth of ice was limited to the present-day coastline of island archipelagos. As it turns out, there is geological evidence against both of these ice sheet ideas.

Weichselian glaciations

Terrestrial geological evidence from Svalbard shows a sequence of ice sheet growth and decay throughout the Weichselian. The Late Weichselian is thought to represent the time of the largest glaciation in Svalbard and the Barents Sea, when one considers that global sea level and solar radiation values were lowest at this time. However, there have been other occasions in the Weichselian when sea level and solar insolation were at low points, and these appear to coincide with geological evidence for ice sheet expansion. The Middle Weichselian glaciation may well have been more extensive in eastern regions such as the Yamal Peninsula than the ice sheet at the LGM (e.g. Svendsen et al., 1999; Forman et al., 1999a).

High productivity zones: conditions for ice sheet growth

Analysis of sea-floor sediment cores suggests that the rapid growth of ice in the Barents Sea around the LGM was associated with enhanced precipitation over this region, fed by a nearby moisture source in the Norwegian–Greenland Sea (Hebbeln et al., 1994). These cores display sections with a high abundance of planktonic foraminifera, including subpolar species which live in relatively warm waters. These forams indicate that the sea surface was likely to be at least seasonally free of sea ice when they were alive. Periods when these foraminifera are present are referred to as 'high productivity' (HP) zones (Figure 11.8). In the periods before and after HP zones, the sea is likely to be perennially ice covered (Dokken and Hald, 1996). In the Arctic Ocean today, summer sea surface temperatures (SSTs) are around 0 to –1°C in the upper sea ice covered layer. This sets the minimum SST for the HP zones. However, the high concentration of the polar species *Neogloboquadrina pachyderma* (sinistral) during HP zones indicates that the maximum temperature must have been less than 5°C (Pflaumann et al., 1996). The interpretation of HP zones comes from studies of sea floor sediment cores on the western Svalbard and Barents margins; a region influenced by the northern end-member of the North Atlantic Drift (Dokken and Hald, 1996). HP zones are associated with an increased flux of the subpolar species *Globigerina quinqueloba*, and suggest an influx of warm Atlantic water on a number of occasions during the Weichselian. It is assumed that during these periods, there was northwards advection of Atlantic water. HP zones across the western Norwegian–Greenland Sea are recorded during periods indicated in Figure 11.8. In the periods between the HP zones it is assumed that oceanographic circulation was modified so that a year-round sea ice cover existed, more in line with the CLIMAP (1981) reconstruction of 'glacial' circulation. Although the HP zones are likely to be linked with an increase in precipitation over the western Eurasian Arctic, it is not yet known which HP zones cause the build-up of the ice sheet, and which are associated with warm conditions during deglaciation.

This palaeoceanographic reconstruction, using inferences based on evidence of HP zones, compares well with several previous studies showing evidence of the influx of Atlantic water to the northern reaches of the Norwegian–Greenland Sea during marine isotope stages (MIS) 5 and early 4 (Baumann et al., 1993). More recent investigations have shown that Atlantic water advected into the Norwegian–Greenland Sea repeatedly during the last 120 000 years (Henrich et al., 1995). Such Atlantic water may have been present along the eastern margin of the Norwegian Sea for as much as 50 per cent of the Weichselian (Dokken and Hald, 1996). In fact, even during the coldest stages, such as MIS 2 and 4, Atlantic water may have reached as far north as the continental margin off western Svalbard (Hebbeln et al., 1994) and the Arctic Ocean (Knies and Stein, 1998). The zones HP1 and HP2 are recorded synchronously across these regions and show very similar high levels of planktonic foraminifera to those recorded in a high-resolution core from the Faeroe–Shetland Channel (FSC) (Rasmussen et al., 1996). A similar correlation is seen between HPs 1–3 and the periods when high carbonate percentages are observed in cores off the northern Barents margin (Knies and Stein, 1998). There are also indications of a correlation between HP0 and HP4 and increased abundance of planktonic foraminifera in the Central Arctic Ocean (Nørgaard-Pedersen et al., 1998). At present, Atlantic water enters the Norwegian Sea through the FSC (at 62°N). The correlation of HP zones from the western Barents margin and the Arctic Ocean suggests that advection of Atlantic water, during these periods, occurred extensively across the polar seas between 62°N and to the north of 80°N (Figure 11.8).

IRD and δ^{18}O records

High amounts of IRD in sea floor sediment cores off western Svalbard reflect periods when the Eurasian High Arctic Ice Sheet extended beyond fjords, on to the continental shelf (Elverhøi et al. 1995) (Figure 11.8). A smaller component of the IRD probably reflects iceberg rafting from the western Scandinavian Ice Sheet, as chalk grains with a source restricted to the North Sea area have been identified in some of the cores (Hebbeln et al., 1994). Large peaks in IRD are likely to be associated with enhanced production of icebergs due to the break up of these marine ice masses, whilst smaller peaks are more likely to reflect short-lived ice sheet fluctuations that may occur over the cycle of ice growth and decay (Dowdeswell et al., 1999). Five major phases of deglaciation are recognised from IRD during the Weichselian over time intervals illustrated in Figure 11.8. The IRD events at around 15 000 calendar years BP and around 55 000–50 000 years BP, reflect the deglaciation of two very large ice sheets when they retreated from the continental shelf edge. The youngest event correlates with the LGM over Svalbard and the Barents Sea (Mangerud et al., 1998) and western Scandinavia (Baumann et al., 1995). The older IRD events correlate to glaciation ‘E’ over Svalbard and the Barents Sea and ‘L’ across western Scandinavia (Figure 11.8). The IRD events are also associated with δ^{18}O depletions, indicative of increased supply of isotopically light melt-water to the ocean. The isotope and IRD signals associated with the two youngest deglaciations can be traced over the entire eastern North Atlantic (Stein et al., 1996) and the Arctic Ocean (Nørgaard-Pedersen et al., 1998). Zones HP6 and HP1, respectively, precede these two deglaciations.

Geological evidence for ice sheet extent

A recent synthesis of marine and terrestrial geological evidence has revealed the timing and extent of the Eurasian Ice Sheet during the Weichselian (Svendsen et al., 1999) (Figures 11.1, 11.7). There are two main ways to summarise this evidence. The first is the maximum extent of the ice sheet when the ice sheet was at the height of a ‘maximum’ phase of expansion (Figure 11.1). The second comprises generalised glaciation curves for Svalbard and Scandinavia (Figure 11.7), based on a large number of glacial sedimentological and geomorphological investigations combined with chronological evidence (e.g., Svendsen et al., 1999). The generalised glaciation curves show that Svalbard was completely covered by ice between 28 000 and 14 000 years ago (Late Weichselian), and around 60 000 years ago (Mid-Weichselian), whereas a more limited ice sheet occurred at 90 000 years ago, and at 110 000 years ago (Landvik et al., 1998) (Figure 11.8). Over western Scandinavia, the Late Weichselian glaciation is clearly recorded (Mangerud et al., 1996). In addition, glacial advances at 40 000 years, 60 000 years, 90 000 years and 110 000 years ago have also been identified (Mangerud et al., 1998). The exact timing of these events is less well known the further back in time they occur, but this geological evidence represents an interpretation of geological data that ice sheet model reconstructions can be compared with.

Previous Late Weichselian ice sheet reconstructions

Proponents of the 'maximum' ice mass idea

Former reconstructions (prior to the 1990s) of the Eurasian Ice Sheet have utilised two interesting features about the geography of the Eurasian High Arctic. The first is that the Eurasian High Arctic sea floor has been eroded by a number of glaciations, carving channels near the continental shelf and depositing sediments across the sea bed and on to the shelf break. The second is that the geography of the Eurasian High Arctic is similar to that of West Antarctica (if the ice were taken off). Therefore it could be argued that the Eurasian Arctic was a prime candidate for a West Antarctic style ice sheet at the LGM. Maximum ice sheet reconstructions, covering the entire Eurasian High Arctic, Scandinavia and British Isles, were established by a number of authors (Grosswald, 1980, Denton and Hughes, 1981), and were adopted by CLIMAP.

Opponents of the 'maximum' ice mass

There are two main problems with the 'maximum' reconstruction. First is that the Barents Sea (mean water depth ~250 metres) would have been submerged throughout the glacial cycle (120–135 metres of sea level fall). Thus, a mechanism (which is unknown) by which a marine-based ice sheet can form, has to be accounted for within any reconstruction/numerical model. Second, geological evidence from Svalbard has been interpreted in a number of different ways, yielding alternative ice sheet scenarios (e.g., Boulton et al., 1982).

A. GEOLOGICAL INFORMATION FOR THE ONSET OF LATE WEICHSELIAN GLACIATION AND THE LGM

In the Eurasian High Arctic, moisture source comes from GIN seas. Therefore, the onset of glacial activity may have occurred predominantly from Svalbard since it is the closest archipelago to the moisture source. Consequently, any geological information regarding the timing and size of the ice sheet from the Svalbard area is very useful.

Onset of glacial activity

Kapp Ekholm is a major sedimentary section, located in central Svalbard. The section is characterised by high rates of proglacial sedimentation indicating interstadials, truncated at the top by an unconformity caused by glacial erosion during glaciation, and a thin till layer deposit. Dates within the interstadial till provide maximum ages for the onset of glacial activity. Kapp Ekholm sediments are dated at 38 ka BP beneath the last till layer. However, the relatively low amino acid values indicate that a long cold terrestrial phase occurred after deposition of sediments. Mangerud and Svendsen (1992) indicate that these data signify glacial onset after 30 ka BP.

Onset within the western Barents Sea

Composite stratigraphy for Spitsbergenbanken (western Barents Sea) indicates that the youngest diamict date is 22.4 ka BP (carbon years). This date equates to about 25 ka BP (calendar years), and marks the last measured date after which glaciation occurred within the Barents Sea. This relatively late phase of ice growth, within the global glaciation cycle, is compatible with the idea that, in order to attain glaciation of the Barents Sea, significant sea level fall would have had to occur in order to help glaciation of the sea region (i.e. the shallower the sea, the more likely glaciation of it becomes). The process by which complete glaciation of a shallow sea is possible is discussed later.

Surrounding ocean conditions

The western limit of the GIN seas, and the southern margin of the Arctic Ocean, were seasonally free of sea ice (as discussed earlier) at several stages during the Weichselian. Importantly, one such stage was immediately prior to the LGM, during ice sheet growth. In addition, evidence, from chalk fragments, that northerly transport of warm waters occurred within two phases between 27–22.5 and 19–14.5 ka ago. This has implications for the source of precipitation to the Eurasian Arctic in that, although glacial activity did not have long to grow (25–18 ka), a high rate of accumulation was available to assist in enhanced ice growth across the west of the region.

LGM ice sheet limits

Svendsen et al. (1999) collected geological evidence from around the Eurasian Arctic to establish the maximum limits of the LGM ice sheet (Figure 11.1). This analysis was based on, among others, the following datasets.

Sedimentary fan systems across the Eurasian continental margin

Glacigenic sediment fans are known to be located at the mouths of bathymetric troughs across the western Eurasian Continent. Specifically, large fans are located in front of the Norwegian Channel, a series of channels west of Norway, the Bear Island Trough and the Storfjorden Trough within the western Barents Sea (Chapter 6). These fan systems, formed by sediments deposited on the margin by ice streams, locate the maximum extent of the ice sheet. Further across the northern margin of the continent, geophysical data are absent. However it is thought that the large trough between Franz Josef Land and Severnaya Zemlya remained free of grounded ice. Therefore, the maximum sized ice sheet that may have been responsible for carving this deep bathymetric trough was probably pre-Weichselian.

Eastern margin – Severnaya Zemlya

The eastern extent of the LGM Eurasian Ice Sheet within the Barents and Kara seas is still under debate. Whilst many suggest a complete ice coverage (e.g. Solheim et al., 1990), some propose a more restricted glaciation with the major parts of the Barents and Kara seas being covered instead by perennial sea ice (Pavlidis, 1992; Velichko et al., 1997). Contradicting interpretations also exist for the extent of glaciation in the Kara Sea (Punkari, 1995), the ice extent on Severnaya Zemlya (Bolshiyanov and Makeyev, 1995; Pavlidis et al., 1997; Svendsen et al., 1999) and the ice advance on to the West Siberian mainland (Tveranger et al., 1995; Astakhov, 1997; Astakhov et al., 1999; Mangerud et al., 1999). Marine geological evidence indicates that the 400 metre deep St Anna Trough, east of Franz Josef Land, was covered by ice at the maximum phase of glaciation, and may have been the site of a large ice stream which drained ice from the central Barents Sea into the Arctic Ocean (Polyak et al., 1997). However, it is not yet clear whether this ice stream was active at the LGM.

In contrast, the Laptev Sea and most parts of the Taymyr Peninsula remained free of ice during the Late Weichselian. Evidence for the lack of ice within the Laptev Sea, located east of Severnaya Zemlya, comes from the occurrence of Late Weichselian permafrost and polygonal ice wedge systems in this shelf area, neither of which can be formed beneath grounded glacier ice (Romanovski, 1993; Kleiber and Niessen, 1999).

Geological evidence for ice-free conditions on the Taymyr Peninsula during the Late Weichselian comes from investigations of permafrost and lacustrine sequences on land and unconsolidated sediment obtained from modern lakes (e.g. Möller et al., 1999). For example, Melles et al. (1996) and Siegert et al. (1999) found continuous, syngenetic permafrost and ice wedge development since Middle Weichselian time at the shore of Labaz Lake, Taymyr Lowland (Figure 11.1), which excludes the presence of Late Weichselian glaciers. Similar findings were made on the northern part of the peninsula, at the shore of Taymyr Lake (Kind and Leonov, 1982; Möller et al., 1999).

Geological evidence for the Late Weichselian glaciation of Severnaya Zemlya is rather sparse. However, it indicates that ice extent over the archipelago was not significantly greater during the Late Weichselian than the present value of 18 300 square kilometres or 50 per cent of the archipelago (Dowdeswell et al., in press). For example, radiocarbon dates on mammoth remains indicate ice-free conditions at several stages during the Late Weichselian (Vasiĺchuk et al., 1997). These mammoth remains were sampled close to the modern ice margins on Severnaya Zemlya (Figure 11.1). This originally led Makeyev et al. (1979) to conclude that the local ice extent at the LGM was in fact smaller than that of today.

In the far east of Russia, toward the Bering Sea, geological evidence suggests that during the LGM, the extent of ice was only a little greater than the present day cover by mountain and valley glaciers across the higher altitude regions such as the Koryak Mountains (Gualtieri et al., 2000). This means that reconstructions involving a large ice sheet across the eastern regions of Russia are probably incorrect.

Southern margin of the ice sheet

The northern region of mainland Russia is characterised by a series of terminal moraines. Grosswald and Hughes (1995) suggested that one of the southernmost moraines marked the edge of a massive Late Weichselian ice sheet. However, as part of PONAM and QUEEN, geological analysis of these features shows that the LGM ice margin was much closer to the coastline. The Markhida moraine in northern Russia was thought by Grosswald and Hughes (1995) to be caused by the readvance of the Barents ice sheet after 9000 14C years ago. However, Tveranger et al. (1995, 1998) show that this moraine consists of a basal till, overlain by lacustrine sediments which date to between 9000 and 10 000 ^{14}C years ago. Thus, the Markhida moraine must have been deposited prior to 9000 years, marking the maximum possible extent of the LGM ice sheet (i.e. if the LGM were further south, this moraine would have been eroded). Importantly, Tveranger et al. (1995) did not rule out the possibility that the Markhida moraine was from mid-Weichselian age, suggesting that the LGM southern margin was further north still. Examination of the sea floor sediments offshore in the Pechora Sea have confirmed that the largest Weichselian glaciation was prior to 40 000 years ago, implying that the Markhida Moraine is indeed mid-Weichselian (Polyak et al., 2000). This work also indicates that the LGM ice sheet could not have reached the present-day shoreline of the Pechora Lowlands. The lacustrine sediments over the Markhida glacial material represents a proglacial lake that developed from the meltwater of the last ice sheet during deglaciation. The northern and eastern limits of the LGM Fennoscandian ice sheet lie to the west of mainland Russia. This ice sheet would have coalesced with the Eurasian Ice Sheet across the southern region of the Barents Sea and not over the mainland.

Recent analysis of the stratigraphy of the Yamal Peninsula in Russia shows that the most recent glacial diamicts are about 40 000 years old (Forman et al., 1999a). This finding indicates that the LGM Eurasian Ice Sheet did not cover the peninsula and, so, must have been restricted to the Barents and Kara seas. It also shows that the Middle Weichselian glaciation could be larger than the Late Weichselian in this area.

Mechanisms for the glacierisation of the Eurasian Arctic seas

The method by which a large continental shelf sea can become covered by a grounded ice sheet has been debated for several years (e.g. Hughes, 1987). If one assumes that iceberg calving at the grounded margin of an ice mass will increase pseudo-linearly with water depth (Brown et al., 1982; Pelto and Warren, 1991; van der Veen, 1996), then as the ice margin migrates to deeper water, so increases the rate of calving. Consequently, migration of the grounded margin into deeper water will be countered by the process of iceberg calving. Thus, any process by which a marine-based ice sheet may form must overcome the problem of enhanced calving rates that are associated with relatively deeper water. Several mechanisms by which this may happen have been proposed.

Denton and Hughes (1981) and Hughes (1987) proposed that the initiation of a marine-based ice mass such as the Late Weichselian Barents Sea Ice Sheet could have occurred because permanent sea ice thickened, through surface accumulation of ice exceeding the basal melt rate, to form an ice shelf which subsequently grounded over the sea floor. Hughes (1987) predicted that the total grounded ice coverage of the Barents Sea could only be achieved after the grounding of such an ice shelf (his 'Marine Ice Transgression Hypothesis'). If no ice shelf formed, then grounded ice was confined to the relatively shallow (<300 metre water depth) western Barents Sea. However, the likelihood of the emplacement of an ice shelf over the Barents Sea has been questioned, since the underlying water convection (and hence the basal melting of the ice shelf) may not have been properly accounted for (e.g. Elverhøi et al., 1993).

A further mechanism that may be at least partly responsible for ice sheet growth within the Barents Sea involves the isostatic behaviour of the crust during the early stages of ice sheet development (Lambeck, 1995; Siegert and Fjeldskaar, 1996; Howell et al., 2000). The theory is that as ice grows initially over the island archipelagos located across the northern edge of the Eurasian continental shelf, their combined isostatic action initiates a forebulge within the central regions of the Barents Sea. This uplift, together with sea level reduction, may aid the growth of grounded ice over the shallow water (either by ice flow from nearby ice sheets, or by grounding of an ice shelf).

B. NUMERICAL ICE SHEETS MODELS

In recent years, as part of the PONAM and QUEEN programmes, a numerical model of the Eurasian Ice Sheet has been used to make predictions about the last glaciation of the Eurasian High Arctic. An informal inverse approach to modelling was adopted, where ice sheet margins were forced to match those derived from recent geological investigations. The only parameter allowed to vary within this inverse procedure was the model's palaeoclimate (surface air temperature and accumulation). Thus, model results indicate the thickness and dynamics of the ice sheet over the eastern

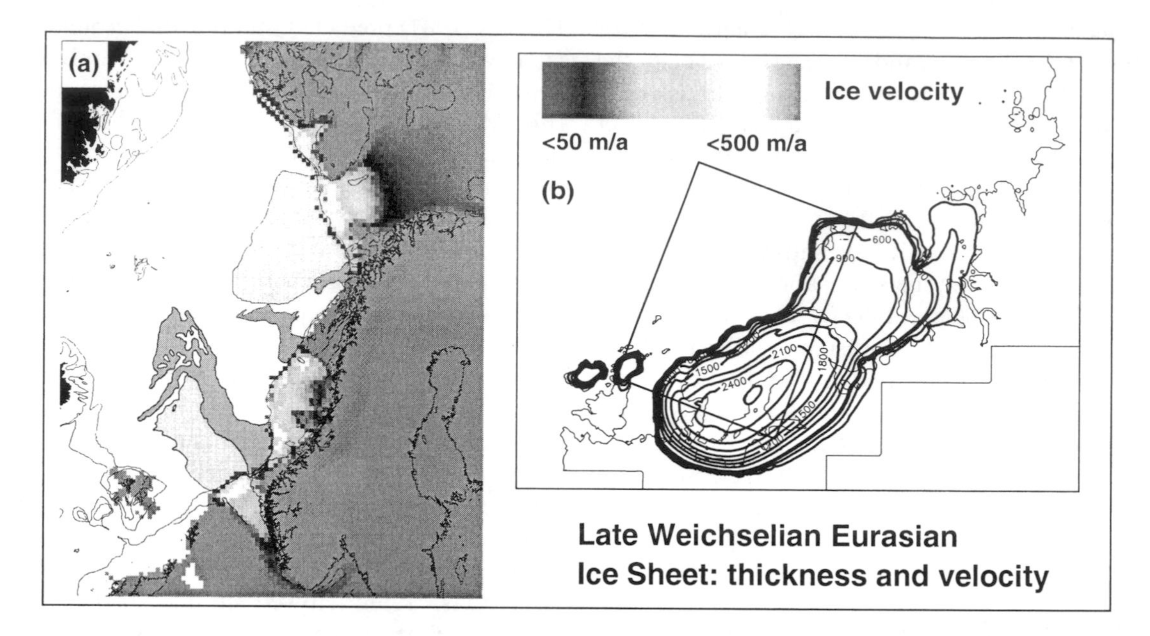

Figure 11.10 Eurasian Ice Sheet model results (a) Ice sheet velocity and the locations of large sedimentary fans (light grey shade) and major sedimentary slides (dark grey shade). (b) Ice sheet thickness (contours in 300 metres). Adapted from Dowdeswell and Siegert (1999) Siegert et al. (1999) and Taylor (1999).

Eurasian High Arctic, and plausible scenarios for the palaeoclimate in this region. The model was forced, initially, by the global sea-level curve and air temperature-induced changes to the ELA, controlled by solar insolation values (Figure 11.8).

Weichselian glaciations

The first point to note about the ice sheet reconstruction is that there are four prominent phases of glaciation, which become progressively larger from Early to Late Weichselian (Figure 11.11). A first, relatively short phases of glaciation at around 110 000 years ago caused small ice caps to form on Svalbard and Scandinavia. These ice caps then decayed by 105 000 years ago. After this time, ice accumulated steadily across Scandinavia and the Arctic archipelagos. The first major period of glaciation (Early Weichselian) predicted in the model was at around 90 000 years ago, when a 1.75 kilometre thick ice cap grew over Scandinavia, and a 0.75 kilometre thick ice mass developed from Novaya Zemlya, covering most of the Barents and Kara seas (Figure 11.12). The deepest regions of these continental seas, such as the Bear Island Trough and the St Anna Trough, were not glaciated at this time (Figure 11.12). The maximum volume of this ice sheet was around 4 000 000 cubic kilometres (Figure 11.11). The ice sheet decayed by 80 000 years ago. This was succeeded by 20 000 years of ice growth, to a maximum at 60 000 years ago (the Middle Weichselian), where the entire northern and western Eurasian continental margins were covered by grounded ice (Figure 11.12). The ice divide was located in the central Barents Sea, where ice thickness was over 1.25 kilometres. Fast-flowing ice streams were activated within topographic troughs, which delivered glacial sediments to the trough mouth regions on the continental shelf edge (Dowdeswell and Siegert,

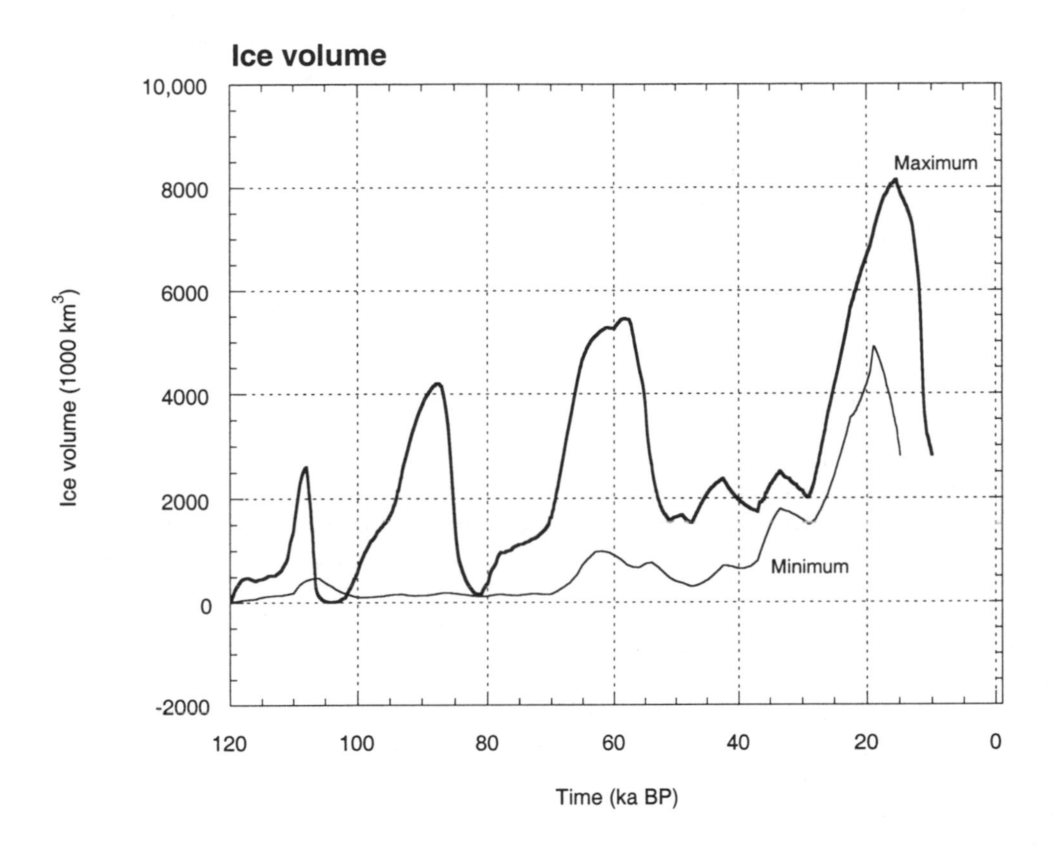

Figure 11.11 Ice-sheet volume determined from numerical modelling. The maximum ice volume relates to reconstructions shown in Figures 11.11 and 11.12. The minimum ice volume curve relates to the reconstruction shown in Figure 11.13. Adapted from Siegert et al. (in press).

1999). To the south, Scandinavia experienced less ice than in the previous glaciation, and the southern Barents and Kara seas were not covered by grounded ice. The maximum volume of this ice sheet was just under 5 500 000 cubic kilometres. Deglaciation began at around 56 000 years ago until 50 000 years, which left a small ice cap over Scandinavia that existed for around 20 000 years.

After 30 000 years ago, the ice sheet grew again to a maximum, Late Weichselian, position (Figure 11.10). Ice thickness over Scandinavia was around 2.75 kilometres, whilst over the Barents Sea it was around 1.75 metres. The ice divide was located over Novaya Zemlya. The entire Barents and Kara seas were covered by grounded ice, with fast-flowing sediment transporting ice streams within the bathymetric troughs. The maximum volume of this ice sheet was around 8 000 000 cubic kilometres. Late Weichselian deglaciation started at around 16 000 years ago and was completed by 10 000 years ago. By a small adjustment to the maximum model, where the accumulation of ice is curtailed across the eastern Kara Sea during the Late Weichselian, ice thickness across the northern Kara Sea is limited to a few hundred metres (due to ice flow from the Barents Sea), and no ice is developed across the Taymyr Peninsula (Siegert et al., 1999).

Late Weichselian ice sheet

Maximum ice sheet

Numerical modelling results indicate that, between 25 000 and 22 000 years ago, several ice domes were initiated and developed within the Eurasian High Arctic. Specifically, 25 000 years ago the ice sheet was characterised by ice domes over Svalbard (400 metres a.s.l., 400 metres ice thickness), Novaya Zemlya (400 metres a.s.l., 400 metres ice thickness) and Scandinavia (1600 metres a.s.l., 1200 metres ice thickness) (Figure 11.13). After 22 000 years ago, ice continued to build up over Scandinavia and the Barents Sea. Ice flowed northwards from Scandinavia into the Barents Sea causing the marine portion of the ice sheet to thicken. By 20 000 years ago the Svalbard–Barents ice mass became the northern component of a Scandinavian–Barents ice sheet divide (Figure 11.13). The increase in ice thickness was moderated by the development of ice streams within the bathymetric troughs on the western and northern Eurasian continental shelves which acted to drain ice from the Barents Sea. Ice also flowed eastwards across Novaya Zemlya, resulting in a continuous ice sheet over the Barents and Kara seas. The ice thickness in the Barents Sea was about 1000 metres, whilst in the south-western Kara Sea it was 900 metres (Figure 11.13).

At 15 000 years ago the maximum ice thickness over Scandinavia was ~2700 metres, whilst over the Barents Sea it was between 1500 and 1800 metres (Figure 11.13e). Maximum ice thickness across the Kara Sea varied from 1200 metres close to Novaya Zemlya to a grounded ice margin along the eastern coast of the Kara Sea. The volume of the Eurasian High Arctic ice sheet 15 000 years ago was calculated at almost 8 000 000 cubic kilometres (Figure 11.11).

Modelling results predict that: (i) ice streams within bathymetric troughs were active by around 25 000 years ago, (ii) ice extended to the shelf break along the both the western Barents Sea margin and the Arctic Ocean margin north of the Barents and Kara seas, and (iii) ice streams draining the Barents and Kara seas were present within most major bathymetric troughs during full-glacial conditions (Figure 11.13).

For the Bear Island Trough ice stream, an ice velocity of about 800 m/yr^{-1} was calculated at 15 000 years ago (Figure 11.13). A sedimentation rate of 2–4 cm/yr^{-1} was predicted along the mouth of the Bear Island Trough between 27 000 and 14 000 years ago (Dowdeswell and Siegert, 1999). This is equivalent to 0.07 to 0.13 cm/yr^{-1} averaged over the fan. Similarly, high delivery rates of 2–6 cm/yr^{-1} of glacial sediments (equivalent to 0.2–0.6 cm/yr^{-1} averaged over the fan) were predicted between 27 000 and 12 000 years ago at the mouth of the Storfjorden Trough (Dowdeswell and Siegert, 1999). The modelled volumes of sediment which accumulate at the continental margin of the Bear Island and Storfjorden troughs (4600 cubic kilometres and 900 cubic kilometres) (Dowdeswell and Siegert, 1999) are similar to the volumes of Late Weichselian sediment measured over the respective fans using seismic methods (4200 cubic kilometres and 700 cubic kilometres). The model also predicts that major glacier-fed fan systems would also have built up on the northern Arctic Ocean margin of the Barents and Kara seas, particularly on the continental slope adjacent to the St Anna and Franz-Victoria troughs (Dowdeswell and Siegert, 1999). In the maximum model, one major cross-shelf trough draining ice to the Arctic Ocean, the St Anna Trough, is predicted to deliver sediments at a rate which produces a fan intermediate in size between those offshore of the Bear Island and Storfjorden troughs (Dowdeswell and Siegert, 1999). These glacigenic sediments are then distributed over the large submarine fans mainly by gravity driven slope processes, and both side-scan sonar and seismic investigations have shown that a series of stacked debris flows make up the major building blocks of these fan systems (e.g. Vorren et al., 1989; Hjelstuen et al., 1996; Dowdeswell et al., 1996a; Elverhøi et al., 1997; Vorren et al., 1998).

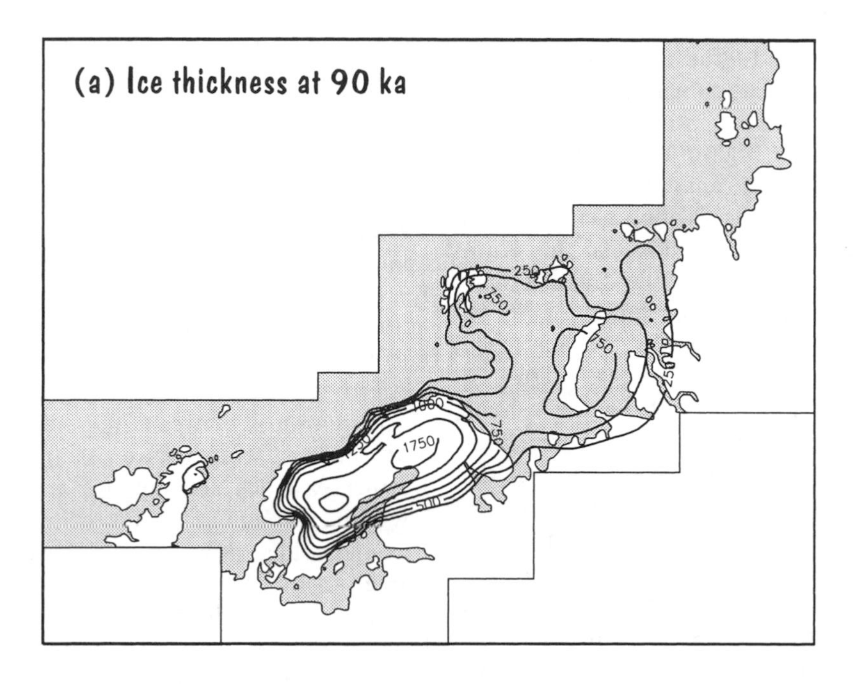

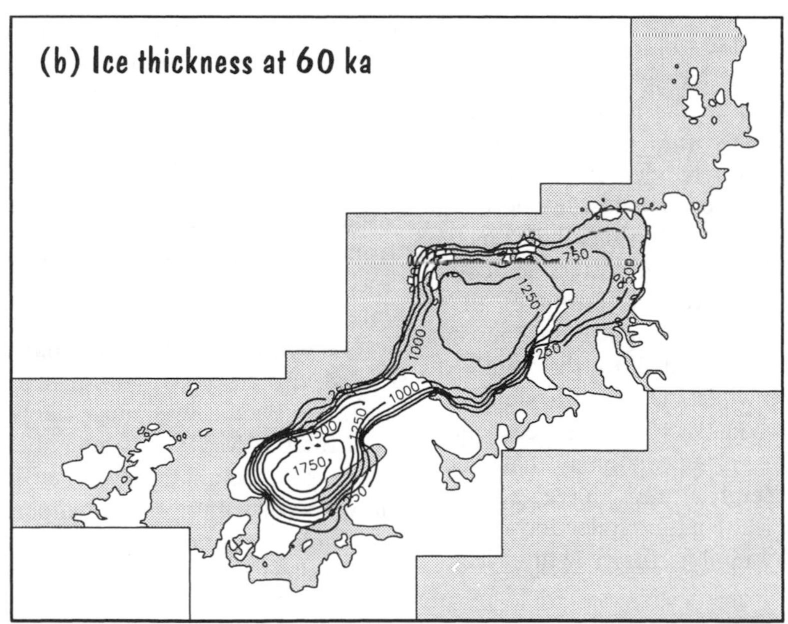

Figure 11.12 Numerical reconstructions of the Eurasian ice sheet at (a) 90 000 years ago and (b) 60 000 years. Adapted from Siegert et al. (in press).

Minimum ice sheet

Further adjustment to the model's palaeoclimate resulted in a situation where accumulation of ice is curtailed across the Kara Sea and Severnaya Zemlya (Figure 11.14). Cold temperatures (–22°C mean annual) do not permit surface melting and so ice is allowed to build up slowly via (a) sea ice thickening over the Kara Sea and (b) ice flow from the Barents Sea. Under these extreme environmental conditions a small ice mass develops over the Kara Sea with a maximum thickness of 300 metres at 15 000 years ago (Figure 11.14). The ice volume across the Kara Sea was 160 000 cubic kilometres, and 5 500 000 cubic kilometres over the entire Eurasian Arctic (Figure 11.11); about 40 per cent and 70 per cent of the maximum model size in these areas. The ice surface of the minimum reconstruction is characterised by an ice divide over the Barents Sea with major ice stream development to the north and west of the continent within bathymetric troughs (Figure 11.14). Ice flows more slowly from the ice divide east over Novaya Zemlya. Large bathymetric troughs to the north of the Kara Sea remained largely free of grounded ice. These regions were surrounded by grounded ice at a depth of 200 metres below modern sea level. It should be noted that instability in the ice margin along the northern Kara Sea could lead to rapid, short-lived glaciation of these trough regions. A maximum ice thickness of only 50 metres is modelled to the south of Severnaya Zemlya whilst in the north, similar to the 'maximum reconstruction', ice-free conditions existed. Thus, it appears likely that much of Severnaya Zemlya may have actually remained free of ice (e.g. nunataks and regions where topography inhibits the growth of ice).

Palaeoclimate reconstruction for the Late Weichselian

One result of matching the ice sheet results to geological information is that a plausible palaeoclimate is established (Siegert et al., 1999). The palaeoclimate associated with this ice sheet at the LGM is characterised by a maritime climate across the western margin of the ice sheet, and polar desert conditions towards the east. Specifically, rates of ice accumulation of over 600 mm/yr^{-1} are calculated across the western margin of Scandinavia. The mean annual surface air temperature in this region is relatively warm at –5°C. There are large gradients of precipitation and air temperature northwards and eastwards such that around the western margin of the Barents Sea the rate of accumulation is between 200 and 100 mm/yr^{-1} and the air temperature is around –20°C. Further east, ice accumulates at an extremely low rate (<40 mm/yr^{-1}) and the air temperature becomes colder (down to –40°C over the eastern Laptev Sea). Across the Taymyr Peninsula, precipitation is reduced to zero so that ice sheet build up here or further east is curtailed.

Marsiat and Valdes (in press) recently employed an Atmospheric General Circulation Model (AGCM) to investigate the role of sea-surface temperature (SSTs) on the LGM climate. Their work was motivated by the apparent discrepancy between recently established SSTs reconstructed for the northern Atlantic and Pacific oceans, and SSTs derived from the CLIMAP reconstruction (1976). Specifically, there is a large marine geological database to show that the North Atlantic was several degrees warmer than that depicted in the CLIMAP reconstruction. Conversely, the northern Pacific was most likely to be colder than the CLIMAP analysis indicates.

The palaeoclimate developed by the AGCM shows a distribution of temperature and ice accumulation similar to that established from the ice sheet model. The major difference is that the AGCM indicates even colder conditions across the Kara Sea (mean annual temperature less than –35°C). However it must be noted that the ICE-4G reconstruction used by the AGCM shows a significant ice sheet over the Kara Sea reaching more than 2000 metres. Although air temperatures simulated by the two models are

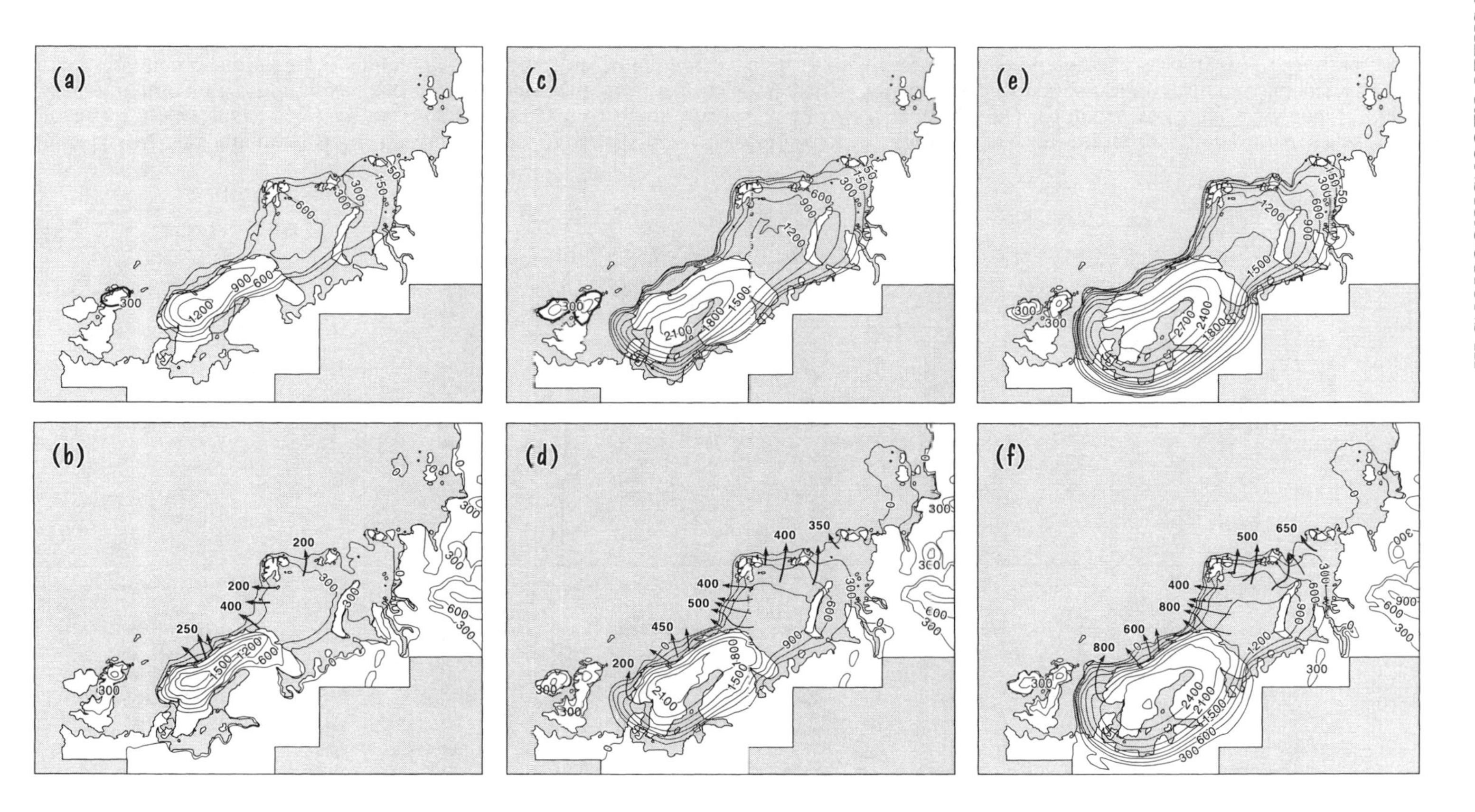

Figure 11.13 The 'maximum-sized' Eurasian High Arctic ice sheet. (a) Ice thickness 25 000 years ago, (b) Surface elevation (metres a.s.l.) 25 000 years ago. (c) Ice thickness 20 000 years ago, (d) Surface elevation (metres a.s.l.) 20 000 years ago. (e) Ice thickness 15 000 years ago, (f) Surface elevation (metres a.s.l.) 15 000 years ago. Ice thickness contours are given at 50 metres, 150 metres, 300 metres and thereafter at 300 metre intervals. Surface elevation contours are given in 300 metres intervals from 0 metres a.s.l. In each case maximum ice stream velocity (m yr^{-1}) is indicated in surface elevation maps. Taken from Siegert et al. (1999), and reproduced by permission of Academic Press.

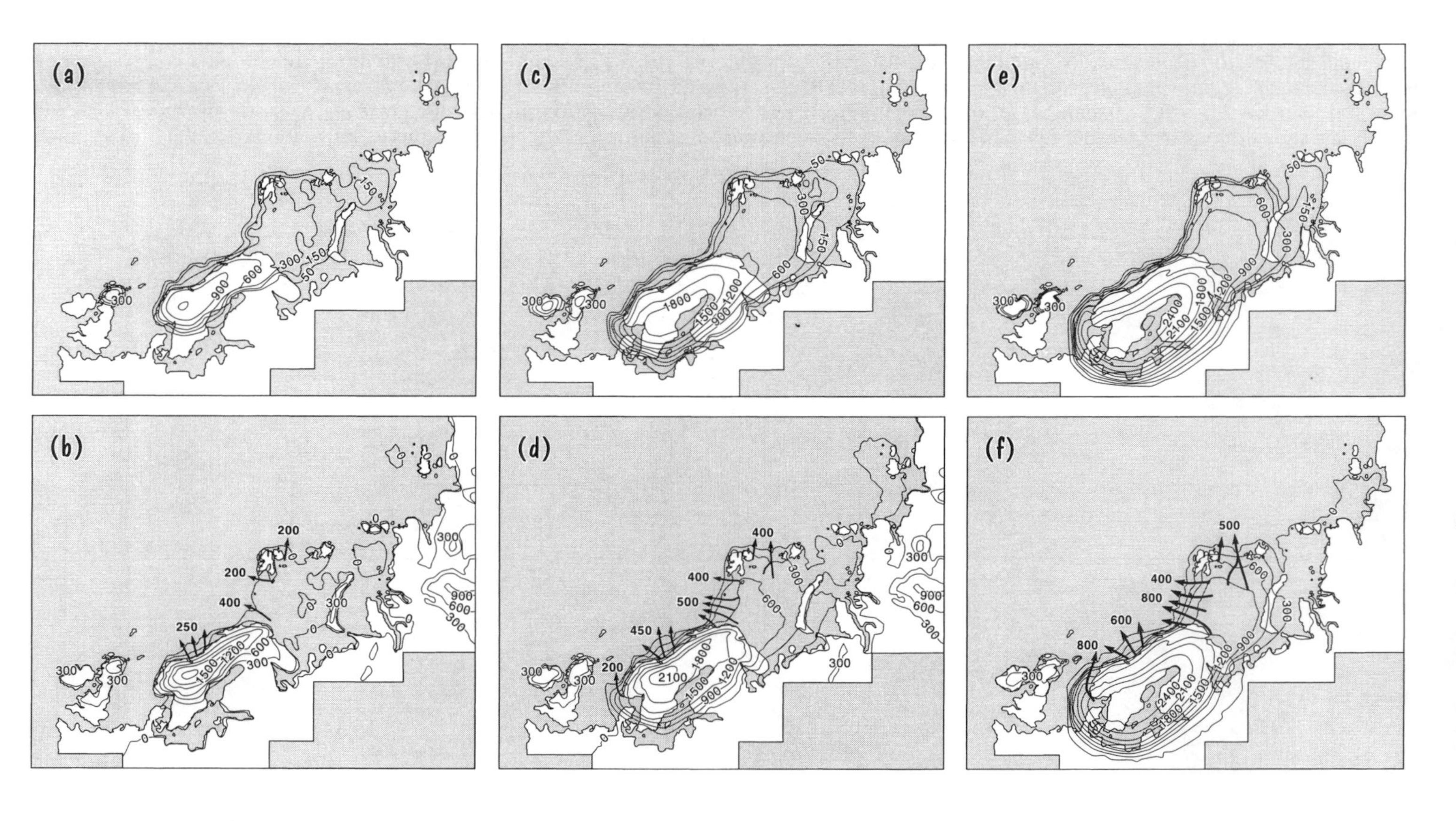

Figure 11.14 The 'minimum-sized' Eurasian High Arctic ice sheet. (a) Ice thickness 25 000 years ago, (b) Surface elevation (metres a.s.l.) 25 000 years ago. (c) Ice thickness 20 000 years ago, (d) Surface elevation (metres a.s.l.) 20 000 years ago. (e) Ice thickness 15 000 years ago, (f) Surface elevation (metres a.s.l.) 15 000 years ago. Ice thickness contours are given at 50 metres, 150 metres, 300 metres and thereafter at 300 metre intervals. Surface elevation contours are given in 300 metre intervals from 0 metres a.s.l. In each case maximum ice stream velocity (m yr^{-1}) is indicated in surface elevation maps. Taken from Siegert et al. (1999), and reproduced by permission of Academic Press.

similar over the south-western part of the Scandinavian Ice Sheet, colder conditions are calculated by the AGCM across the north-eastern half of the ice sheet. Mass balance calculations are in notable agreement with that established from the ice sheet model. High net accumulation rates (over 800 millimetres of water equivalent per year) are simulated along the south-western tip of the ice sheet, extending northward along the Atlantic coast in good correlation with the sea ice margin and storm track position as observed in the sensitivity experiments. Accumulation rates decrease rapidly when moving north-eastward to values close to 300 milllimetres of water equivalent per year over the centre of the ice sheet. Dryer conditions are simulated over the Kara Sea, not exceeding 150 millimetres of water equivalent per year.

AGCM results show that a polar desert forms across the Taymyr Peninsula because of a steady atmospheric circulation cell above it. According to the AGCM, this cyclonic system is related to cold SSTs over the northern Pacific Ocean (Marsiat and Valdes, in press). This atmospheric cell acts to stop warm moist air derived from the North Atlantic flowing across the eastern Eurasian High Arctic. The ice sheet model reconstruction demonstrates that this cell must be responsible for near-zero rates of precipitation.

Isostatic modelling

Recent isostatic modelling of the LGM ice sheet has provided an ice sheet scenario similar to the ice sheet modelling reconstructions (Lambeck et al., 1995). This model indicated that, if ice caps existed over Scandinavia and the archipelagos that surround the Barents Sea, then the resulting isostatic forebulge within the Barents Sea may have been enough, when combined with eustatic sea level fall, to sub-aerially expose some of the sea floor. Lambeck (1995) suggested that this mechanism may have been instrumental in the formation of grounded ice within the Barents Sea.

However, the global model of LGM ice sheets, ICE-4G, developed by Peltier (1994), driven by isostatic measurements shows an ice sheet that is over 2 kilometres thick across Novaya Zemlya. The ice sheet limits of these studies have been fixed to recent geological evidence and so provide results that are in some way compatible with the QUEEEN and PONAM geological datasets. However, the main problem with these reconstructions remains across the Kara Sea, where ice thickness may have been considerably less than indicated in ICE-4G.

Geological data – deglaciation

There are several independent geological datasets that record the Late Weichselian deglaciation of the Eurasian Arctic. They include palaeoceanic oxygen isotope records from the Fram Strait, ice marginal moraines left during brief halts in deglaciation and post glacial isostatic uplift.

Analysis of the $\delta^{18}O$ content of sea floor sediments within the Fram Strait shows a substantial meltwater spike at around 15 ka BP (e.g. Jones and Keigwin, 1988). These records show the break up of the Barents/Eurasian ice mass early within the last deglaciation. Thus, the Eurasian Arctic Ice Sheet may have been one of the first ice sheets to decay and, in doing so, could have been instrumental in starting sea level rise which lead to (i) Southern Hemisphere ice decay and (ii) break up of other Northern Hemisphere ice sheets.

Moraines left by the retreating ice front across the floor of the Barents Sea have been measured through sonar and seismic imaging (Landvik, 1998). These indicate that the deeper sea regions of the ice sheet broke up first. At 15 ka, the Bear Island Trough, and the deepest regions at the front of other bathymetric troughs, were deglaciated leaving a series of open ocean embayments surrounded by calving ice walls. By 12 ka, the ice sheets had decayed further such that they were limited to the archipelagos and shallow seas that surrounded them.

The pattern of ice decay within the Barents Sea is recorded within isostatic uplift curves. Carbon 14-dated raised beaches across Svalbard and Franz Josef Land, are measured to wrap around these northern archipelagos, indicating that the last phase of deglaciation would have been limited to these islands and the surrounding sea (Figure 5.2). This fits with the overall idea that deeper sea represents a region where grounded ice is difficult to 'grow', and easy to decay.

Ice sheet modelling – deglaciation

Siegert et al. (submitted) were able to model the deglaciation of the Eurasian Ice Sheet by matching model results to the geological evidence detailing the pattern of deglaciation (Figure 11.15). The model indicates the break up of the ice sheet in terms of iceberg calving and surface melting. The model shows that iceberg calving was the dominant ablation mechanism across the Barents Sea. These icebergs were issued into the Norwegian–Greenland Sea and the Arctic Ocean. Surface melting affected the southern margin of the ice sheet, allowing water to build up into proglacial lakes across northern Russia.

Deglaciation began through a dramatic increase in the rate of iceberg calving at 15 000 years, which reached a maximum by 14 500 years ago. During this time, the ice sheet decayed by 30 per cent, and 600 000 cubic kilometres of icebergs were released into the Norwegian–Greenland Sea and the Arctic Ocean. This phase of deglaciation caused the deepest marine portions of the ice sheet to break up. Another pulse of icebergs at 12 500 years was concentrated west of Scandinavia and the Bear Island Trough, when a further 400 000 cubic kilometres of icebergs were calved into the Norwegian–Greenland Sea. This resulted in the ice sheet margins receding to the shorelines of archipelagos and land masses. In total the model indicates that 7 500 000 cubic kilometres worth of icebergs were calved from the Late Weichselian ice sheet.

The model predicts that the volume of water released through surface melting during deglaciation was much less than that caused by iceberg calving. However, distinct melting events were predicted across the ice sheet as the ELA was elevated due to post LGM air temperature rise. The first episode of meltwater production was at 14 000 years ago due to the decay of the ice sheet over Ireland. The British Isles ice sheet melted next at 13 000 years, and Scandinavia and the southern section of the ice sheet melted from 11 000 years ago. In total the model calculates that 200 000 cubic kilometres of ice was melted from the ice sheet during the Late Weichselian. It should be noted that the model did not account for rapid flow of ice across the southern margin of the ice sheet and, so, may underpredict the rate of melting over Scandinavia.

Isostatic modelling of ice sheet decay

Isostatic modelling of ice sheet break up shows a history of deglaciation comparable to the ice sheet modelling results across the Eurasian Arctic. The most recent of such modelling studies was by Lambeck (1995) who calculated that in order for the model to match measured rates of isostatic uplift, the marine sections of the ice sheet have to decay first, leaving behind smaller sized ice caps across island archipelagos. The most recent uplift curves from the Russian High Arctic (Forman et al., 1995, 1997, 1999b) have yet to be included in isostatic models. However, interpretation from these uplift curves appears to be consistent with Lambeck's earlier model results (Forman et al., 1997).

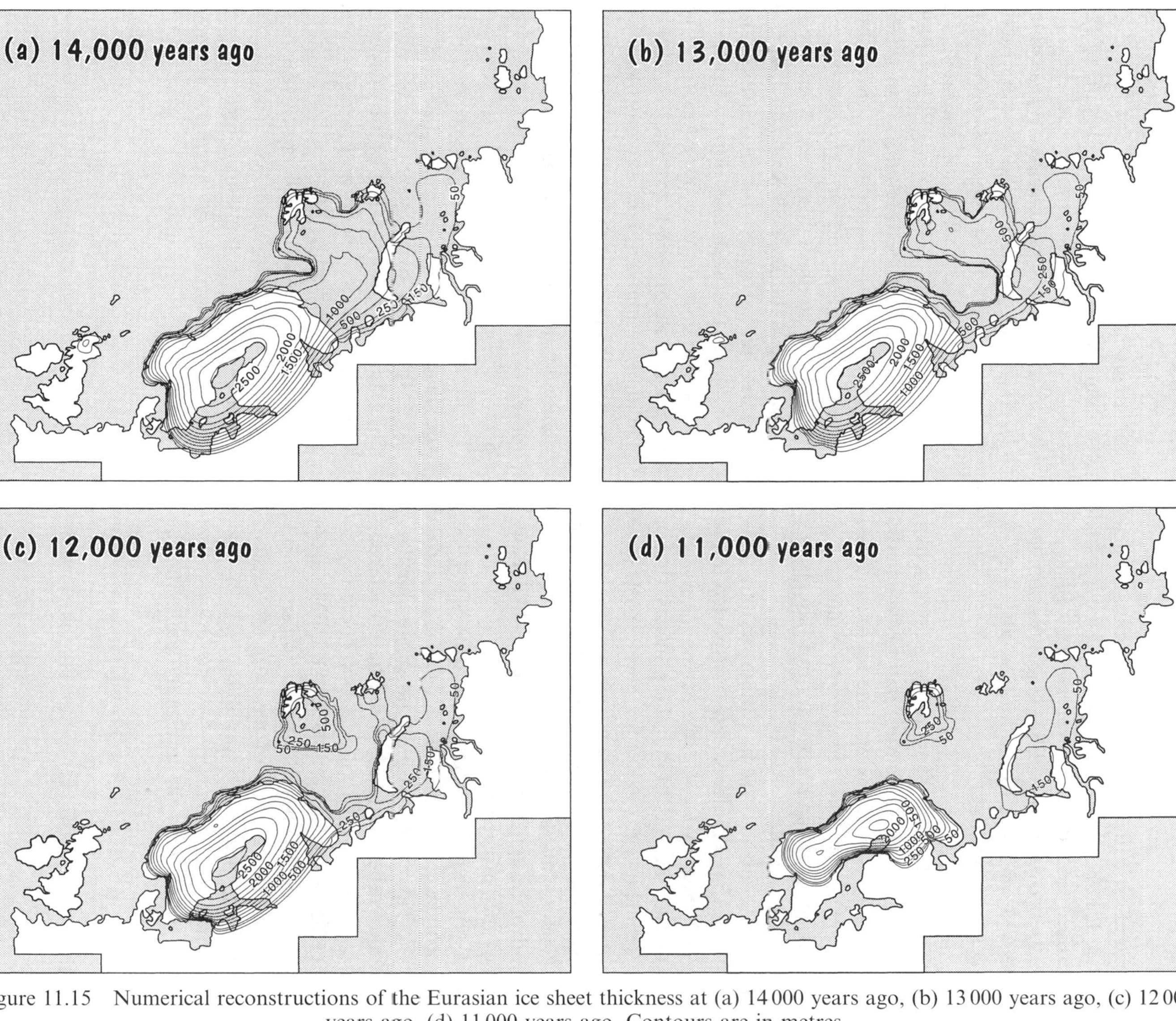

Figure 11.15 Numerical reconstructions of the Eurasian ice sheet thickness at (a) 14000 years ago, (b) 13000 years ago, (c) 12000 years ago, (d) 11000 years ago. Contours are in metres.

C. SUMMARY OF THE LATE QUATERNARY GLACIATION OF THE EURASIAN ARCTIC

Our understanding of the LGM Eurasian High Arctic Ice Sheet during the last glacial cycle has improved considerably over the last 10 years due to two successive ESF field programmes, PONAM and QUEEN, and to the use of numerical modelling.

- The most recent phase of ice growth across the Eurasian Arctic was relatively late in the glacial cycle, allowing time for interstadial sediments at Kapp Ekholm to build up and marine sediments in the central Barents Sea to accumulate.
- Measurements from trough mouth fans show that sediments were transported to the shelf edge by ice streams across the western margin at the LGM. This means that a significant grounded ice sheet must have existed across the Barents Sea.
- Moraine features near the northern coast of Russia show that the LGM ice sheet did not advance significantly southwards like many of the former ice sheets.
- Lacustrine sediments and permafrost sequences demonstrate a lack of ice across the Taymyr Peninsula, and suggest that ice across the Kara Sea would have been relatively thin (<0.5 kilometres). There is also evidence for lack of ice at the LGM across the Yamal Peninsula further to the south.
- The rapid growth of ice over the western Barents Sea was assisted by high precipitation rates, caused by the evaporation of water from the unusually warm surface of the eastern Norwegian Sea, where seasonally open ocean conditions prevailed. Evidence for these conditions comes from sub-polar forams found within Late Weichselian sea floor sediments.
- In contrast to the maritime climate of the west, the Kara Sea and the Taymyr Peninsula were characterised by a polar desert climate inhibiting the growth of ice.
- The central, deepest regions of the Barents Sea were the last to become glaciated. Ice growth in these regions may have been assisted by isostatic forebulges, caused by ice loads across Scandinavia, Svalbard, Franz Josef Land and Novaya Zemlya.
- The LGM ice sheet was over 1 kilometre thick across the Barents Sea, and had fast flowing ice streams within bathymetric troughs to the west and north. However, ice thickness may have been less than 0.5 kilometres across the Kara Sea.
- The deepest marine-based regions of the ice sheet broke up early in the last deglaciation. The high ^{16}O content in the iceberg meltwater has been observed in sea floor sediments within the Fram Strait, dated at about 15 ka BP.
- The terrestrial portions of the ice sheet were the last to decay, allowing the isostatic uplift patterns over Svalbard and Franz Josef Land to reflect a last phase of the ice sheet limited to these archipelagos and neighbouring shallow seas.
- The decay of ice was by iceberg calving over the marine sections of the ice sheet, and by surface melting in the south, forming proglacial lakes across the northern Russian mainland.

CHAPTER 12

Ice Sheet Reconstructions IV – North America, Laurentide Ice Sheet, Cordilleran Ice Sheet, Innuitian Ice Sheet

A. GEOLOGICAL EVIDENCE

Introduction

The glaciation of North America is one of the most well studied ice age glacial phenomena. The extent of the ice sheet at the last glacial maximum, and during deglaciation, has been relatively easy to determine because of the well preserved moraines and proglacial lake sequences. However, it is only in the last few years that information regarding the time-dependent 'instability' of the ice sheet been discussed, and how this behaviour may be linked with rapid oceanic and climate change. At the LGM, the Laurentide Ice Sheet was the second largest (to Antarctica) in the world. It, therefore, contributed most to global sea level fall during the last ice age. Moreover, in its coalescence with the Cordilleran Ice Sheet, the ice cover over North America constituted more than one-third of the world's glacial cover. In North America, the period of the last glacial–interglacial is called the Wisconsin, and the last glacial itself is referred to as the Late Wisconsin.

Present day ice masses in North America

A number of ice caps exist within the Canadian High Arctic, including those on Devon Island (representing 16 200 square kilometres of glacierised area), Ellesmere Island (80 500 square kilometres), and Axel Heiberg Island (11 700 square kilometres) (Dowdeswell, 1995). The bulk of glacier ice is found in the islands of the eastern Canadian Arctic. This region is relatively close to the moisture source provided by the seasonally sea ice-free waters of Davis Strait, Baffin Bay and Nares Strait, and to the North Water, which is an area where open water is found year-round. More westerly locations within the High Canadian Arctic have lower precipitation and little glacier ice, with the exception of the mountainous terrain of the Yukon Territory. A few glaciers and ice caps in the Canadian Arctic islands have been investigated in some detail, but relatively little is known about the ice dynamics and mass balance of many of these ice masses. For example, the form and flow of the Barnes Ice Cap on Baffin Island has been studied in some detail (e.g. Holdsworth, 1977), the mass balances of the Devon and Meighen ice caps, and of the White Glacier on Axel Heiberg Island, have been monitored over a relatively long period (e.g. Cogley et al., 1995; Dowdeswell et al., 1997), and ice cores have been taken from several ice caps for the purposes of palaeoclimatic reconstruction (e.g. Koerner, 1977, 1997).

Just as the distribution of modern ice caps in North America is controlled by the spatial distribution of air temperature and precipitation (Figures 12.1a,b), so the growth of ice during the last glaciation would have been

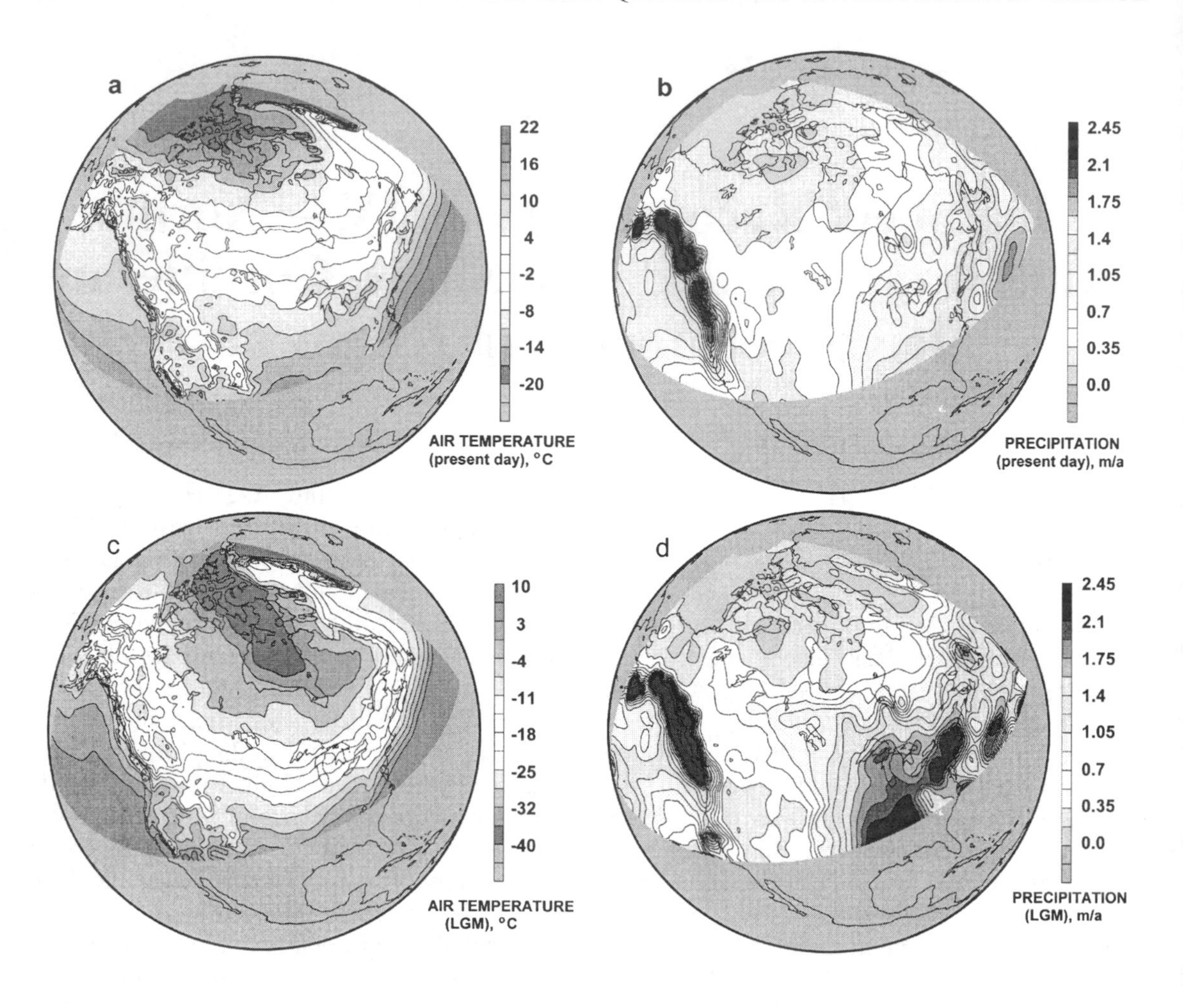

Figure 12.1 The present day and LGM climate of North America derived from GCM modelling (Vettoretti et al., 1999). (a) Modern values of mean annual surface air temperature. (b) Modern rates of mean annual precipitation. (c) LGM values of mean annual surface air temperature. (d) LGM rates of mean annual precipitation. Taken from Marshall et al. (2000) and reprinted from the Canadian Journal of Earth Sciences with permission from NRC Research Press.

controlled to a first order by these climate parameters. Moisture sources from the Pacific and North Atlantic provide precipitation to the west and east of the continent, respectively. At the LGM, rates of precipitation are thought to have increased across the south-east, and decreased in areas to the north of the continent (Figures 12.1c,d). The mean annual air temperature was up to 35°C colder at the LGM (Figure 12.1).

Onset of ice growth (Late Wisconsin)

The Late Wisconsin glacial stage in North America is between 35–10 000 years ago. It is not known whether growth occurred over ice-free land or a pre-existing ice sheet from earlier in the Wisconsin. There were two main ice sheets in North America, the Laurentide and the Cordilleran. At the LGM, in confluence, these ice sheets represented 16 000 000 square kilometres of glacial coverage. The onset of North American glaciation at 35 000 years ago is presumed to have started by Milankovitch forcing (with associated feedback processes). This makes the ice sheet a typical, text-book, land-based large Northern Hemisphere ice sheet. As the solar insolation values decreased in the summer, so did summer air temperatures resulting in a reduction in the summer melting season. This allowed glacier growth from the northern latitudes and highlands southwards to the mid North American continent and the ablation zone.

Maximum extent of ice sheets

Cordilleran Ice Sheet and the northwest of North America

The LGM ice sheet extended from the coast ranges of Alaska, over to the east of the Rocky Mountains (Figure 12.2). The precipitation source for the ice sheet accumulation was the Gulf of Alaska, resulting in a maritime climate in the west of the ice sheet. Because of the precipitation source and topography, the Pacific coastline of the Cordilleran Ice Sheet was dominated by tidewater glaciers and ice shelves. To the east of the Rocky Mountains, an accumulation shadow caused restriction of the eastward flow of ice over the mountains. To the north and west, the Aleutian Islands were partly glaciated and a small ice cap also existed, independent of the Cordilleran Ice Sheet, over the Brooks Range in Alaska.

Laurentide Ice Sheet

The Laurentide Ice Sheet covered most of Canada and the very northern edge of the United States (Figure 12.3). The western extent of the ice sheet was the east of the Rocky Mountains, where there was very little precipitation, and the ice sheet coalesced with the Cordilleran Ice Sheet. From here the ice sheet extended to the north-west Atlantic Ocean. North of 60°N, the ice sheet (a sector known as the Innuitian Ice Sheet) spread over Ellesmere Island and was connected to the north-west Greenland Ice Sheet via the Smith Sound ice stream (Chapter 10). The southern margin of the Laurentide Ice Sheet, during its maximum stage and during deglaciation, is characterised by a series of ice sheet 'lobes'. These lobes represent advances or surges of part of the ice sheet, which left a series of complex moraines to the south of the ice sheet (e.g. Des Moines Lobe; James Lobe) which advanced in excess of 500 kilometres southwards from the parent ice sheet. The ice sheet also left behind huge glacial geological features (including drumlins, mega-flutes and other mega-scale lineation features). Some of these glacial features are so large that it is only until fairly recently that they have been mapped and interpreted using satellite remote sensing techniques (Clark, 1993).

Extent of ice during ice sheet decay

Ice sheet decay and the formation of lobes

The southern margin of the ice sheet flowed over unconsolidated sediments. Whilst the ice sheet base was cold, these sediments would have been frozen and would not have affected the flow of ice. However, as the ice sheet built up, and the ice base became warm, the geotechnical properties of these sediments would have changed rapidly from frozen and strong, to water saturated and weak. In this situation the shear stress at the ice sheet base would have far exceeded the strength of the material and it would have deformed. This basal deforming sediment would have resulted in rapid ice sheet flow above and the formation of an ice sheet lobe. Many of these lobe surges caused the ice extent to flow southwards by 500 kilometres, where a moraine was deposited. However, the sudden southward advance of ice took the ice front into the deep ablation zone. Once the ice surge had ceased, the supply of ice could not meet the rate of ablation and so the ice front quickly returned to the parent ice mass, where ice sheet build-up could

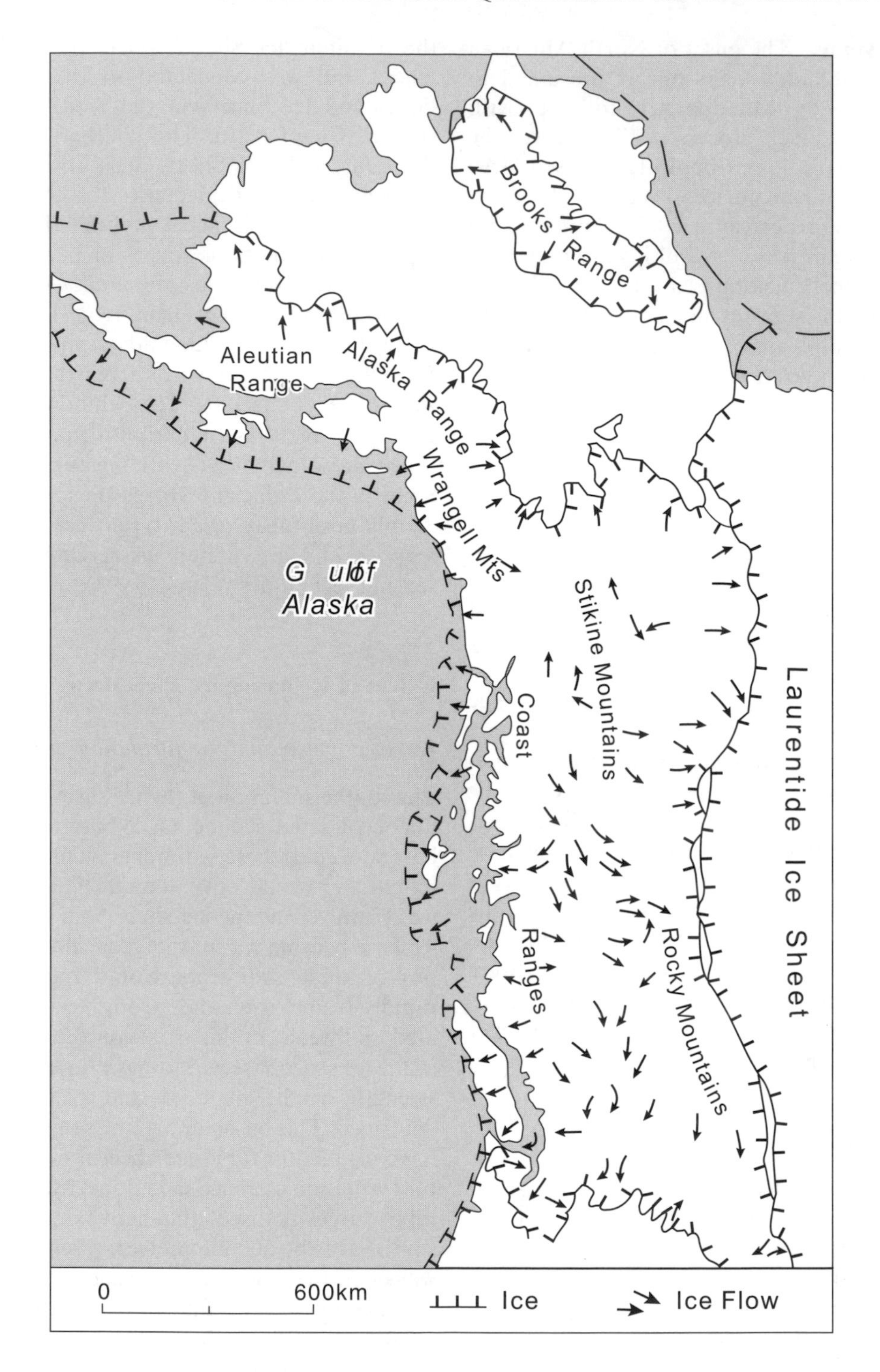

Figure 12.2 Glaciation of the Rocky Mountains and Alaska: The Cordilleran Ice Sheet. The ice limits and flow directions have been established from a variety of geological data. Taken from Flint (1971).

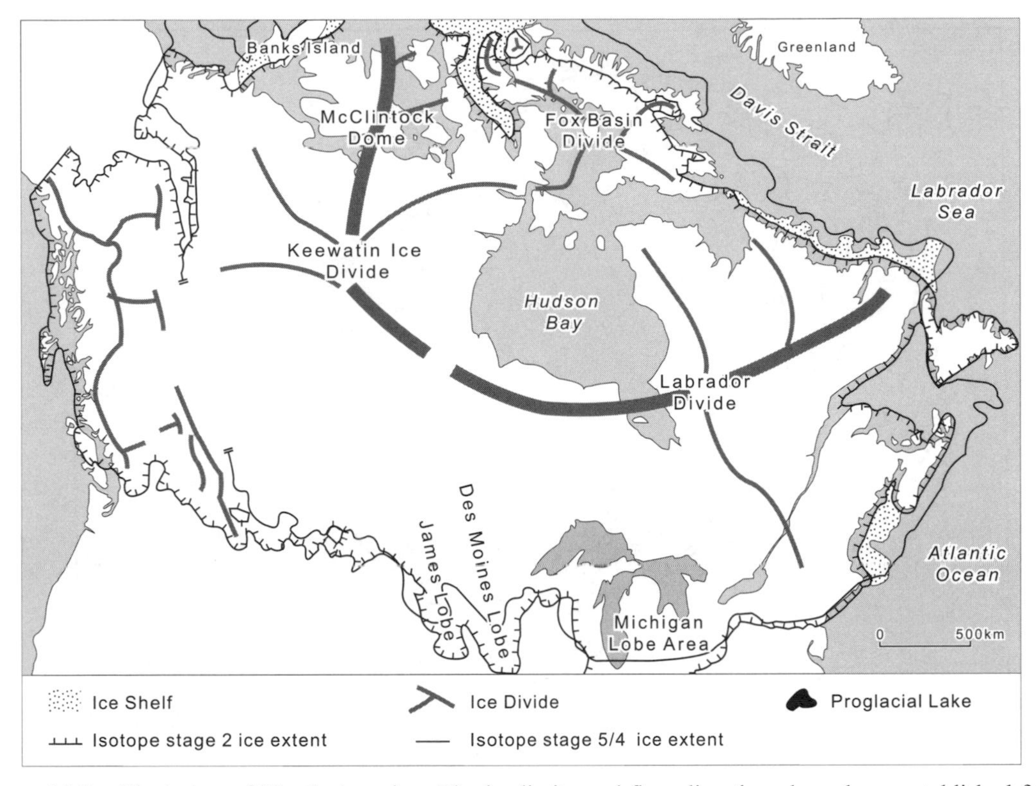

Figure 12.3 Glaciation of North America. The ice limits and flow directions have been established from a variety of geological data. Taken from Dyke and Prest (1987).

begin again as the first stage of the next ice lobe advance. The lobes of the Laurentide Ice Sheet often spread into the Great Lakes region. In summary, the unstable behaviour of the southern Laurentide Ice Sheet was induced by internal ice sheet dynamics rather than by climate change.

Ice sheet decay and the formation of proglacial lakes

The southern margin of the Laurentide Ice Sheet was an ablation zone. During the main phase of deglaciation, after 13 000 years ago, the surface meltwater issued by the ice sheet ran off the ice surface to the proglacial zone where some of it could be routed through existing rivers such as the Mississippi. However, the topography at the margin of the ice sheet was influenced heavily by isostatic depression such that water collected around the ice margin as a series of proglacial lakes. Also, the formation of lobes often dammed the course of rivers allowing the further build up of lakes. The existence of former proglacial lakes are easily identifiable from the glaci-lacustrine geology that is left behind. As the ice lobe decayed, some lakes merged together, through rapid flooding events, to form bigger lakes. For example, by 12 000 years ago, lakes Whittesley, Chicago and Saginaw, coalesced into a new lake, Lake Algonquin. Further east, stagnation and retreat of the Ontario ice lobe resulted in the formation of Lake Iroquois (supplied by discharge from Lake Algonquin). By 11 000 years ago, further deglaciation had resulted in the formation of the larger post Algonquin Lake, and overflow into the Champlain Sea. For a fuller description of the history of ice lobe advances, read Dyke and Prest (1987). The glacial sequences left by

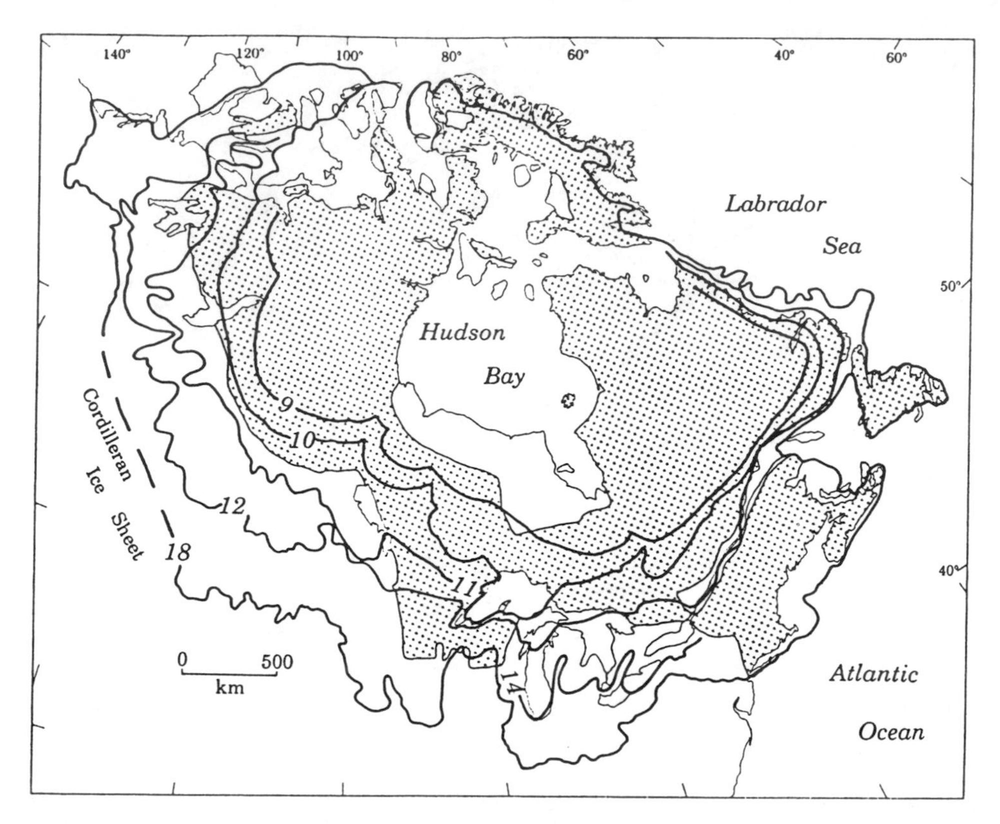

Figure 12.4 Decay of the Laurentide Ice Sheet determined from geological evidence. Also shown is the area of the ice sheet base resting on rock (shaded region) and soft deformable sediment (white region). Taken from Dyke and Prest (1987).

proglacial lakes and moraines mark the edge of the ice sheet at the LGM and several stages during deglaciation. These data have been used to map the retreat of the ice sheet across Canada during deglaciation (Figures 12.4 and 12.5).

The chronology of proglacial lakes Agassiz, Algonquin and Ojibway

The progressive retreat and thinning of the Red River lobe eventually caused, after 12 000 years, the formation of Lake Agassiz. This massive lake fed fresh water into the Mississippi River, as did lakes within the Great Lakes region between 12 500 and 11 000 (as is evident in the Gulf of Mexico isotope record). However, after 11 000 years, the lack of ^{16}O within the Gulf of Mexico sediments reflects a re-routing of lake waters through the Great Lakes into the St Lawrence (e.g. Clark et al., 1996).

During its 4000 year history, Lake Agassiz played a significant role in the deglaciation of the Laurentide Ice Sheet (Figure 12.5). The lake reached maximum extent between 9900 and 9500 years ago when it occupied an area of 350 000 square kilometres, within large regions of Ontario and Manitoba. As the Lake Superior ice lobe decayed, a large lake was left behind; Lake Ojibway. At 8400 years ago, Hudson Bay and James Bay lowlands experienced glacial

Figure 12.5 Decay of the Laurentide Ice Sheet according to interpretation of the geological record. Taken from Dyke and Prest (1987).

14000 years before present

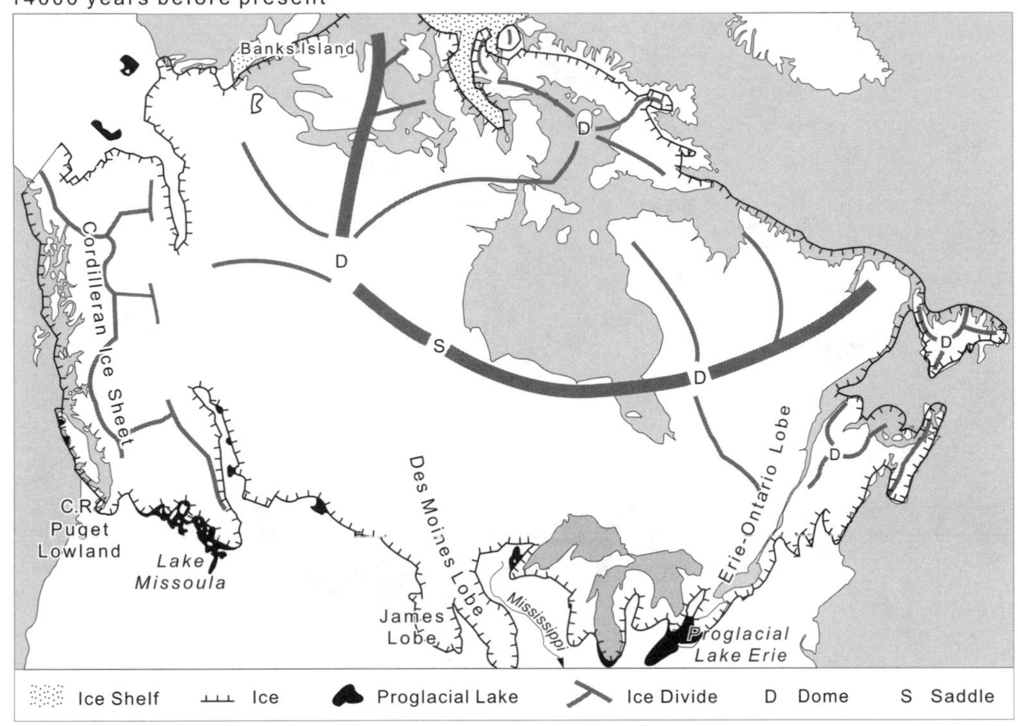

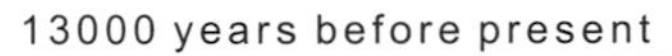
13000 years before present

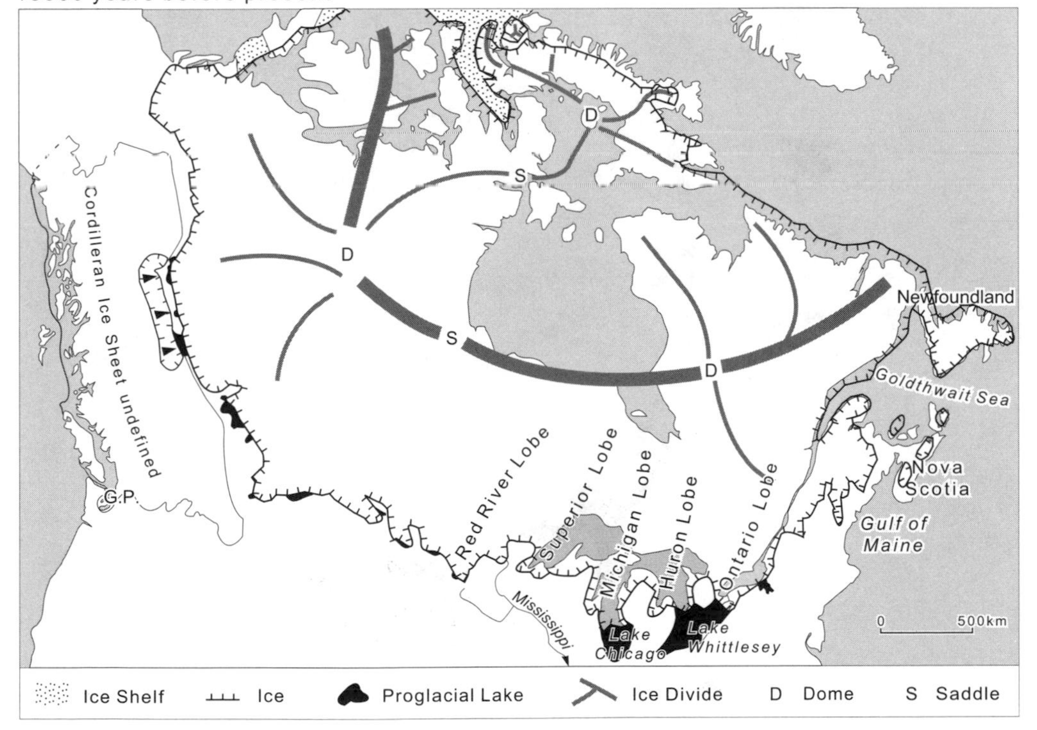

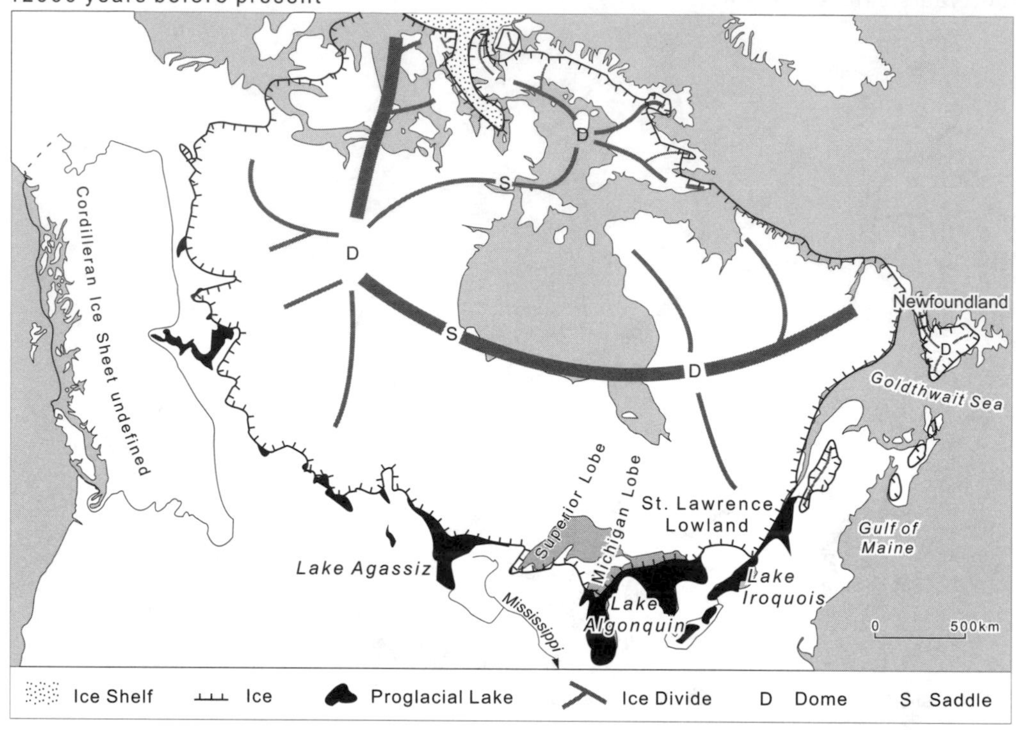
12000 years before present
Cordilleran Ice Sheet undefined
D
S
D
S
D
Newfoundland
D
Goldthwait Sea
Superior Lobe
Michigan Lobe
St. Lawrence Lowland
Gulf of Maine
Lake Agassiz
Lake Algonquin
Lake Iroquois
Mississippi
0 500km
Ice Shelf
Ice
Proglacial Lake
Ice Divide
D Dome
S Saddle

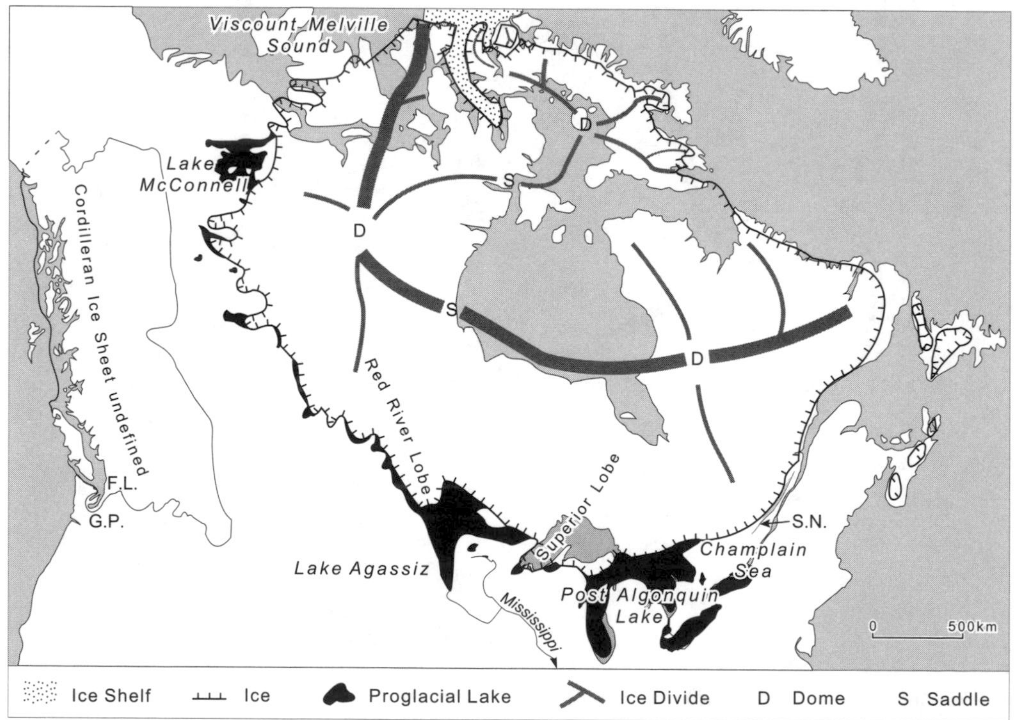
11000 years before present
Viscount Melville Sound
Lake McConnell
Cordilleran Ice Sheet undefined
D
S
D
S
D
Red River Lobe
Superior Lobe
F.L.
G.P.
S.N.
Champlain Sea
Lake Agassiz
Post Algonquin Lake
Mississippi
0 500km
Ice Shelf
Ice
Proglacial Lake
Ice Divide
D Dome
S Saddle

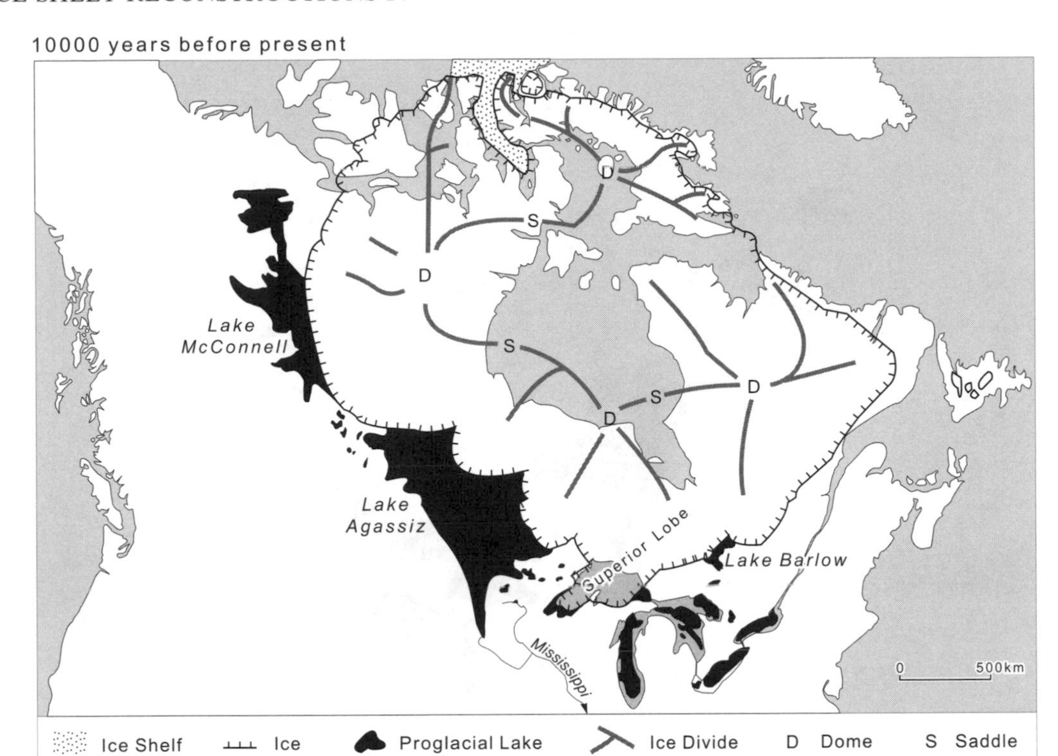
10000 years before present
Lake McConnell
Lake Agassiz
Superior Lobe
Lake Barlow
Mississippi
D
S
0 500km
Ice Shelf
Ice
Proglacial Lake
Ice Divide
D Dome
S Saddle

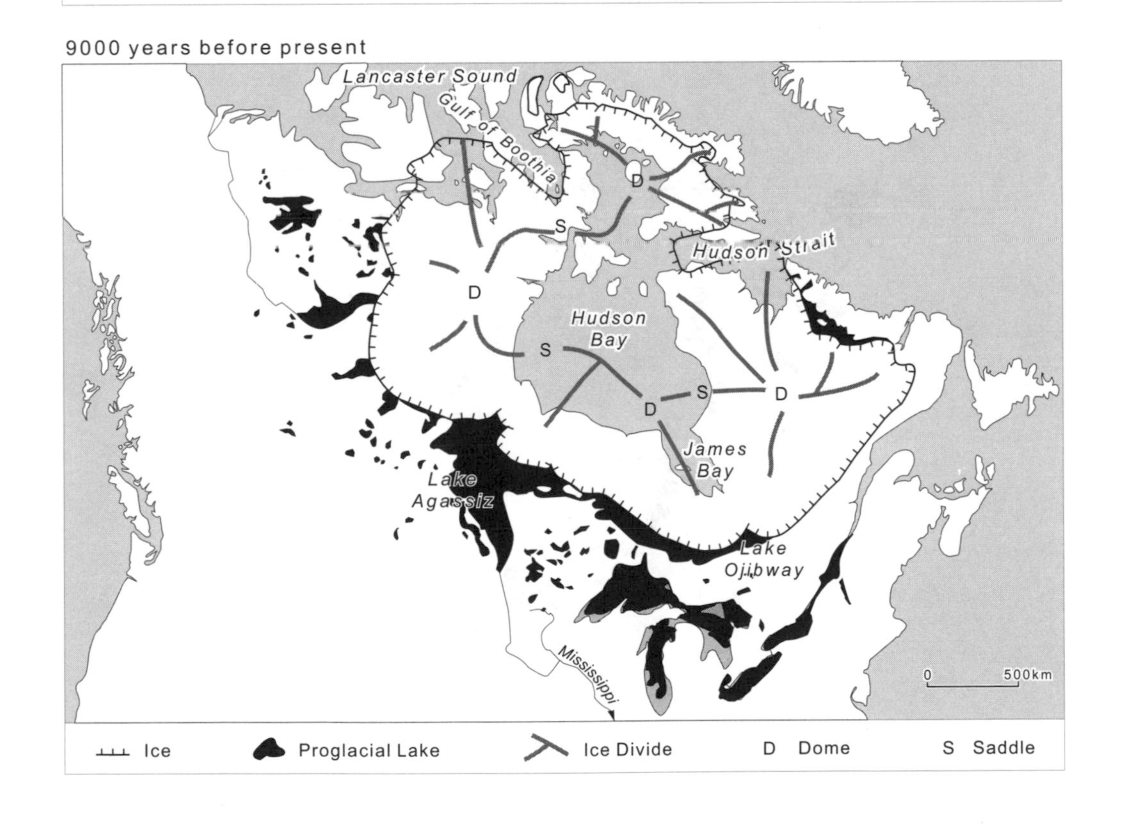
9000 years before present
Lancaster Sound
Gulf of Boothia
Hudson Strait
Hudson Bay
James Bay
Lake Agassiz
Lake Ojibway
Mississippi
D
S
0 500km
Ice
Proglacial Lake
Ice Divide
D Dome
S Saddle

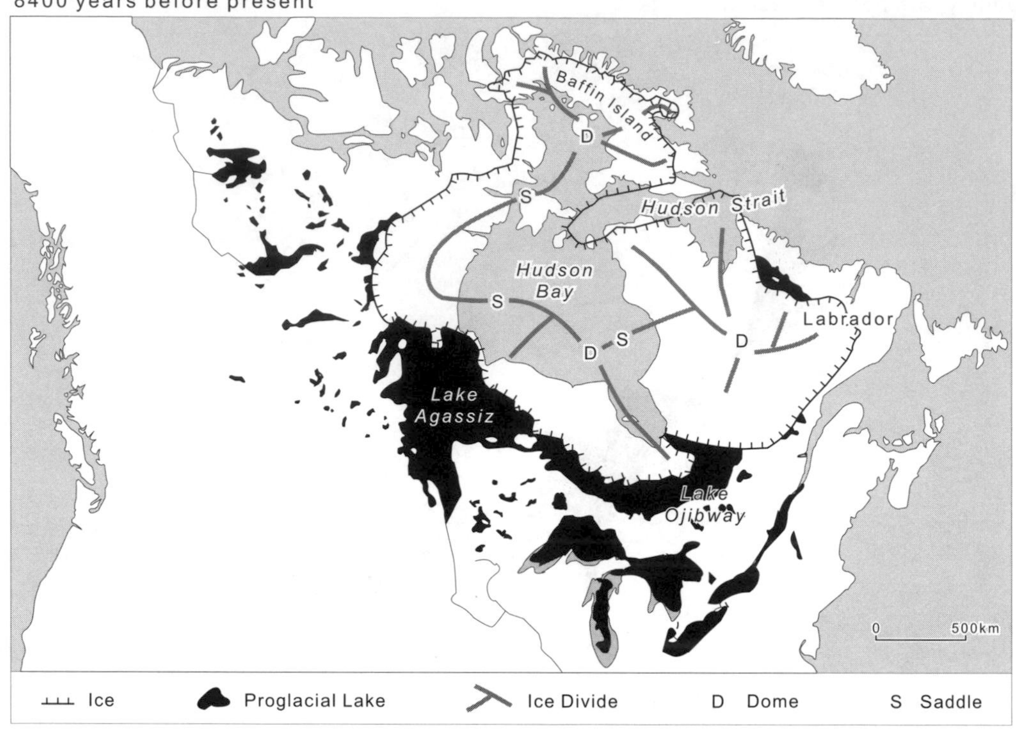
8400 years before present
Baffin Island
D
S
Hudson Strait
Hudson Bay
S
D
S
D
Labrador
Lake Agassiz
Lake Ojibway
0 500km
Ice
Proglacial Lake
Ice Divide
D Dome
S Saddle

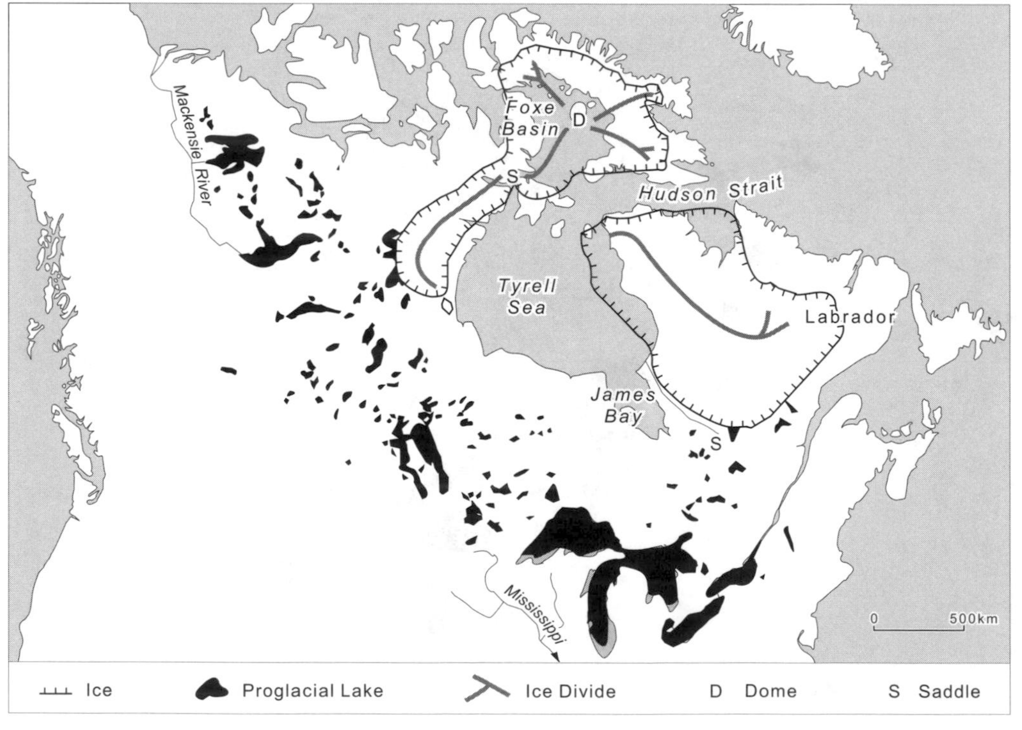
8000 years before present
Mackensie River
Foxe Basin
D
S
Hudson Strait
Tyrell Sea
Labrador
James Bay
S
Mississippi
0 500km
Ice
Proglacial Lake
Ice Divide
D Dome
S Saddle

surges known as the Cochrane ice advances. At this time, ice pushed into Lake Ojibway such that the southern and south-western ice margin was connected to a 3100 kilometre lake margin. The Cochrane ice advances were succeeded by the disintegration of ice in Hudson Bay.

At 8200 years ago, ice sheet stagnation was accompanied by catastrophic drainage of Lake Ojibway and Lake Agassiz into the Hudson Bay. Thus, the drainage path of the meltwater was dramatically altered. Estimates put the volume of water involved in the instantaneous discharge at between 70 000 and 150 000 cubic kilometres. This very sudden influx of fresh water into the North Atlantic is bound to have affected the salinity of surface waters in the North Atlantic, NADW production and, hence the climate at this time (Chapter 8). Estimates indicate that the draining of Lake Ojibway and Lake Agassiz caused an increase in global sea level of 0.2–0.4 metres within two days. Barber et al. (1999) match the sudden drainage of these lakes through the Hudson Strait to a widespread cool period found in the Greenland ice cores at 8200 calendar years ago (Figure 12.6). The alteration to the loading of the Earth's surface caused by this virtually instantaneous increase in sea level has yet to be investigated. However, such catastrophic drainage must have caused some seismic and earthquake activity as the Earth's crusts adjusted to this new loading distribution. By 7000 years BP, only small remnants of the Laurentide Ice Sheet, and the proglacial lakes that it was responsible for, remained.

The effect of proglacial lakes on climate and ice sheet dynamics

The occurrence of large proglacial lakes in North America such as Lake Agassiz would have had an influence on the climate of the continent. Hostetler et al. (2000) investigate this influence by numerical modelling of the late glacial climate across North America, taking into account the position of the proglacial lakes and ice sheets. They found that when the lake level is high, moisture supply to the ice sheet is restricted compared to when the lake is at a lower level. Because of this, when the lake is at a high level, the ice sheet is likely to retreat, opening up a pathway for the lake water to the North Atlantic. However, when the lake was at a low level, ice sheet advance could have closed this water route, stopping the supply of water to the North Atlantic. Hostetler et al. (2000) conclude that the climatological effects of Lake Agassiz, and other proglacial lakes in North America, may have influenced climate, ice sheet dynamics and, in doing so, their own rapid discharge to the North Atlantic.

Brief summary of proglacial lake chronology

At 13 000 years ago the first large lakes formed at the margin of ice sheet. By 12 000 years ago lakes Chicago, Saginaw and Whittlesey merged to form Lake Algonquin. Also, the Red River lobe decayed allowing the formation of Lake Agassiz, water from which was routed through the Mississippi. At 11 000 years ago drainage of Lake Agassiz caused water flow east through the St Lawrence. Also, Lake Algonquin drained into the Champlain Sea. By 9000 years ago, as the Superior ice lobe decayed, Lake Ojibway was formed. At around 8200 years ago lakes Ojibway and Agassiz decanted their water into the Atlantic Ocean within a few days, resulting in catastrophic flooding, drainage and climate change.

Raised beaches and post glacial isostatic uplift

The ice sheet reconstruction of the CLIMAP programme involved a huge, single domed ice sheet centred over the western side of Hudson Bay. However, this model was discredited as a result of mapping the post-glacial emergence of the Laurentide Ice Sheet. If a single-domed ice sheet existed at the LGM, its decay would have caused concentric isostatic uplift lines around the zone of maximum ice thickness (i.e. where maximum displacement of the crust occurred).

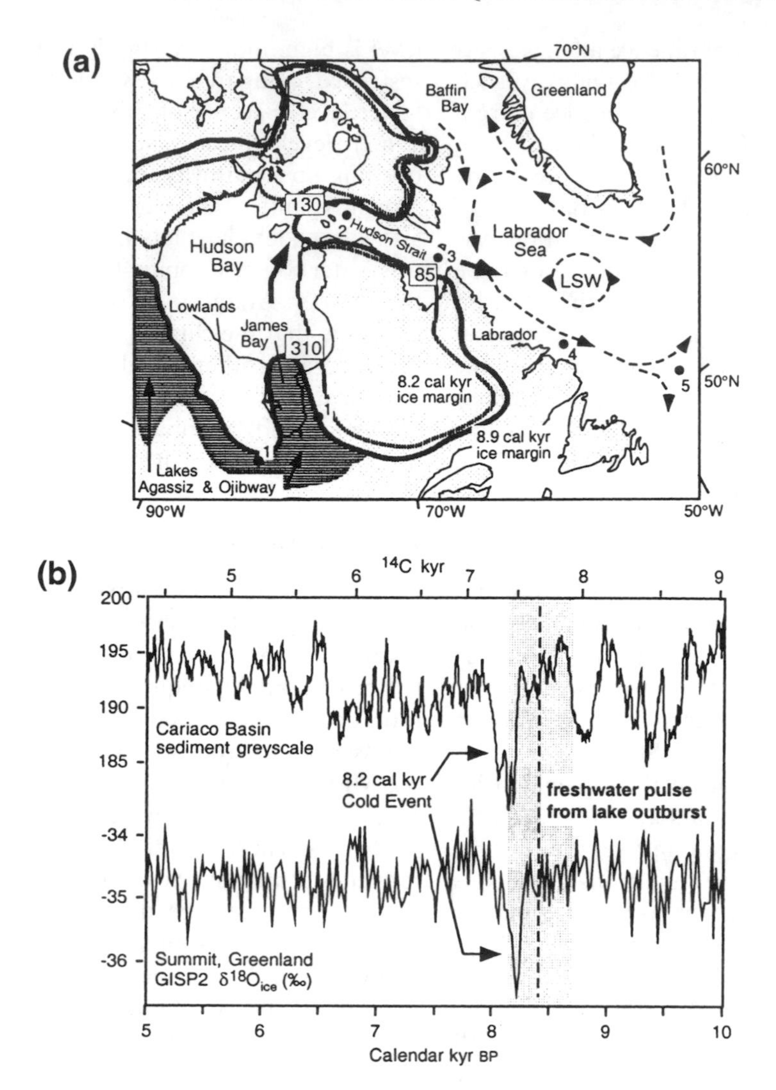

Figure 12.6 Decanting of Lake Agassiz into the North Atlantic and the effect on ocean circulation and climate. Barber et al. (1999) contend that this process was responsible for the 'cold event' observed in the Greenland ice core record at 8200 years ago. (a) routes of glacial meltwater at 8200 years ago. (b) proxy records of the 8200 year 'cold event' from the colour of sediment transported by the meltwater and $\delta^{18}O$ record from the GISP2 ice core. Reprinted with permission from *Nature* vol 400, pp. 344–348, copyright (1999) Macmillan Magazines Limited.

However, Dyke found that the actual post-glacial uplift is organised into a series of concentric circles around North America, each denoting a region of former maximum ice loading (Figure 12.7). Thus, from consideration of the uplift patterns, it has been concluded that a multi-domed ice sheet is more likely at the LGM than the CLIMAP singled domed reconstruction. These uplift data have been used as input to an Earth model (Peltier, 1994) to quantify the ice sheet dimensions responsible for the measured isostatic response. It is now thought by many that there were three main components comprising the Laurentide Ice

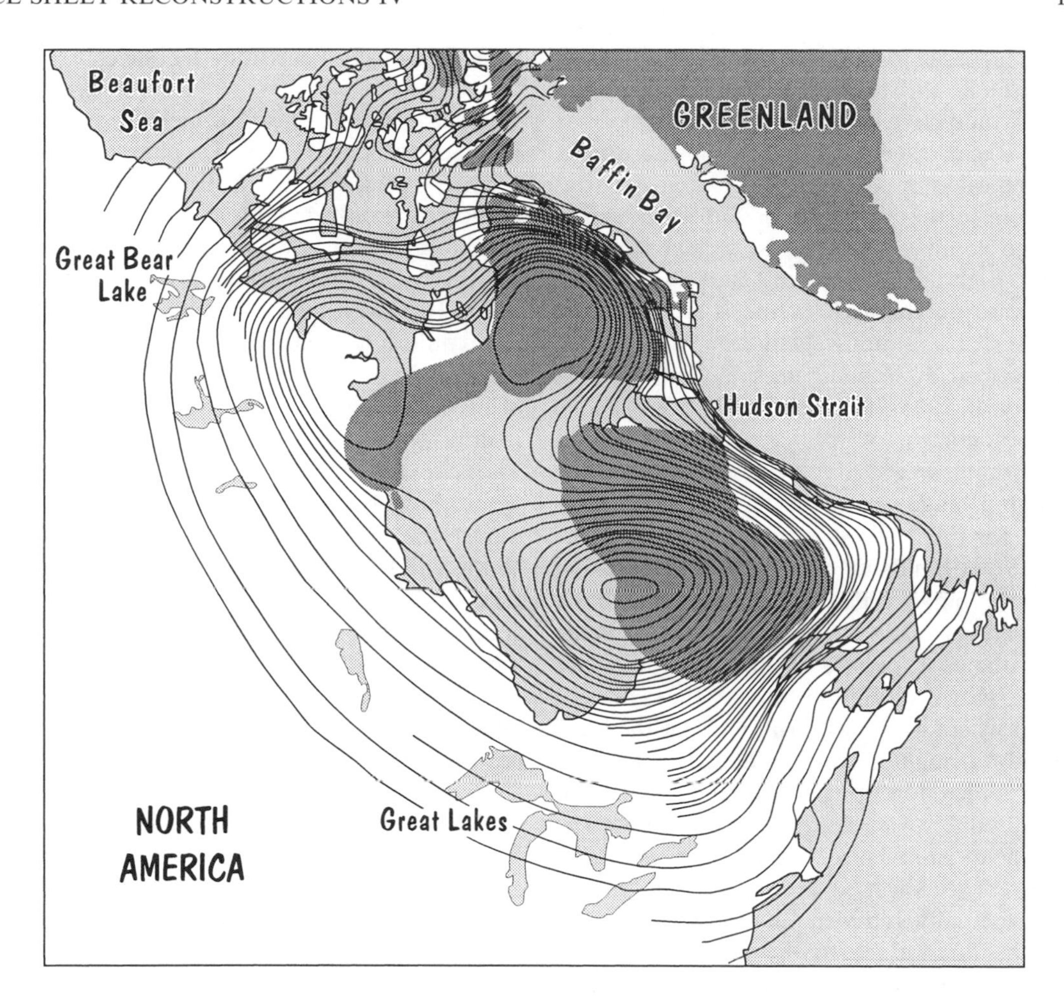

Figure 12.7 Patterns of isostatic uplift across North America at 8000 years ago. The isolines have been interpolated from dated raised beach and ancient proglacial lake shorelines. The uplift clearly indicates that there were multiple ice domes across North America at the last ice age. This uplift is at odds with the notion of the large single domed ice sheet predicted by the CLIMAP group. Adapted from information supplied by C. O'Cofaigh from research by Art Dyke.

Sheet; the Labrador, Baffin and Keewatin sectors, to the east, north and west of Hudson Bay, respectively (e.g. Clark et al., 2000).

The rapid decay of ice across North America caused several hundred metres of uplift and would have induced enormous alterations to the stress field of the North American Plate. Such changes to the Earth's crust are likely to have resulted in significant seismic activity during deglaciation. Wu and Johnston (2000) suggest from numerical modelling that deglacial isostasy may be responsible for the modern relatively large earthquakes occurring across supposedly stable intra-plate regions near the former ice sheet margin.

Production of icebergs from the Laurentide Ice Sheet

The north-western margin of the ice sheet terminated as a calving ice wall along the Hudson

Strait. Icebergs formed and would have drifted out into the North Atlantic where their sediments would have dropped to the ocean floor as IRD. Recent sedimentological investigations have revealed a series of IRD events, corresponding to the 7000 year periodic production of huge volumes of icebergs, named Heinrich events (Figure 12.8; see also Chapters 6 and 8). The cause of Heinrich events is still debated. However, a commonly held view is that they were formed by periodic unstable ice dynamics (MacAyeal, 1993). In much the same way as the southern margin of the ice sheet experienced lobe surging, so too may the north-eastern side, through a binge–purge mechanism. The explanation for this process is quite simple. The ice sheet slowly builds up over an essentially frozen base (the binge phase). Eventually the basal temperatures reach the pressure melting value that initiates rapid basal motion (the purge phase). However, unlike across the southern margin where ice lobes form, most of the ice is drained through a single outlet, the Hudson Strait. As ice flows out of the Hudson Strait it calves, and flows across the North Atlantic. The drainage of so much ice depletes the reserves in the parent ice sheet so that it becomes thinner and, eventually refrozen to the base at which time the flux of ice to the Hudson Strait is decreased, and iceberg volumes are reduced. This is then followed by ice re-growth. The cycle of binge–purge has been modelled by MacAyeal (1993), at around 7000 years, a remarkably similar periodicity to that measured from IRD in the North Atlantic (Figure 12.8). It should be noted that the purge (surging) phase lasts for only a short time (around 700 years).

Climate and the Laurentide Ice Sheet

There are several ways in which the Laurentide Ice Sheet was responsible for climate change at a variety of scales. First, the atmospheric circulation was affected by the ice sheet. At its maximum, two eastward, upper atmosphere jet streams were split north and south of the ice sheet. However, as the ice sheet decayed these jets would have coalesced. Second, the thermohaline-driven ocean conveyor system may have been influenced by meltwater and icebergs issuing from the ice sheet during its decay. The climate change that may have resulted would have been at a hemispheric level. Third, during the end of the last glacial, a cold north–south wind crossed the Laurentide Ice Sheet, was undercut by warm south–north winds south of the ice sheet. Thus, a very steep temperature gradient was set up in the proglacial area at the margin between these two winds, as is evident from proxy temperature records from lakes in the northern USA (Levesque et al., 1997).

The LGM climate model of Marsiat and Valdes (in press) accounts for cooler Pacific SSTs than predicted under the CLIMAP scenario. These conditions lead to cooler winds across the Laurentide Ice Sheet and, hence, a reduction in precipitation over the ice sheet. This cool, dry climate would have had consequences for the ice thickness at the LGM, and puts into question the possibility of an ice sheet over 3 kilometres thick.

Deglaciation of the Cordilleran Ice Sheet

The timing of deglaciation began sometime after 15 000 years ago. Shortly after this first phase of

Figure 12.8 The occurrence of Heinrich layers from North Atlantic sea floor cores. (a) Location of cores where Heinrich layers have been identified. Filled circles are where carbonate deposits are found in all Heinrich layers. Half filled circles show where carbonates are found in some Heinrich layers and open boxes are where no carbonate material is found. (b) Details of the Heinrich event record within several North Atlantic cores. Note that a map of Heinrich layer 1 is provided in Figure 6.6. Reprinted with permission from *Nature* vol 360, pp. 245–249, copyright (1999) Macmillan Magazines Limited.

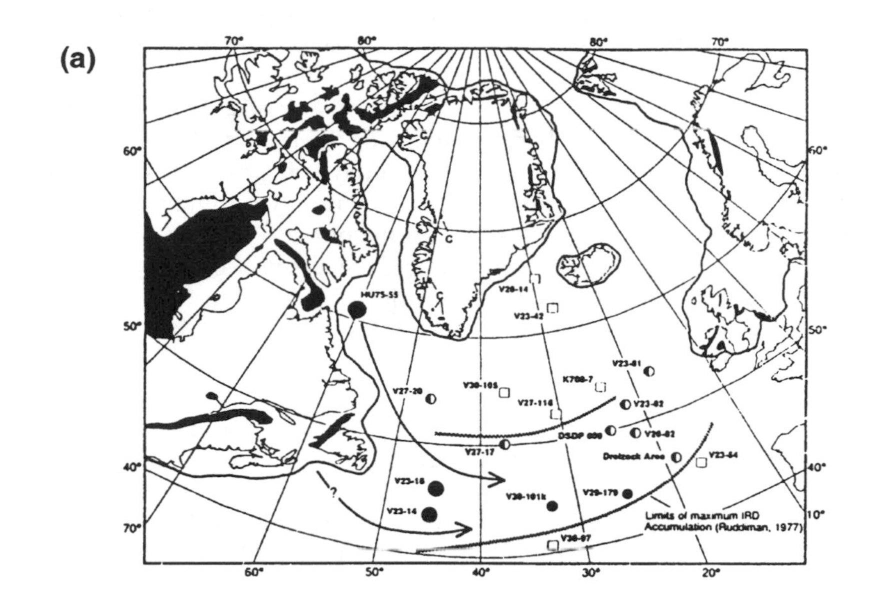
(a)
HU75-55
DSDP 609
Dreizack Area
Limits of maximum IRD
Accumulation (Ruddiman, 1977)

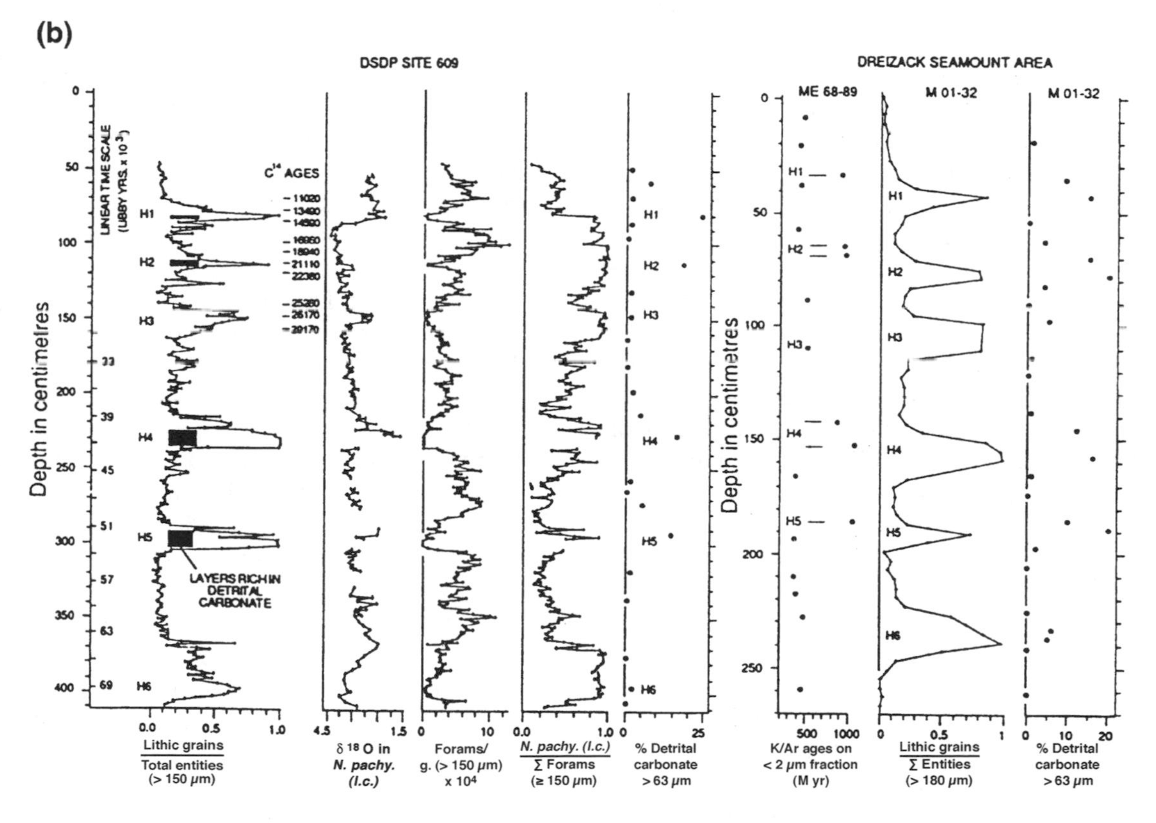
(b)
DSDP SITE 609
DREIZACK SEAMOUNT AREA
ME 68-89
M 01-32
M 01-32
Depth in centimetres
LINEAR TIME SCALE (LIBBY YRS. x 10^-3)
C^14 AGES
H1
H2
H3
H4
H5
H6
LAYERS RICH IN DETRITAL CARBONATE
Lithic grains / Total entities (> 150 μm)
δ 18 O in N. pachy. (l.c.)
Forams/ g. (> 150 μm) x 10^4
N. pachy. (l.c.) / Σ Forams (≥ 150 μm)
% Detrital carbonate > 63 μm
K/Ar ages on < 2 μm fraction (M yr)
Lithic grains / Σ Entities (> 180 μm)
% Detrital carbonate > 63 μm

deglaciation an ice-free corridor was established between the Laurentide and Cordilleran ice sheets. This may have very important implications for the migration of animals into the North American continent from Asia via the Bering Strait. During deglaciation in Montana, several pro-glacial lakes were formed, the largest of which was Lake Missoula; a periodical lake which grew whenever ice advanced to dam the Clark Ford River. The withdrawing of this dam caused several Jökulhlaups of around 2000 cubic kilometres along the river.

Possibility of subglacial lake outbursts during deglaciation

There have been several recent theories regarding the nature of large-scale ice sheet stability and subglacial conditions during the deglaciation of the Laurentide ice sheet. One rather controversial idea, that is worthy of note, is that subglacial water escaped from the ice sheet base during deglaciation, causing a series of fluvial drumlins and other erosional landforms. For example, Shaw et al. (1984) calculated that the Livingstone Lake drumlin field (Saskatchewan, Canada) was caused by a flood involving 84 000 cubic kilometres of water. This calculation assumes that the drumlins were formed within subglacial hollows melted out by the flood water. However, as Shoemaker (1995) points out, if the action of erosion by suspended sediment on the sub-ice hollows is accounted for, then an outburst volume of around 8000 cubic kilometres is required. Shoemaker (1992) postulated that a huge 200 000 cubic kilometres subglacial lake, fed by supra glacial meltwater and proglacial lakes, built up in Hudson Bay after 9000 years. This lake is unlikely to have outburst completely in any single event. Instead, Shoemaker (1992) suggested that outburst volumes would have been limited to much less than 60 000 cubic kilometres (Shoemaker, 1992), and that a series of episodic outbursts lifted the ice sheet off its bed causing ice sheet 'surging' during the final phase of ice sheet decay. This suggests that ice sheet dynamics were influential in the decay of the Laurentide Ice Sheet. Shoemaker (1999) also suggests that water sheet floods from other regions of the decaying Laurentide Ice Sheet may have been influential in the dynamics of the ice sheet, and the formation of drumlins. It should be noted that the physical basis of water sheet outbursts, where the ice sheet base is uplifted by a 1 metre depth of water has been hotly disputed (Walder, 1982, 1994; Shoemaker, 1994).

Further inspection of the Laurentide Ice Sheet glacial geology and unstable flow of ice

Most recent geological work on the Laurentide ice sheet has focused on the nature of deformable sediment at the ice sheet margin during deglaciation, and the consequences for the unstable flow of ice. Alley (1991) and Clark (1994) suggest from analysis of the widespread uniform till sheets within the southern margin of the Laurentide Ice Sheet, that the surging of the southern margin would have been due to deformation of this sediment at the base of the ice sheet. Because of this, the dynamics of the ice sheet are driven internally rather than by climate change. This work is supported by Hicock and Dreimanis (1992), who suggest that basal till deformation within the Great Lakes region was a major influence on ice sheet dynamics.

All of these studies were a result of a seminal paper by Fisher et al. (1985), in which the Laurentide Ice Sheet was modelled using a simple yield stress technique. This form of modelling is not time dependent and cannot be used to examine the ice sheet response to environmental change. However, such modelling is useful because it can provide a quick and simple method by which ice sheet profiles can be established if the extent of the former ice sheet is known (say from terminal moraines). The model assumes that ice behaves as a perfectly plastic material. That is, ice does not deform until a yield stress is reached. At and above this yield stress, ice deformation can occur easily. The model also assumes that the yield stress is equal to the basal shear-stress. If basal shear-stress is replaced by yield stress in Equation 1 (Chapter 3), and the surface slope (α) is rewritten as the change in ice thickness over horizontal distance (dh/dx), then the equation can be integrated, and used to determine the ice thickness along a profile from the ice sheet centre to the specified ice sheet margin.

Fisher et al. (1985) found that only when very low yield stresses were placed across the Hudson Bay, Great Lakes and Prairies areas (indicative of weak water saturated sub-glacial sediment) was the geologically-based reconstruction of Dyke et al. (1982) able to be calculated. The Laurentide Ice Sheet model of Fisher et al. (1995) was effectively 'upgraded' by Clark et al. (1996), using a time-dependent ice sheet model from which it was again concluded that the presence of subglacial water-saturated sediments would have had a fundamental control on the surface topography of the ice sheet.

Local oxygen isotope records, indicating meltwater production, show that the Barents (Eurasian) ice sheets decayed at 15 000 years ago. The same methods have shown the Laurentide Ice Sheet receded at around 13 000 years ago (Jones, 1991b).

B. NUMERICAL MODELS OF THE WISCONSIN NORTH AMERICAN ICE SHEETS

By far the most comprehensive series of numerical ice sheet modelling experiments involving the North American ice sheets has been undertaken by Shawn Marshall (Department of Geography, University of Calgary). In the last few years Marshall has produced a number of papers detailing the configuration, dynamics, growth and decay, mass balance components and run off of the Laurentide and Cordilleran ice sheets. When put together, this folio of work comprises a fully quantitative reconstruction of the Wisconsin glaciation of North America (Figures 12.9, 12.10 and 12.11). Prior to this work of Marshall, the most accepted model of the LGM ice sheet in North America was developed by Peltier (1994), named ICE-4G, through experiments with a solid Earth model forced to match ice sheet topographies with measured raised shorelines.

LGM ice sheet configuration

The ICE-4G model depicts an ice sheet which is considerably thinner than produced by the CLIMAP group in the 1970s (Figure 10.10). Whereas the CLIMAP LGM model involved a huge 4 kilometre thick ice sheet with one central ice dome across the whole of North America, ICE-4G predicted a multi-domed ice sheet that was up to 1500 metres thinner (Figure 10.10, Clark et al 1999). The ICE-4G model calculated ice domes across the Cordilleran Ice Sheet, and two main domes over the eastern and western Laurentide Ice Sheet. With an acknowledgement that numerical ice sheet models tend to over predict the ICE-4G Laurentide ice thickness (due possibly to the standardised treatment of the deformation of ice in models), Marshall and Clarke (1999) modelled the overall topographic appearance of the ICE-4G reconstruction reasonably well in a time-dependent investigation of ice volume and mass balance (Figure 12.9). The total volume of their modelled ice sheet was estimated to be about 34 per cent greater than the ICE-4G reconstruction. Despite this, Marshall and Clarke (1999) showed that, at the LGM, three large supraglacial lakes would have formed across the ice sheet over (a) the confluence between the Laurentide and Cordilleran ice sheets, (b) the centre of the Laurentide Ice Sheet and (c) a smaller lake over the north of the ice sheet, south of the Amundsen Gulf. The limits of the ICE-4G LGM ice sheet was forced to match the well mapped terminal moraines to the south, and the coastlines to the north, east and west. Marshall and Clarke predicted that proglacial

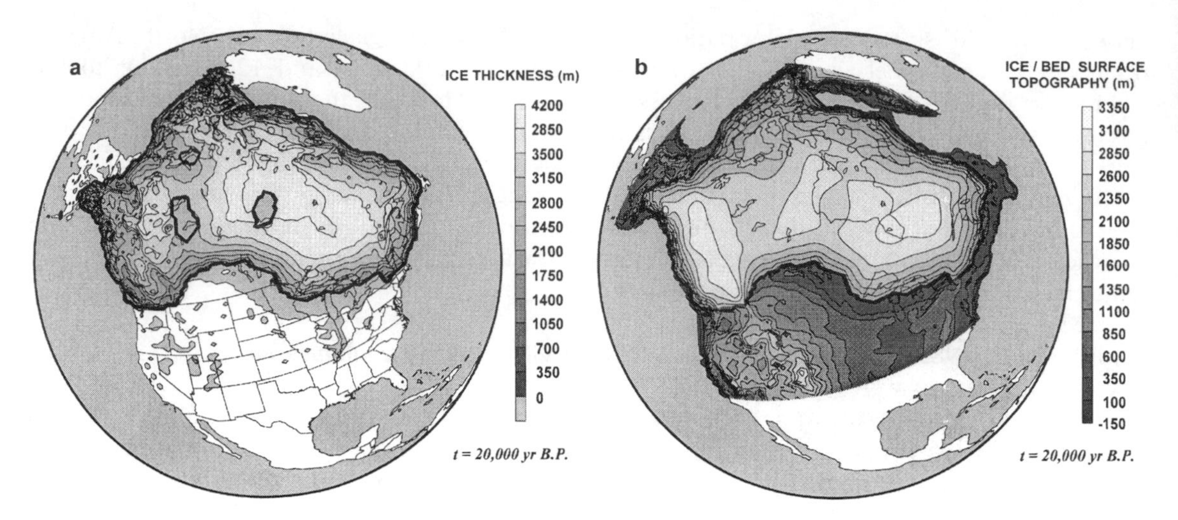

Figure 12.9 The Laurentide Ice Sheet at the LGM reconstructed from numerical modelling. (a) Ice thickness. (b) Surface elevation. Taken from Marshall and Clarke (1999), and reproduced by permission of Academic Press.

lakes would dominate the southern margin of the ice sheet at the LGM (Figure 12.9).

Bearing in mind the over-prediction of ice thickness by the Marshall and Clarke model, they calculated that the Laurentide Ice Sheet would have contained enough ice to lower the sea level by 87 metres. Only 17 per cent of this ice was stored within the Cordilleran Section of the ice sheet.

Time-dependent growth of ice

The model of Marshall and Clarke (1999) calculated the growth of ice throughout a full Wisconsin glacial cycle. The model was forced by air temperature conditions (and related ELA adjustments) derived from the GRIP ice core and global eustatic sea level fall (see Chapter 2). Although the surface mass balance of the ice sheet reflected the high frequency variability within the GRIP record, the interaction between ELA adjustment and sea level resulted in the time-dependent ice sheet development oscillating with Milankovitch-type frequencies (e.g. 20 000 and 40 000 and 100 000 years) (Figure 12.12). The ice sheet model grew from a zero state at 120 000 years ago, and produced four main maximum stages at 107 000, 84 000, 60 000 and 20 000 years ago. The ice sheet volume and surface area at 60 000 years ago is predicted to be almost as great as at the LGM. If this part of the model is realistic, it means that the decrease in global sea level measured in the $\delta^{18}O$ sea floor sediment signal, and low latitude coral reefs, must have been caused by the growth of ice outside of North America (e.g. Eurasia, Chapter 11). Marshall and Clarke's model coupled an ice sheet component with a surface run-off model. They were therefore able to predict the growth and decay of proglacial lakes during the glacial cycle. Their results indicated that proglacial lakes existed at the southern margin of the ice sheet throughout most of the glacial cycle, because surface ablation across this region of the ice sheet fed the proglacial region with water.

Geological and topographic control on ice sheet dynamics

Marshall et al. (1996) modelled the flow of ice from an ICE-4G type ice sheet. They used a

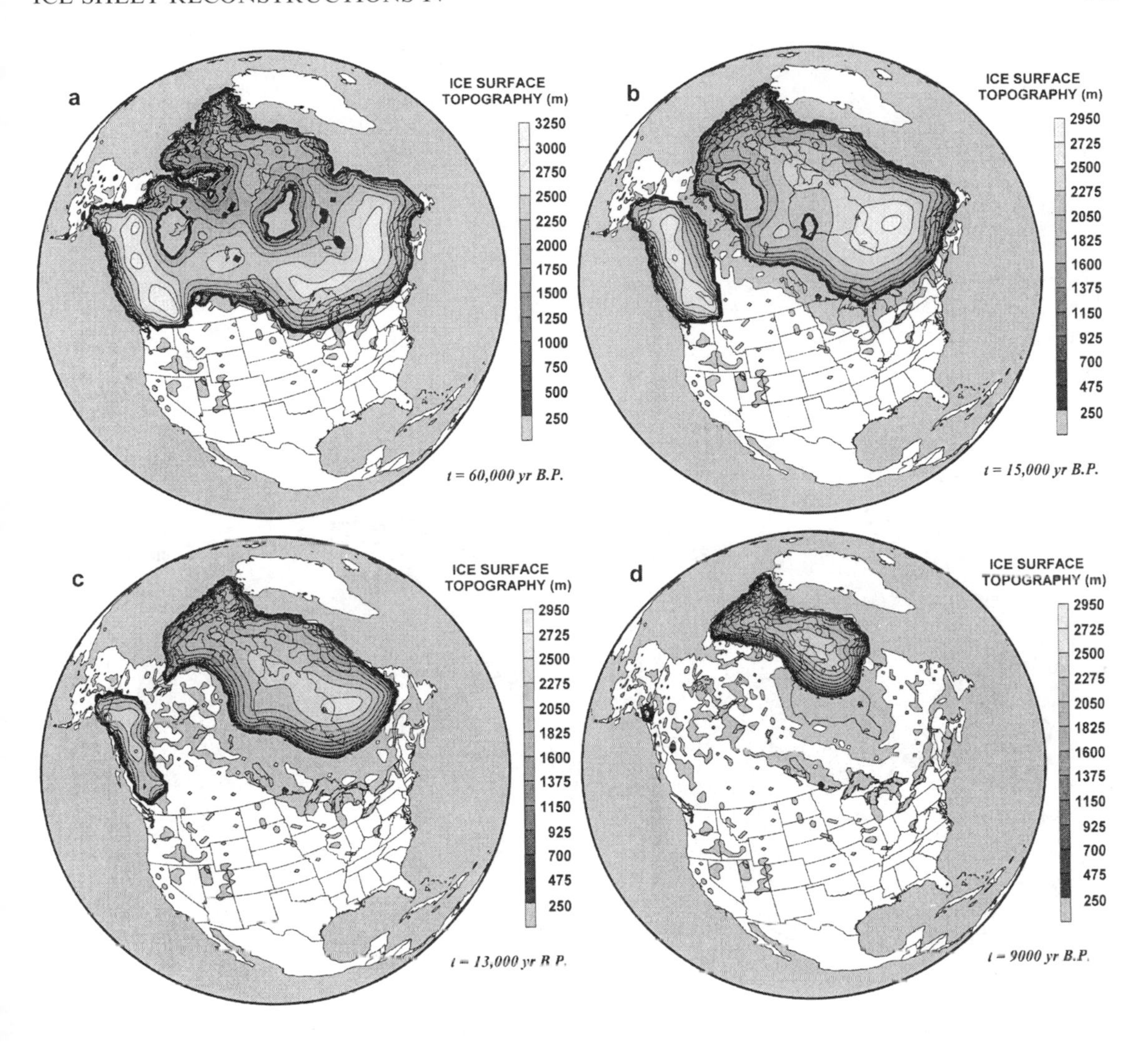

Figure 12.10 Surface of Laurentide Ice Sheet for the last 60 000 years from numerical modelling. (a) Ice surface at 60 000 years ago. (b) Ice surface at 15 000 years ago. (c) Ice surface at 13 000 years ago. (d) Ice surface at 9000 years ago. Taken from Marshall and Clarke (1999), and reproduced by permission of Academic Press.

realistic bedrock topography as input, which included an appreciation of subglacial geology. Utilising the fact that there is plenty of geological data to determine the flowlines and extent of ice streams, they modelled the Laurentide Ice Sheet with an aim to establishing the subglacial topographic and geological controls on ice stream development in North America. The analysis of surface geology revealed that large regions of the southern margin of the ice sheet are characterised by thick (up to 20 metres), continuous deposits of sediment. They concluded that the southern plains and continental shelf regions are 'predisposed' to rapid basal motion of ice, whether it be caused by deformation of subglacial material or by subglacial sliding. This work was in strong agreement with the earlier investigation of Fisher et al. (1995).

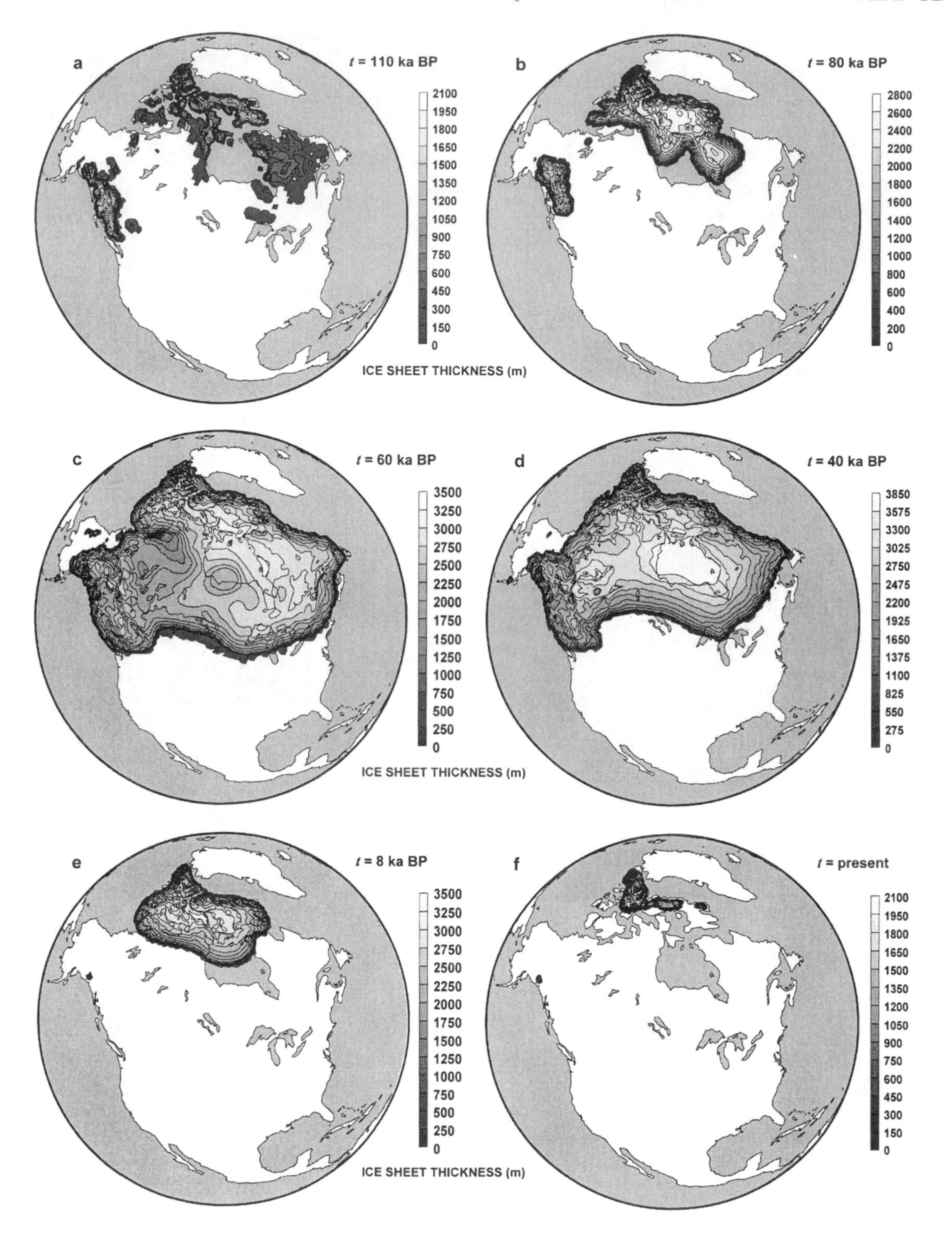
a
t = 110 ka BP
2100
1950
1800
1650
1500
1350
1200
1050
900
750
600
450
300
150
0
ICE SHEET THICKNESS (m)
b
t = 80 ka BP
2800
2600
2400
2200
2000
1800
1600
1400
1200
1000
800
600
400
200
0
c
t = 60 ka BP
3500
3250
3000
2750
2500
2250
2000
1750
1500
1250
1000
750
500
250
0
ICE SHEET THICKNESS (m)
d
t = 40 ka BP
3850
3575
3300
3025
2750
2475
2200
1925
1650
1375
1100
825
550
275
0
e
t = 8 ka BP
3500
3250
3000
2750
2500
2250
2000
1750
1500
1250
1000
750
500
250
0
ICE SHEET THICKNESS (m)
f
t = present
2100
1950
1800
1650
1500
1350
1200
1050
900
750
600
450
300
150
0

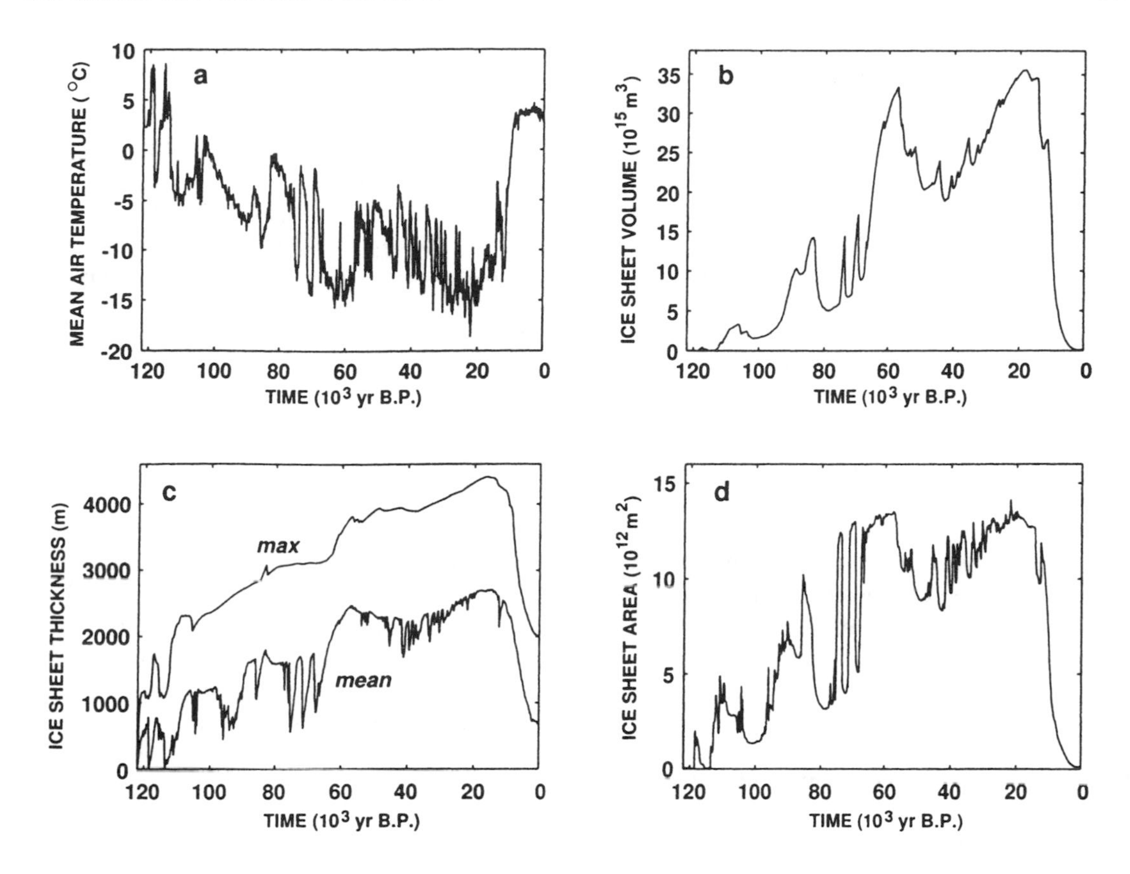

Figure 12.12 Time-dependent ice sheet model results. (a) mean annual air temperature over the ice sheet. (b) Ice sheet volume. (c) Ice Sheet thickness and (d) Ice sheet area. Taken from Marshall and Clarke (1999), and reproduced by permission of Academic Press.

Unstable ice flow and ice stream thermo-mechanics

Payne (1995) and Payne and Donglemans (1997) showed how the build up and decay of ice has a strong control on the subglacial thermal regime of an ice sheet. Under relatively stable external forcing conditions, an ice sheet may oscillate between periods of ice growth when the ice is cold-based, and ice decay caused by the attainment of warm-based thermal conditions and the consequent initiation of rapid basal motion. This mechanism is, effectively, the Binge–Purge idea that MacAyeal (1993) put forward for the cause of iceberg pulses from the Hudson Strait.

Marshall and Clarke (1997a,b), used a similar thermo-mechanical ice sheet model to Payne (1995) to predict the oscillations forced by the thermal regulation of the ice sheet. They found that the periodicity of warm–cold cycles yielding iceberg fluxes from the Hudson Strait

Figure 12.11 Thickness of the Laurentide Ice Sheet during the last 110 000 years from numerical modelling. (a) 110 000 years ago. (b) 80 000 years ago. (c) 60 000 years ago. (d) 40 000 years ago. (e) 8000 years ago. (f) present day ice cover. Taken from Marshall et al. (2000) and reprinted from the Canadian Journal of Earth Sciences with permission from NRC Research Press.

was comparable to the 7000 year Heinrich layer signal (Figure 12.13). However, the ice stream velocity associated with, and the total volume of ice evacuated by, a single surge event was much less (an order of magnitude) than predicted by MacAyeal. Marshall and Clarke (1997b) calculated that sea level would rise by between 0.04 to 0.6 of a metre because of one oscillation of the 'binge–purge' mechanism. Further, they suggested that the amount of iceberg production would have been significantly less than proposed by Dowdeswell et al. (1995). Because of this, Marshall and Clarke (1997b) indicate that an efficient means by which sediment is entrained within the ice is required to match their results with measured thickness of Heinrich layer IRD in the North Atlantic. This would mean that Heinrich layer icebergs would be much dirtier than icebergs of the present day.

Time-dependent numerical studies of deglaciation

A great deal is known about the deglaciation of North America from numerical modelling investigations. However, time-dependent fluxes of meltwater, surface run off and icebergs during deglaciation are not well known, despite being critical to the circulation of the ocean and, therefore, global climate. Marshall and Clarke (1999) ran an ice sheet model through a full Wisconsin cycle, and matched their results to geological evidence. The numerical results indicated the time- and spatially-dependent changes in the flux of fresh water to the oceans caused by (i) icebergs, (ii) run off through rivers and (iii) catastrophic release of water from proglacial lakes. These results, detailed below, compare reasonably well with the available geological evidence for meltwater pulses in the Gulf of Mexico and the North Atlantic.

The novel part of Marshall and Clarke's model was the division of North America into surface run-off drainage systems. From these systems it was possible to predict where and when the meltwaters would have been routed during the last deglaciation. The deglaciation of Late Wisconsin North America was predicted to have been controlled predominately by melting at the southern margin, causing the ice sheet limit to retreat northwards leaving behind proglacial lakes and uncovering formerly glaciated landscapes which became involved in the routing of water (Figure 12.10). The model suggests that, between 14 000–10 000 years ago, a 'moat' of meltwater was located in the area of lakes Agassiz and Ojibway, caused by the topographic depression from former ice loading. This moat grew northwards as deglaciation continued. Although this model is unrealistic in terms of the exact size and location of proglacial lakes established from the geological record, on a continental scale it appears to be quite a good general approximation. The model also accounted for run off from supraglacial lakes as the ice sheet decayed, run-off of surface rainfall and the flux of icebergs to the ocean. At the time of maximum extent of proglacial lakes, the volume of water held within proglacial lakes was large enough to raise sea level by a few metres.

The interaction between available drainage paths and sources of meltwater during the last deglaciation formed a number of distinct changes to the output of freshwater from North America into the ocean. Prior to 15 000 years ago, the dominant outlet was through the Mississippi River. However, after this time, the amount of water through the Mississippi River was reduced, and increased through the St Lawrence River. By 12 000 years, the switch from water routed through the Mississippi, to water routed through the St Lawrence, was completed. It was not, however, until 10 000 years ago that significant volumes of water were transported through the Hudson Strait or to the Arctic. The rapid change between waters issued through the Mississippi to the St Lawrence ties in nicely with measurements of sea floor ^{16}O measurements at around 12 000 years ago (Clark et al. 1996).

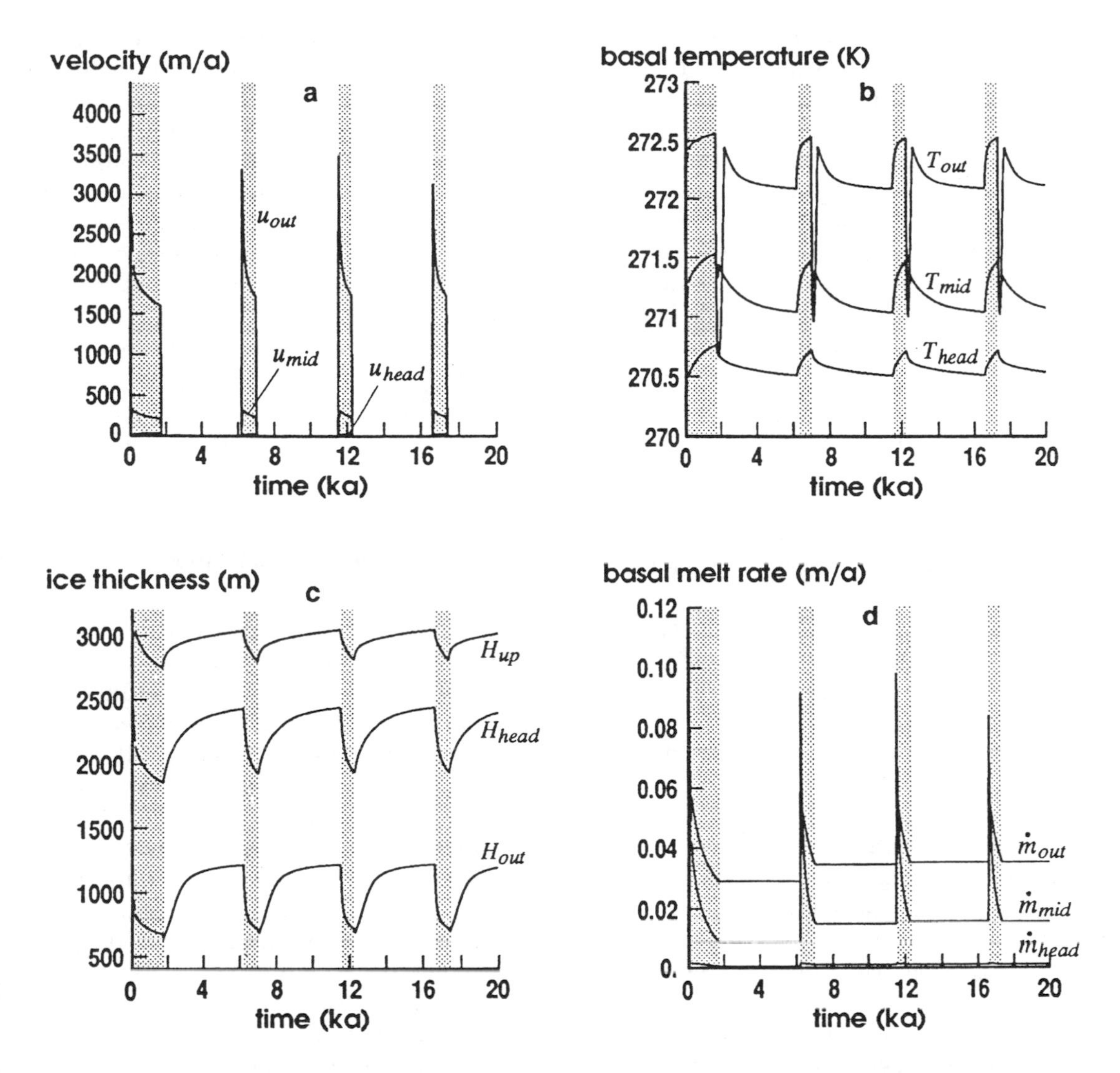

Figure 12.13 Thermo-mechanical ice sheet modelling results showing ice sheet instability at a frequency comparable to the measured occurrence of Heinrich layers. The graphs show the time-dependent behaviour of a number of glaciological characteristics at three locations within a Laurentide Ice Sheet ice stream during the last glaciation. (a) Ice velocity. (b) Basal temperature. (c) Ice thickness. (d) Rate of basal melting. Grey shading refers to the periods during when Heinrich layers have been recorded. Taken from Marshall and Clarke (1997), reprinted from *Journal of Geophysical Research*, with permission of American Geophysical Union.

Ice-Ocean interactions: Heinrich events

One problem associated with the notion of binge–purge oscillations being responsible for Heinrich layers is the method of entrainment of material within the ice sheet during the surge phase when most icebergs are produced. The problem is that a surge-type situation will involve the ice sheet effectively decoupled from the glacier bed which may not allow the take-up of material into the basal ice layers. However, Alley and MacAyeal (1994) established a method that allows both entrainment and surging to take place, which results in the production of sediment-laden icebergs. The model assumes that the start of the surge phase involves enough frictional heat to melt the ice base, lubricating the sediments and causing them to deform. However, as soon as the purge phase starts, the ice stream begins to thin. This thinning results in colder temperatures at the ice base and so water and sediment freeze on to the base of the ice stream. However, beneath the refrozen basal ice the sediment is still warm and water saturated and so the ice stream is still active. When all the water-soaked sediment has frozen on to the ice stream base it becomes 'cold-based' and the purge phase ends. This soft-bed model produces, in theory, a volume of sediment within the ice sheet that is larger than required to account for the Heinrich layers. However, a hard-bed model would produce far less. So, the presence of both hard- and soft-bed configurations would seem to make the model fit the measurements. Although this model remains unproven, Alley and MacAyeal (1994) suggest a number of good reasons as to why alternative entrainment mechanisms are less likely. For example, to account for enough debris to be incorporated due to ice tectonics such as folding and cavitation, would require a rate of uptake of material far in excess of that calculated and measured in modern ice sheets and glaciers. Further, a pressure-induced basal freezing process results in 2 orders of magnitude less sediment than is required. Therefore, the soft-bed freezing model appears a likely mechanism to explain the entrainment of material into the ice sheet during the surge phase cycle of the binge–purge model.

Dowdeswell et al. (1995) showed how the thickness distribution of Heinrich layers 1 and 2 (at 14.5 ka and 21.1 ka) is similar. This suggests that the volume and size of icebergs responsible for their formation, the proportion of sediment within these icebergs and the ocean currents were similar on both occasions (Figure 6.6). In addition, Dowdeswell et al. (1999) showed that there appeared to be a lack of correlation between Heinrich events and IRD features from the Fennoscandian Ice Sheet. This suggests that the cause of Heinrich events was particular to the Laurentide Ice Sheet, precluding the likeliness of a climate-induced mechanism that would affect all ice sheets in the Northern Hemisphere. Thus, there is a balance of data in support of the oscillatory binge–purge mechanism for Heinrich layers.

However, it should be noted that some argue for the involvement of the European ice sheets in at least the first phase of Heinrich layer depisition (Chapter 11; Grousset et al., 2000).

Implications for oceanic circulation

The production of icebergs from the Laurentide Ice Sheet in 'binge' mode into the North Atlantic would have injected vast quantities of fresh water to the ocean over very short periods of time (thought to be less than 1000 years). The thermohaline-driven circulation of the ocean is very likely to be affected by such freshwater. Put simply, the freshwater would reduce the amount of dense water formation and, therefore, the development of NADW thus reducing the northward drift of warm mid-Atlantic surface water, so initiating cooler climate conditions around Greenland and Europe (a D-O event).

One important event at the end of the last deglaciation is the 1000-year return to colder conditions during the Younger Dryas. Several authors have suggested that the Younger Dryas may have been caused by the release of meltwaters and icebergs into the North Atlantic. This episode would then have been both a D-O and Heinrich event (Alley and Clark, 1999). Climate change during the Younger Dryas led to the growth of many Northern Hemisphere glaciers. In addition to the Younger Dryas, the drainage of water from huge proglacial Lake Agassiz into the North Atlantic may also have had significant consequences for ocean

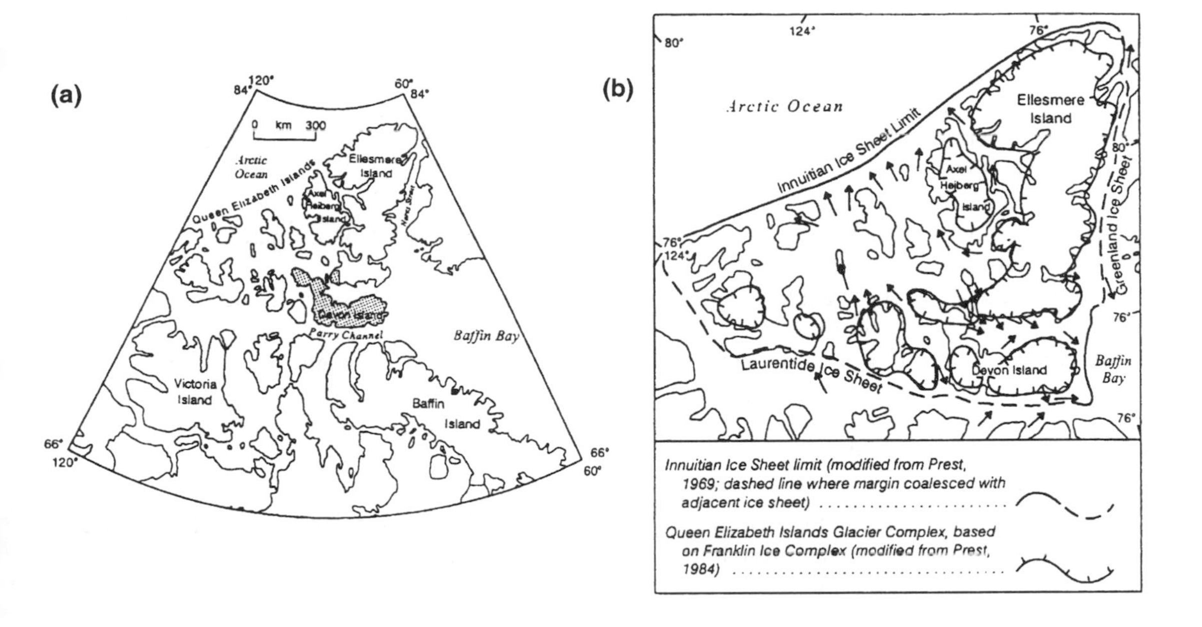

Figure 12.14 Location and glaciological reconstruction of the Innuitian Ice Sheet. Reprinted from *Quaternary Science Reviews*, 18, A. Dyke, Last Glacial Maximum and deglaciation of Devon Island, Arctic Canada: support of an Innuitian ice sheet, pp. 393–420, copyright (1999), with permission from Elsevier Science.

circulation (see Chapter 8). The GRIP and GISP2 ice-core records show a complex history of rapid (decadal scale), significant (mean annual temperatures changing by several degrees centigrade) climate change, and that these changes are coincident with the input of freshwater from the Northern Hemisphere ice sheets.

The Late Wisconsinan Innuitian Ice Sheet

One region where there has been a great deal of debate as to the extent of the ice sheet in North America is the Innuitian sector of the Laurentide Ice Sheet (Figure 12.14). This region covers the present island archipelagos from north-eastern Canada to the Nares Strait off the northern edge of Greenland. Recent geological evidence appears to favour the notion of a large ice sheet, occupying the entire area and linking the Greenland and Laurentide ice sheets (e.g. England, 1999).

Evidence from Cl dating of glacially-polished bedrock within the Nares Strait shows that the last deglaciation occurred at 10 000 years ago (Zreda et al., 1999). This evidences supports the notion of a large ice stream within the Nares Strait at the LGM, blocking the flow of water between the Arctic and Atlantic oceans (see Chapter 10).

Dyke (1999) provides an excellent summary of geological evidence, and the interpretation of this evidence spanning more than 20 years. He indicates that the glacial geology of Devon Island reflects the cover of a large ice sheet at the LGM. However, although the ice sheet may have connected with the Laurentide Ice Sheet to the south, Dyke (1999) suggests that local ice divides over these islands meant that the flow of ice was controlled by local parameters rather than by a large overriding ice sheet from the south (as in the CLIMAP scenario). In support of a large Innuitian ice cover, O'Cofaigh et al. (2000) report evidence for relatively thick ice within the Eureka Sound, between Ellesmere and Axel Heiberg islands. Geological indicators of former ice flow show that ice flowed into the sound from both islands, suggesting that an ice divide was located on each.

Ice advance occurred quite late in the glacial cycle, at around 23 000 years. Deglaciation, exposing the present coastline took place at 10 000 years ago, and was complete by 8000 years ago. There is geological evidence across Ellesmere Island, in the form of striated bedrock, erratics and proglacial sediments, to suggest that the ice sheet was warm-based during deglaciation (O'Cofaigh et al., 1999). This means that the ice cover must have been relatively thick and dynamic to induce subglacial pressure melting.

Additionally, there is dated sedimentological evidence from a fjord in Baffin Island showing a significant expansion of ice at the LGM (Marsella et al., 2000). This adds weight to the evidence from Ellesmere Island to suggest a large Innuitian Ice Sheet at the LGM.

However, it should be noted, the geological record is not entirely consistent with the 'full-cover' ice sheet reconstruction. Wolfe and King (1999) showed that ice cover across Devon Island was limited, and that ice free conditions existed within Jones Sound (thought by some previously to be an outlet of the Innuitian Ice Sheet), between Devon and Ellesmere Islands at the LGM. Their evidence comes from limnological data from a pre-Holocene lake across the north-east of Devon Island. This shows that a lacustrine system was active in Jones Sound during the Late Wisconsin, ruling out the possibility that the region was covered by grounded ice.

Clearly, all geological data need to be integrated before an unequivocal model is established for the Innuitian Ice Sheet. So far numerical modelling has yet to be employed in this area. It is therefore possible that in the next years, ice sheet modelling may be targeted to help solve the problem of the Innuitian Ice Sheet once and for all.

C. GLACIATION OF OTHER PARTS OF NORTH AMERICA

Alaska and the Brooks Range

The northern extension of the Cordilleran Ice Sheet was located across the western margin of Alaska. Today, the region is characterised by a series of alpine glaciers, often terminating within fjords. At the LGM, these ice masses grew in response to ELA depression, so that they coalesced across the marine margin of Alaska, where they terminated as calving ice walls. As precipitation decreases to the north of Alaska, so the LGM ice sheets reduce in size (Hamilton, 1986). For example, a small mountain-based ice cap was located across the Brooks Range, and another isolated ice mass to the east of Bristol Bay at the northern edge of the state. It is difficult to estimate the volume and thickness of these ice caps because they represent a complex coalescence between alpine, piedmont and valley glaciers. In other words, the bedrock morphology beneath the ice caps is complex and strongly controls the flow of ice above.

The Quaternary history of the region shows that Alaska experienced two, or possibly three, major glacial advances in the Wisconsin. The glaciated regions of Alaska are typified by a glacial drift which marks the extent of former ice cover. The youngest part of the drift correlates to the LGM. At this time, palaeosol data from the Seward Peninsula show that the mean annual air temperature was below about 6–8°C (Höfle et al., 2000). This would have caused a reduction in the ELA by several hundred metres, leading to the growth of ice. The major debate concerning the former glaciation of Alaska relates to how far offshore the ice caps were able to flow in the west. For instance, offshore investigations have revealed that much of the continental shelf was deglaciated by about 13 000 years ago. However, some of these continental shelf areas remained free of ice throughout the last glacial. Deglaciation of the land-based regions occurred around 13 000 years ago, quickly across the whole region, leaving behind a series of alpine glaciers and ice caps that are today in recession from the LIA cool period ending about 100 years ago.

Aleutian Islands

The westernmost extent of ice from the Cordilleran Ice Sheet was situated across the eastern Aleutian Islands in the North Pacific. It is thought that the islands were glaciated by small ice caps similar to the glacier cover across the Russian High Arctic today, at the LGM. These ice caps were probably connected by floating ice shelves to the east, most likely similar to ice over Franz Josef Land, Russian High Arctic. This ice cover was connected to the ice sheet across Alaska and, in turn, the northern Cordilleran Ice Sheet. To the east, relatively small ice masses were confined to the islands, in a manner which could be similar to those on Severnaya Zemlya today.

Rocky Mountain National Parks

The Rocky Mountains of the northern USA mark the border between the Cordilleran Ice Sheet to the West and the Laurentide Ice Sheet to the East. In the regions now known as Glacier, Yellowstone, Grand Teton and Rocky Mountain national parks, many small cirque-type glaciers exist in today's climate. However, at the LGM when the ELA was lower, these glaciers grew down to lower elevations and merged to form individual ice masses across these regions, punctuated by nunataks (Elias, 1996). The glacial deposits formed by these Late Wisconsin and former ice masses are dateable because of the volcanic ash deposited over them by eruptions of volcanoes like Mount St Helens.

Sierra Nevada

Widespread glacial deposits denote the growth of a 15 000 square kilometre ice cap across the Sierra Nevada mountains (Fullerton, 1986). This relatively small ice cap would have grown as a result of ELA lowering, and was thin enough to be punctuated by a series of nunataks. It should be noted that the region has been glaciated several times during the Quaternary and precise correlation between the various moraine systems is difficult.

Glacial History of Mexico and Central America

Several of the volcanoes in central southern Mexico were glaciated at the LGM. One of these volcanoes, Iztaccihuatl, is covered today by a 12 square kilometre ice field. At the LGM, this glacier extended to around 52 square kilometres. This form of small-scale mountain glaciation appears to typify several of the other volcanoes in central Mexico. However, the Iztaccihuatl ice cap is thought to have been the largest (White, 1986). Additionally, the highest elevations of central America were likely to have been glaciated to some extent at the LGM, as the ELA lowered due to a fall in the mean annual air temperature. For example, Orvis and Horn (2000) use geological information and numerical modelling to show how LGM glaciers could exist above about 3300 metres a.s.l. in Costa Rica.

D. SUMMARY OF THE GLACIATION OF NORTH AMERICA

The glaciation of North America involved two major ice masses; the Cordilleran and the Laurentide ice sheets. The Cordilleran Ice Sheet was located over the Rocky Mountains and was fed by moisture-laden westerly winds from the Pacific Ocean (Figure 12.1). Consequently, across the lee side of the mountains, precipitation was curtailed, resulting in an ice sheet margin to the east of the Rocky Mountains. The Laurentide Ice Sheet occupied most of the rest of Canada and the northern borders of the United States. At the LGM the ice sheet coalesced with the Cordilleran Ice Sheet to the west and the Greenland Ice Sheet to the north east (connected by a section of the Laurentide Ice Sheet named the Innuitian Ice Sheet). The ice sheet experienced surface ablation along the southern margin, and iceberg calving to the east and north. These ablation processes played a critical role in the decay of ice in North America.

- The Wisconsin glaciation of North America occurred between 120 000 and 8000 years ago. Although the ice sheet waxed and waned several times during this period a substantial ice cover existed between 50 000 years ago and the LGM. This makes the ice sheets different to other Northern Hemisphere ice sheets that all but decayed prior to the Late Wisconsin advance.
- Ice growth occurred as a result of ELA lowering, leading to the expansion of ice from the high latitudes and high elevations generally southwards across most of Canada.
- The LGM ice extent in North America is extremely well known from geological evidence. Terminal moraines and proglacial lacustrine features have been extensively mapped. They show that the margin of the ice sheet was characterised by a series of lobe advances, and a multitude of proglacial lakes. The process controlling the ice sheet lobes is thought to be similar to a glacier 'surging' mechanism, associated with deformable subglacial sediment.
- The maximum thickness of the Laurentide Ice Sheet is thought to be around 2500 metres. The surface topography of the ice sheet was divided into three major ice domes across the Cordilleran Ice Sheet, the eastern Laurentide Ice Sheet and over the northern side of the Laurentide Ice Sheet. This relatively new reconstruction contrasts markedly with that proposed earlier by CLIMAP, where a massive 4 kilometre thick, single-domed ice mass was predicted.
- During deglaciation, as the ice sheet retreated northwards, meltwater from the ice sheet was routed into proglacial lakes causing a massive build up of meltwater at the southern ice sheet margin. These lakes were subject to episodes of out-bursting as the ice walls that acted as dams decayed themselves. The outbursting of the largest of these lakes, Lake Agassiz at around 8200 years ago, may have resulted in 20 centimetres of global sea level rise within a few days, and has been correlated to a widespread 'cold event' measured in Greenland ice cores.
- The periodic issuing of large volumes of icebergs from the Hudson Strait into the North Atlantic is recorded in anomalous layers of IRD on the ocean floor. These so-called Heinrich layers display a periodicity of around 7000 years. There have been several explanations for the production of icebergs (Heinrich events). The most widely accepted is that the Laurentide Ice Sheet underwent surge-type oscillations, termed the 'binge–purge' model. During the binge phase, the ice sheet builds up over an essentially frozen but warming base until the temperature reaches the pressure melting value. Once this is attained the dynamics of the ice sheet are altered, and the 'purge' begins, involving an increase in ice velocity and the transfer of ice from the central regions to the iceberg calving front. The purge ends when so much ice has been lost that the ice sheet regains a frozen base.
- The decay of ice in North America resulted in over 60 metres of global sea-level rise. Most of this was routed into the Atlantic Ocean either as icebergs or runoff. A significant input of fresh water into the North Atlantic would have effected the production of saline North Atlantic Deep Water and, consequently, the ocean circulation. Palaeo-

climate records from the Greenland Ice Sheet (Chapter 8) appear to indicate that air temperatures were highly changeable over the North Atlantic during deglaciation (Dansgaard-Oeschger events, see Chapter 8). This has led some to link climate variability with the production of glacier meltwater from North America.

CHAPTER 13

Ice Sheet Reconstructions V – Remaining LGM Ice Cover. Iceland, South America: Patagonia and the Andes, South Island (New Zealand) and Tasmania, Mainland Europe, Tibet

INTRODUCTION

In addition to the development of continental-scale ice sheets at the LGM, outlined in Chapters 9–12, many other land-based regions experienced Late Quaternary glaciation. This chapter summarises briefly what is currently understood about these relatively small ice masses, and concludes with a simple calculation of the volume of ice held in ice sheets at the LGM, which is compared to calculations of global ice volume from $\delta^{18}O$ analysis (Chapter 2).

ICELAND

Modern ice cover across Iceland is unique in two respects. First, its location within the middle of the Norwegian–Greenland Sea means it is influenced by an extreme maritime environment. Any change in North Atlantic Ocean conditions is bound to affect the mass balance and dynamics of glaciers in this region. Second, being situated over a geothermal hotspot leads Icelandic glaciers to possess an unusual thermal regime, associated with highly active volcanoes and hot springs. Currently, there are four major ice caps on Iceland, totalling 11 000 square kilometres of ice, the largest being the Vatnajokull ice cap (8000 square kilometres) located over the Grimsvotn volcanic caldera. In 1996 this volcano issued hot material to the underside of the ice cap causing subglacial melting and the production of a subglacial lake. As melting continued, and the lake became larger, the ice that was acting to dam the water failed and a large outburst of water, known as a 'jökulhlaup' resulted. This association between glaciers and volcanic activity is observed elsewhere in Iceland. For instance, the Myrdalsjökull ice cap is located beneath the Katla volcano, which has a tendency to erupt every 50 years or so, causing jökulhlaup activity. Volcano-induced jökulhlaups are a serious geo-hazard in Iceland. They leave behind a very visible sequence of glacifluvially reworked material (often of a glacial origin). Jökulhlaups are also very likely to have been highly active during the Late Quaternary, when ice cover over Iceland was much greater than at present.

Ship-borne geophysical surveys of the seas surrounding Iceland have revealed several moraine-type features at the continental shelf break which have been used by some to infer a corresponding LGM ice expansion (e.g. Egloff and Johnson, 1979; Boulton et al., 1988).

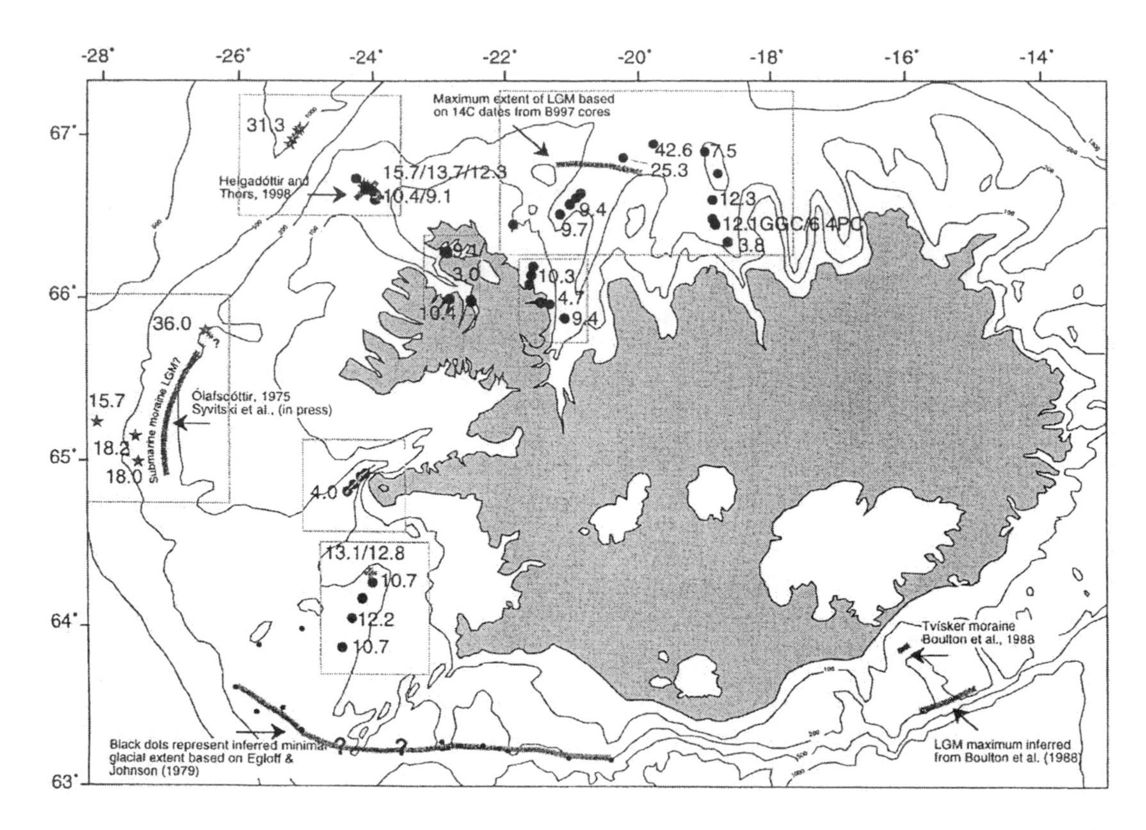

Figure 13.1 The location of Icelandic LGM ice sheet terminal moraines. Also shown are the locations of radiocarbon dates from sea floor sediments used to date the moraines. Reprinted from *Quaternary Science Reviews*, 19, J. Andrews et al., The N and W Iceland Shelf: insights into the Last Glacial Maximum ice extent and deglaciation based on acoustic stratigraphy and basal radiocarbon AMS dates, pp. 619–631, copyright (2000), with permission from Elsevier Science.

Andrews et al. (2000) produced a number of AMS radiocarbon dates from the continental shelf in an attempt to provide some chronological order to the interpretation of these sediments (Figure 13.1). Although their work was limited to the north and west of the Iceland shelf, the results provide an important new insight into the LGM configuration of the Iceland ice sheet. The main conclusions from this work were that the south-western part of the former ice sheet extended to the continental margin, but that the northern side may have been restricted to the mid-shelf regions. To date there have been no numerical modelling investigations of the Iceland ice sheet, possibly because the complex relation between ice, climate and geothermal heat does not lend itself easily to traditional numerical analysis. However, bearing in mind that some material within Heinrich layers is thought to have originated from Iceland, it would be rather important to the understanding of the North Atlantic ice-ocean-atmosphere interaction at the LGM if the extent, thickness and dynamics of the former ice cover in this region is known.

It should be noted here also that the small island of Jan Mayen, located 300 kilometres north of Iceland in the central Greenland Sea, is also likely to have been at least in part covered by ice at the LGM.

SOUTH AMERICA

Patagonian ice cap

The glaciation of the Northern Hemisphere at the LGM is quite easy to conceptualise given that it was probably forced by solar insolation changes to this half of the planet. However, there is plenty of evidence to suggest that several regions of the Southern Hemisphere also experienced glaciation. The reconstructions of these ice masses provide important clues as to the climate of the Southern Hemisphere during the Late Quaternary.

The largest former Southern Hemisphere ice cap outside Antarctica was located over Patagonia. Patagonia is located within the Andean Cordillera, which provides the topographic constraint to the former ice cover. At present there are a number of small valley glaciers here, which are generally in a phase of retreat, and often terminate in proglacial lakes formed as the ice margin migrates past the formerly over-deepened bedrock. The climate of Patagonia is dominated by westerly winds off the adjacent Pacific Ocean. The mountains exert a strong control on the precipitation gradient from west to east. There is also currently a noticeable precipitation gradient from south (50°S) to north (40°S), due to the occurrence or otherwise of westerly storm fronts. There is quite an abundance of geological evidence of former ice growth across Patagonia. However, the relative inaccessibility of the region has meant that this evidence has yet to be surveyed in the same detail as in many other formerly glaciated regions. Nevertheless, enough geological evidence has been collected to infer an LGM ice limit (Hollin and Schilling, 1981) (Figure 13.2).

Hulton et al. (1994) used these LGM ice limits to reconstruct the former ice cover from a numerical ice sheet model. They made an assumption that the ice sheet would have reached an equilibrium with the climate by the time of the LGM. By matching the ice sheet to the geologically-derived LGM limits, they were also able to reconstruct the climate of the region (in an inverse-type procedure). Hulton et al. (1994) calculated that the LGM ice cap grew along a 1800 kilometre length of the Andes (Figure 13.3), and had a volume of over 400 000 cubic kilometres. In the model, the ice sheet grew up as a familiar dome shape. However, in reality, the mountain peaks that today attain elevations of over 3000 metres would have manifested as nunataks at the LGM. Hulton et al. calculated that their ice sheet reconstruction required the ELA to have been depressed by between 160 and 560 metres from the present position, depending on latitude. The ELA rises from west to east by as much as 4 metres per kilometre, causing an ablating ice margin to the east of the ice sheet. In order to lower the ELA by this amount, a temperature depression of around 3°C is required. This conclusion is important in the context of the LGM climate in the Southern Hemisphere. Hulton et al. (1994) indicate that this temperature is within the lower range of the estimated global temperature reduction of around 3–5°C. The cause of this relatively small air temperature fall is assigned to the 'modest' effect that the Antarctic Ice Sheet, and the surrounding seas, may have had on moderating climate to the north.

However, analysis of sediment landform assemblages along the south-eastern side of the Patagonian Ice Sheet (in the Strait of Magellan) show characteristics similar to those found at the margins of modern sub-polar type glaciers (such as thrust moraines and kame topography). The formation of these landforms is possible only if, at the LGM, permafrost occurred close to sea level in the southern extreme of South America. This implies that temperatures were ~7–8°C lower than at present; significantly below the temperatures calculated through, for example, Hulton et al.'s numerical modelling (Benn and Clapperton, 2000). Clearly the establishment of a comprehensive glacial geological record for Patagonia is warranted, since it would allow the numerical ice sheet models to be better constrained.

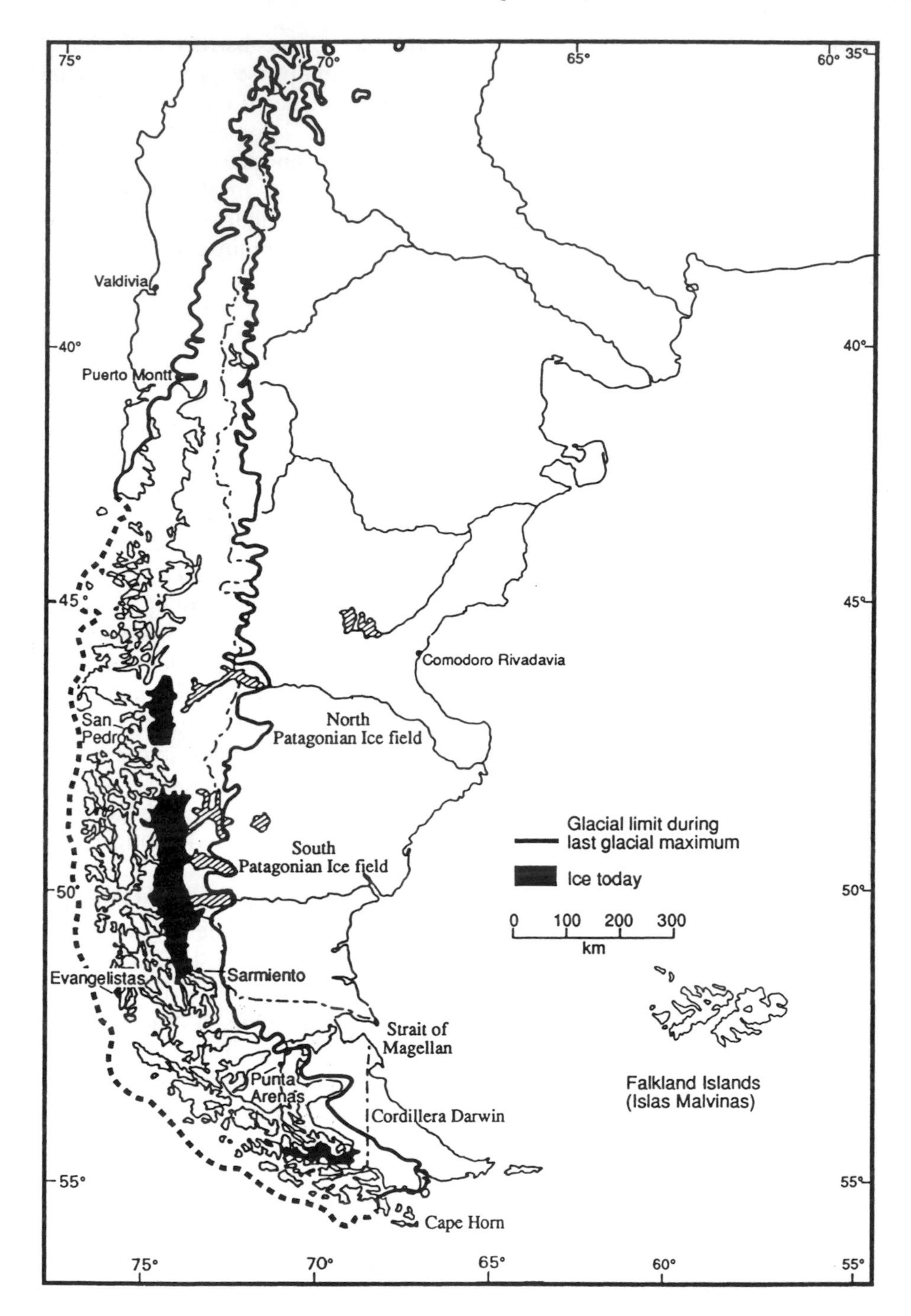

Figure 13.2 Maximum extent of the late Quaternary ice sheet across Patagonia from geological evidence. Taken from Hulton et al. (1994), after Hollin and Shilling (1981), and reproduced by permission of Academic Press.

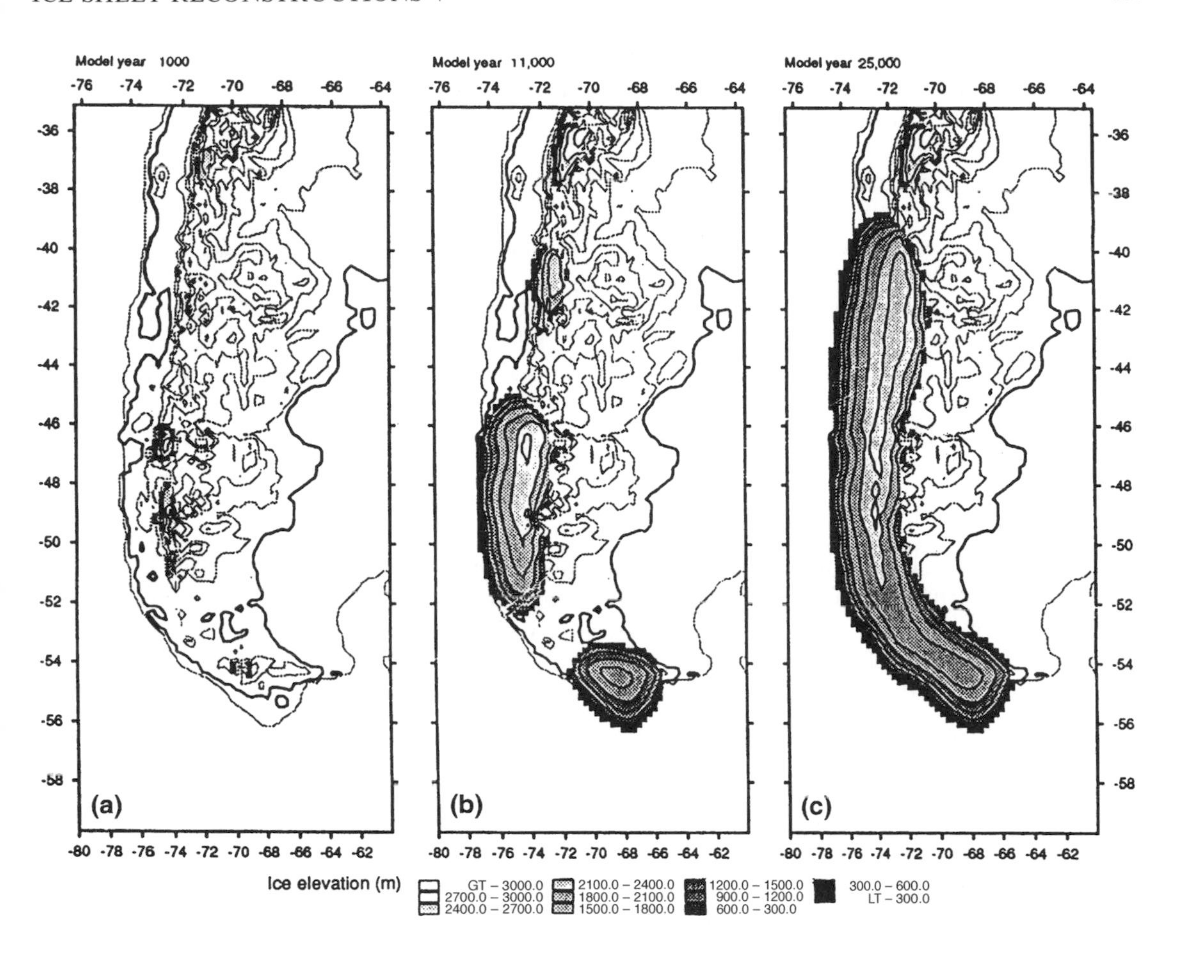

Figure 13.3 The growth of ice across Patagonia from numerical modelling experiments. Maps show the thickness of ice at 1000, 11 000 and 25 000 years worth of model running time. Taken from Hulton et al. (1994) and reproduced by permission of Academic Press.

Equatorial Andes

Clapperton (1983) established a geological framework for the development of an 800 square kilometre ice cap across the Potrerillos Plateau, Eastern Cordillera of Ecuador. The mountain range has an altitude of 4200 metres and there is currently no ice in the area. Existing ice cover is restricted to the very highest elevations (e.g. The Volcan Antisana at 5758 metres a.s.l.). At the LGM, however, glaciers feeding off this ice cap terminated at elevations of around 3850 metres a.s.l. The growth of this type of relatively small ice cap may have characterised the entire northern highlands of South America at the LGM. However, much more geological work is required before a full understanding of the past glaciation can be determined. These ice caps would have formed as a result of ELA lowering by a few hundred metres, as calculated for the Patagonian ice cover further south.

One interesting feature concerning the geological evidence for former ice caps in the Equatorial Andes is that there appears to be some data to support the notion that glaciers here may have advanced again during the Younger Dryas. The possibility of this happening has great significance to the understanding of the Younger Dryas event, since it is thought by many to be confined to the Northern Hemisphere. The problem of identifying Younger Dryas signals in Southern Hemisphere records is compounded by the fact that some data support the idea (see Chapter 8), but most indicate no Younger Dryas feature. The problem regarding the prevalence of the Younger Dryas in the Southern Hemisphere remains to be resolved.

NEW ZEALAND AND AUSTRALIA

South Island

Because of the lack of landmass at mid-latitude locations in the Southern Hemisphere compared to the Northern, there are few sites for potential ice sheet growth. One of these regions is South Island, New Zealand, where a high mountain chain is currently situated toward the western side. Just as in Patagonia, the prevailing wind direction is from the west, and a precipitation shadow occurs to the east of the mountains. Currently, there are a number of glaciers in South Island, showing that the ELA is currently below the elevation of the mountains at several locations. The entire Pacific basin experienced a depression of the mean annual air temperature which led to an ELA depression of about 1 kilometre. The glacial geology of South Island shows a clear indication of the effects of lowering the ELA, namely widespread glaciation. Porter (1975) identified that the ELA was depressed by about 875 metres across the Southern Alps. Radiocarbon dates on moraines and glacial outwash show that the ice extent was at a maximum between 22 000 and 14 000 years ago. The maximum ice sheet extent has been mapped from moraine sequences (Figure 13.4, Broecker and Denton, 1990). The ELA at this time is estimated to have been around 500 metres lower than at present. Deglaciation began at about 14 000 years ago, causing the input of glacial sediments across the eastern side of South Island (the Canterbury Plains) to the ocean where this signal has been measured in a nearby deep sea core.

Tasmania

Similar to South Island New Zealand, Calhoun (1985) established that the ELA across the West Coast Range, Tasmania, was reduced by about 1 kilometre at the LGM. This would have caused the growth of glaciers over the mountains of this small island, located at a latitude of 42°S. The LGM ice cover extended over 108 square kilometres, and have a maximum ice thickness of about 250 metres. It is interesting to note that the LGM ice sheet is much smaller than a previous glaciation, assumed to date from before the last interglacial (isotope stage 6), which involved a 1000 square kilometre ice sheet with a maximum thickness of 300–400 metres (Calhoun, 1985).

GLACIATION ON MAUNA KEA, HAWAII

The highest volcano on Hawaii is the only known region of these islands to have been glaciated. However, it is important to reconstructions of past climate and ocean conditions given the location of Hawaii in the central Pacific Ocean. A small ice cap of around 70 square kilometres is thought to have grown over the mountain during the LGM. Indeed this ice cap may have developed at least four times during the Quaternary. These glaciations leave a series of drift deposits that can be dated from isotopic studies on the volcanic lava that separates drift events (Porter, 1986).

EUROPE

The southern margin of the Eurasian Ice Sheet crossed northern Europe a number of times during the Quaternary (see Chapter 11). However, further south, there is evidence of

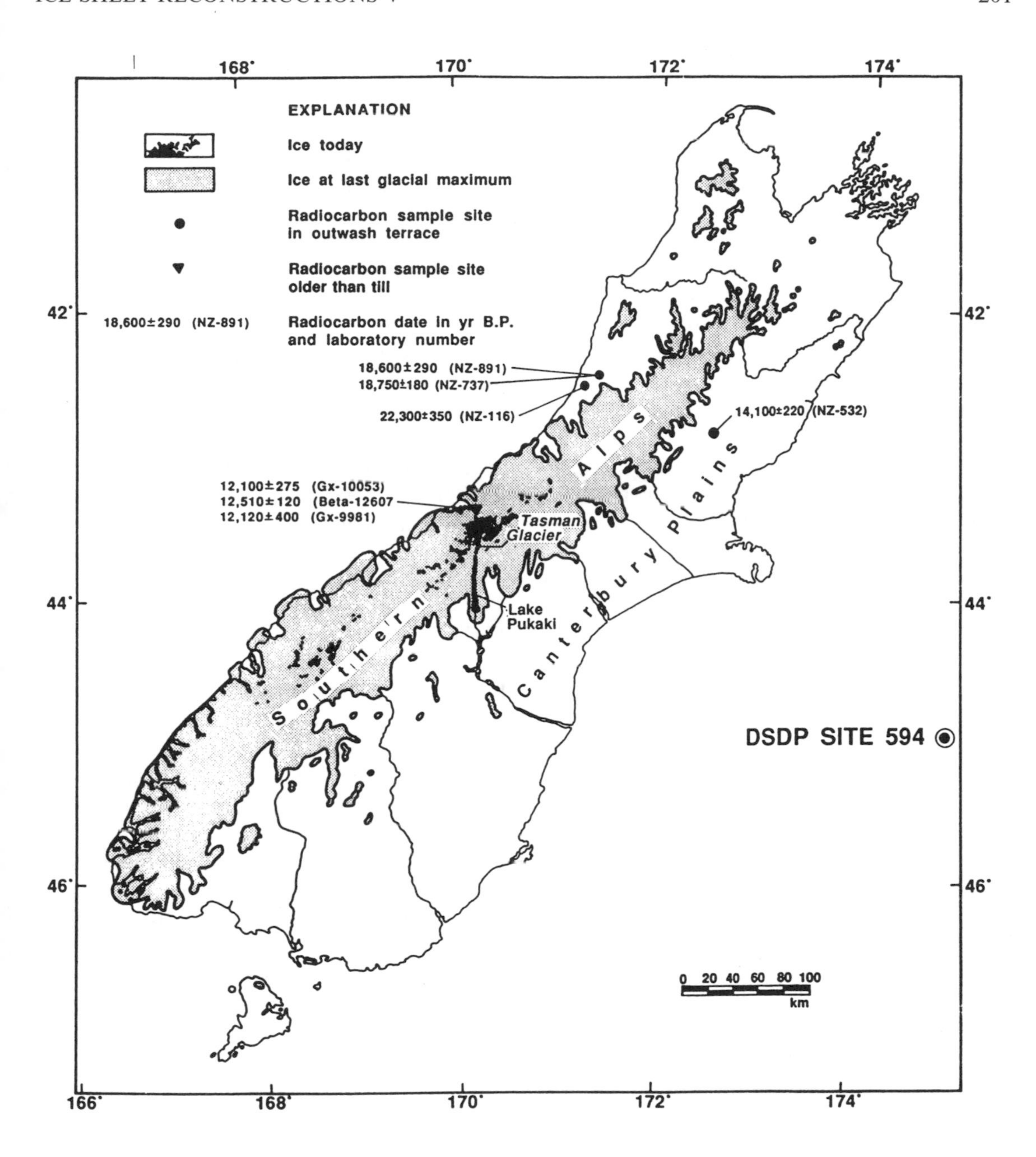

Figure 13.4 Geological evidence of the LGM ice cover on South Island, New Zealand. Also shown is a time-dependent record of the ELA, and a sea floor sediment record from DSDP SITE 594. Reprinted from *Quaternary Science Reviews*, 9, W. Broecker and G. Denton, pp. 305–341, copyright (1990) with permission of Elsevier Science.

glaciation, independent of the Eurasian Ice Sheet, across several regions of central mainland Europe. Very brief details of these ice masses are now given.

Today the Alps, the principal mountain range in Europe covering an area of about 250 000 square kilometres, houses dozens of valley glaciers, some of which are even located on southerly facing slopes. These are testament to the fact that the high elevation of this mountain region (between 2.5 and 4.5 kilometres a.s.l.), and the reduction in mean annual air temperature associated with it, is conducive to permanent ice cover. There is obvious evidence in the presence of U-shaped valleys, to conclude that, in the past, the ice cover across the Alps was far more extensive that at present. The outer end moraines that occur across the northern foreland of the Alps indicate that the last glaciation involved the complete cover of this mountain region by ice (Schlüchter, 1986). However, the exact date of this glacial event has not been identified. This glaciation could have been an LGM event. However, if the 'maximum' ice cover occurred prior to the LGM, then a slightly more restricted glaciation of the Alps must be inferred. Whichever is the case, the LGM Alps would have been covered by far more ice than at present. Such ice growth would have resulted from the lowering of the ELA due to air temperature depression that occurred at the LGM.

Glacial geology shows that the maximum glaciation over the Jura Mountains is likely to have been by a sizeable ice cap, 100 kilometres long, 40 kilometres wide, independent of any ice cover across the Alps (Campy, 1986). However, there is still some uncertainty as to whether this 'maximum' ice cover existed at the LGM or whether more limited glaciation (restricted to small cirque and valley glaciers) occurred at this time.

The Pyrenees, which reach an altitude of over 3000 metres a.s.l., experienced less glaciation than the Alps because of its southerly latitude of 42°N (Hérail, et al., 1986). Instead, glaciation was restricted to large valley glaciers and small ice caps, of which the remnants of many can still be seen today. Similarly, the volcanic mountains of the Massif Central, reaching elevations in excess of 1500 metres a.s.l., have a number of cirque and glacially-eroded valleys which were most probably occupied by glaciers at the LGM (Veyret, 1986). There is also evidence from the Bavarian region in Germany for the growth of valley glaciers. Further east, there is widespread evidence for a relatively large ice cover across both the Puturana Mountains and the Ural Mountains, Eastern Russian Arctic. It is interesting to note that one of the most southerly glaciations in Europe was on the Mediterranean island of Corsica. Despite being in a southerly latitude (41–43°N), the mountains that reach over 2000 metres a.s.l. experienced glaciation by cirque and small valley glaciers at the LGM (Conchon, 1986).

THE GLACIATION OF TIBET

Today, the Tibetan Plateau covers an area of about 2 000 000 square kilometres, with an average elevation of 5000 metres a.s.l. of which around 4 per cent is glacierised. The mountains were formed due to the tectonic collision of the Asian and Indian plates. This process initiated some 20 million years ago and continues today, resulting in an uplift rate of about 10 mm/yr^{-1}. The current ELA is estimated at around 4500 to 6000 metres a.s.l. Crucial to the development of a Late Quaternary ice cap across the Tibetan Plateau is a mechanism by which large-scale glaciation can occur. Bush (2000) used a numerical climate model to propose a positive feedback between ice growth and surface albedo for the Himalayas. In the model, as snowfall induces a broad increase in the albedo, descending cold air is calculated across the Himalayas in the summer time, and the moisture content of monsoon winds from north-western India increases leading to enhanced rates of snowfall. This in turn leads to an increase in the surface albedo, which due to the positive feedback causes yet more snowfall.

To date, there is no consensus regarding the glaciation of this area at the LGM. There are two main schools of thought regarding the former ice cover across Tibet. The first is based on a reduction in the ELA of around 1000 metres. Gupte et al. (1992) estimated from $\delta^{18}O$ records from the Dunde ice cap (at 37°N, 96°E) that the LGM temperature was 4–6°C lower than at present (Kaufmann and Lambeck, 1997). Assuming an adiabatic lapse rate of 1 kilometre per 5°C, Gupte et al.'s estimates are compatible with an ELA reduction of about 1000 metres. In this situation, a large portion of the plateau would be placed above the ELA, causing a substantial ice sheet to build. Estimates on the ice thickness and contribution to LGM eustatic sea level fall of this 'maximum' form of ice cover range from about 250 metres ice thickness and 1.3 metres of sea level (Gupte et al., 1992), to a huge 2 kilometres thick ice sheet responsible for 5.2 m of sea level (e.g. Kuhle, 1988). Gupte et al. (1992) favoured the former, smaller ice thickness because they noted that the decay of such an ice mass would have resulted in a noticeable reduction in $\delta^{18}O$ in sea floor sediments within the Bay of Bengal, and that such a signal is absent from known records. The minimum-type reconstruction is based on the interpretation of glacial geology. Derbyshire et al. (1991) proposed that the glaciation of the Tibetan Plateau was non-uniform (unlike that suggested in the 'maximum' form of glaciation). They indicated that the LGM glaciation was characterised by ice caps over mountain tops, and a series of interlinked piedmont glaciers. The corresponding ELA reduction necessary for this form of glaciation is rather more complex than uniform depressions assumed by Kuhle (1998) and Gupte et al. (1992). Derbyshire et al. (1991) suggested that the LGM ELA depression ranged from 1000 metres over the southern region, to 500 metres in the interior and 300 metres to the north. This 'minimum' form ice sheet at the LGM is supported by glacial geological evidence from the central and eastern Tibetan Plateau (Kehmkuhl, 1998), where a series of cirques, U-shaped valleys and moraine systems mark the proposed extent of ice at the LGM.

There have been no attempts at reconstructing the LGM ice cover over Tibet using numerical ice sheet modelling. However, Kaufmann and Lambeck (1997) have used a solid Earth model to calculate the isostatic effects that potential LGM ice loads would have had on the post-glacial uplift of the region. They showed that, under a maximum ice sheet scenario, the decay of ice could have led to present day uplift rates of around 7 mm/yr^{-1}; more than half the total uplift at present. Therefore, determining an accurate understanding of Late Quaternary ice sheet history in Tibet is fundamental to understanding the modern uplift and tectonics of the region.

CALCULATIONS OF GLOBAL SEA LEVEL AT THE LGM

Global sea level was between 120–135 metres lower at the LGM than it is today (Yokoyama et al., 2000). This change in sea level was caused by a transfer of water from the oceans to the ice sheets. Therefore, the respective volumes of the former ice sheets should tally to that required to lower sea level by the amount measured. Thus, one might expect the simple addition of all the LGM ice, divided by the surface area of the ocean to equate to ~120–135 metres or so. There are two reasons why such a calculation is problematic. First, as has been demonstrated in this book, the volumes of former ice sheets are often not known with any confidence. Usually a 'maximum' and 'minimum' value is predicted. Second, the relationship between sea level and ice volume is a complicated one, considering glacial, tectonic and geoidal effects (Figure 5.1). Andrews (1992) suggested that much work has still to be done in terms of tying in former ice sheet volumes with measurements of sea level fall. He indicated that, the Antarctic Ice Sheet is only expected to contribute a few metres to global sea level fall at the LGM. The

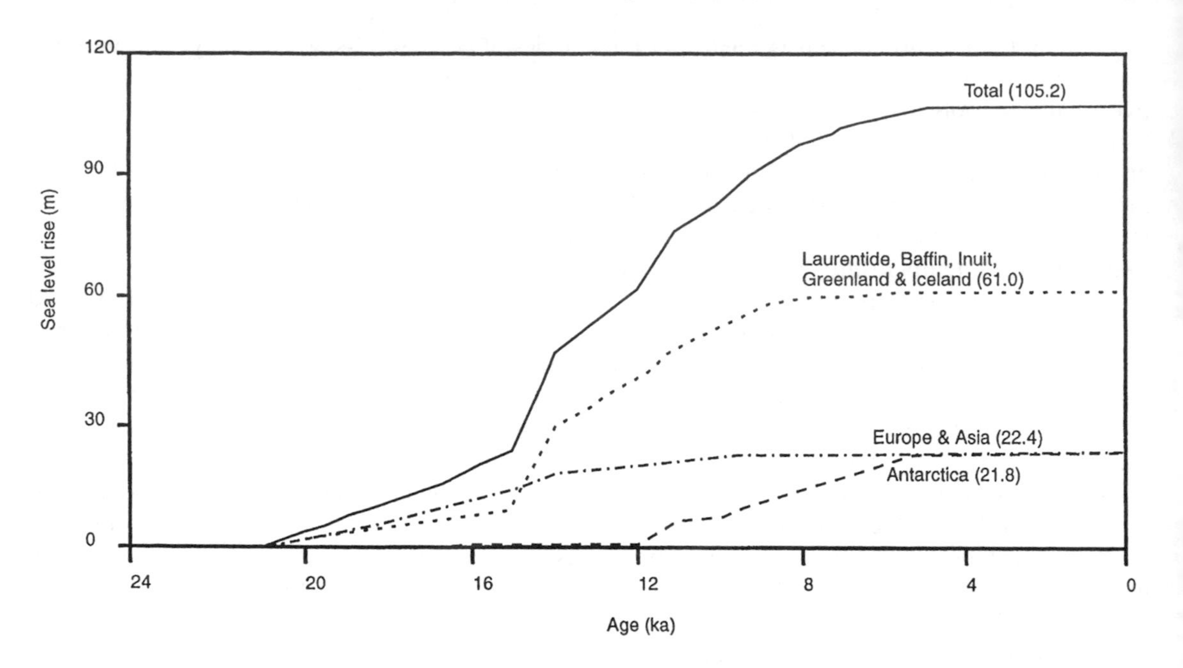

Figure 13.5 Time-dependent contribution to sea level rise caused by the decay of LGM ice sheets calculated by the ICE-4G model of Peltier (1994). Reprinted from *Quaternary Science Reviews*, 18, M. Bentley, Volume of Antarctic ice at the last glacial maximum, and its impact on global sea level change, pp. 1569–1595, copyright (1999), with permission from Elsevier Science.

implication of this is a short fall in global ice volume at the LGM. Denton and Hughes (1981) provided the first examination of global ice volume at the LGM in terms of sea level equivalent amounts. They showed that under their 'minimum' ice reconstructions around 127 metres of sea level fall was expected at the LGM. However, this calculation is problematic because it includes an over-prediction of the sea level fall caused by the LGM expansion of the Antarctic Ice Sheet. In a recent commentary on the work of Yokoyama et al., Clark and Mix (2000) offer few solutions to the problem. They suggest that one way forward is to determine the boundaries of the former ice sheets and use numerical ice sheet modelling to account for the appropriate volume of ice held in these ice sheets.

One highly regarded reconstruction of global ice volume at the LGM was calculated by the ICE-4G isostasy model of Peltier (1994). In this work, the contribution of the Laurentide, Greenland and Icelandic ice sheets was calculated at 61 metres of sea level fall, whilst for Europe and Antarctic it was 22.4 and 21.8 metres respectively (Figure 13.5). The total reduction in sea level at the LGM due to the ice sheets is thus calculated at 105.2 metres, some 15 metres below that measured in New Guinea terraces and $\delta^{18}O$ data from sea floor sediments (Figure 1.4), and as much as 30 metres below that calculated by Yokoyama et al. (2000).

As pointed out by Bentley (1999), there are good geological reasons to estimate the contribution of the Antarctic Ice Sheet to LGM sea level lowering at between 6.1 and 13.1 metres (Chapter 9). Further, recent geological evidence from the Russian High Arctic (Chapter 11) shows that the Eurasian Ice Sheet would have been smaller than that depicted in the results of ICE-4G. Numerical modelling suggests a maximum contribution of only 16 metres of sea level fall from the Eurasian Ice Sheet at the LGM (Chapter 11). Thus, with these calculations of ice volumes in mind, the total volume of the LGM ice sheets, as illustrated in this book, would have lowered sea level by less than 100 metres. Thus, there is a clear discrepancy between the measured sea level reduction at the LGM and that calculated from the most recent reconstructions of Late Quaternary ice sheets. Andrews (1992) sites this problem as a 'case of missing water'. As it stands now, there is no clear indication of where 20 per cent of the water displaced from the oceans was stored at the LGM.

References

Ackert Jr., R.P., Barclay, D.J., Borns Jr., H.W., Calkin, P.E., Kurz, M.D., Fastook, J.L. and Steig, E.J. 1999. Measurements of past ice sheet elevations in interior West Antarctica. *Science*, 286, 276–280.

Adkins, J.F., Boyle, E.A., Keigwin, L. and Cortijo, E. 1997. Variability of the North Atlantic thermohaline circulation during the last interglacial period. *Nature*, 390, 150–154.

Alley R.B. and MacAyeal, D. 1994. Ice-rafted debris associated with binge/purge oscillations of the Laurentide Ice Sheet. *Palaeoceanography*, 9, 503–511.

Alley, R.B., Blankenship, D.D., Rooney, S.T. and Bentley, C.R. 1989. Sedimentation beneath ice shelves – the view from Ice Stream B. *Marine Geology*, 85, 101–120.

Alley, R.B. 1990. Multiple steady states in ice-water-till systems. *Annals of Glaciology*, 14, 1–5.

Alley, R.B. 1991. Deforming bed origin for southern Laurentide till sheets? *Journal of Glaciology*, 37, 67–76.

Alley, R.B. and Clark, P.U. 1999. The deglaciation of the Northern Hemisphere: a global perspective. *Annual Reviews of Earth and Planetary Science*, 27, 149–182.

Alley, R.B., Meese, D.A., Shuman, C.A., Gow, A.J., Taylor, K.C., Grootes, P.M., White, J.W.C., Ram, M., Waddington, E.D., Mayewski, P.A and Zielinski, G.A. 1993. Abrupt increase in Greenland snow accumulation at the end of the Younger Dryas event. *Nature*, 362, 527–529.

Andersen, E.S., Dokken, T.M., Elverhøi, A., Solheim, A. and Fossen, I. 1996. Late Quaternary sedimentation and glacial history of the western Svalbard continental margin. *Marine Geology*, 133, 123–156.

Anderson, B.G. 1981. Late Weichselian ice sheets in Eurasia and Greenland. In: Denton, G.H. and Hughes, T.J. (eds) *The Last Great Ice Sheets*, John Wiley, New York, pp. 1–65.

Andrews, J.T. 1970. A geomorphological study of post-glacial uplift with particular reference to Arctic Canada. Institute of British Geographers, Special Publication, 2, 156pp.

Andrews, J.T. 1992. A case of missing water. *Nature*, 358, 281.

Andrews, J.R., Hardardóttir, J., Helgatóttir, G., Jennings, A.E., Geirsdóttir, A., Sveinbjörnsdóttir, Á.E., Schoolfield, S., Kritjánsdóttir, G.B., Smith, L.M., Thors, K. and Syvitski, J.P.M. 2000. The N and W Iceland Shelf: insights into the Last Glacial Maximum ice extent and deglaciation based on acoustic stratigraphy and basal radiocarbon AMS dates. *Quaternary Science Reviews*, 19, 619–631.

Appenzeller, C., Stocker, T.F. and Anklin, M. 1998. North Atlantic oscillation dynamics recorded in Greenland ice cores. *Science*, 282, 446–449.

Archer, D., Winguth, A., Lea, D. and Mahowald, N. 2000. What caused the glacial/interglacial pCO_2 cycles? *Reviews of Geophysics*, 38, 159–189.

Armand, L.K. 2000. An ocean of ice – advances in the estimation of past sea ice in the southern ocean. *GSA Today*, 10, 1–7.

Astakhov, V.I. 1997. Late glacial events in the central Russian Arctic. *Quaternary International*, 41/42, 17–25.

Astakhov, V.I., Svendsen, J.I., Matiouchkov, A., Mangerud, J., Maslenikova, O. and Tveranger, J. 1999. Marginal formations of the last Kara and Barents ice sheets in northern European Russia. *Boreas*, 28, 23–45.

Ballantyne, C.K. and McCarroll, D. 1995. The vertical dimensions of Late Devensian glaciation on the mountains of Harris and southeast Lewis, Outer Hebrides, Scotland. *Journal of Quaternary Science*, 10, 211–223.

Bamber, J.L., Vaughan, D.G. and Joughin, I. 2000a. Widespread complex flow in the interior of the Antarctic Ice Sheet. *Science*, 287, 1248–1250.

Bamber, J.L., Hardy, R.J. and Joughin, I. 2000b. An analysis of balance velocities over the Greenland ice sheet and comparison with synthetic aperture radar interferometry. *Journal of Glaciology*, 46, 67–74.

Barber, D.C., Dyke, A., Hillaire-Marcel, C., Jennings, A.E., Andrews, J.T., Kerwin, M.W., Bilodeau, G., McNeely, R., Southon, J., Morehead, M.D. and Gagnon, J.-M. 1999. Forcing of the cold event of 8,200 years ago by catastrophic drainage of Laurentide lakes. *Nature*, 400, 344–348.

Bard, E., Hamelin, B. and Fairbanks, R.G. 1990. U-Th ages obtained by mass spectrometry in corals from Barbados: sea level during the past 130,000 years. *Nature*, 346, 456–458.

Bard, E., Rostek, F. and Sonzogni, C. 1997. Interhemispheric synchrony of the last deglaciation inferred from alkenone palaeothermometry. *Nature*, 385, 707–708.

Barnola, J.M., Raynaud, D., Korotkevich, Y.S. and Lorius, C. 1987. Vostok ice core provides 160,000-year record of atmospheric CO_2. *Nature*, 329, 408–414.

Bauch H.A. and Weinelt M.S. 1997. Surface water changes in the Norwegian sea during last deglacial and Holocene times. *Quaternary Science Reviews*, 16, 1115–1124.

Baumann, K.H., Lackschewitz, K.S., Erlenkeuser, H., Henrich, R., Jünger, B. 1993. Late Quaternary calcium carbonate sedimentation and terrigenous input along the east Greenland continental margin. *Marine Geology*, 114, 13–36.

Baumann, K.H., Lackschewitz, K.S., Mangerud, J., Spielhagen, R.F., Wolf-Welling, T.C.W., Henrich, R. and Kassens, H. 1995. Reflection of Scandinavian Ice Sheet Fluctuations in Norwegian Sea Sediments during the past 150,000 Years. *Quaternary Research*, 43, 185–197.

Benn, D.I. and Clapperton, C.M. 2000. Glacial sediment-landform associations and paleoclimate during the last glaciation, Strait of Magellan, Chile. *Quaternary Research*, 54, 13–23.

Benn, D.I. and Evans, D. 1998. *Glaciers and Glaciation*. Arnold, London.

Bennett, M.R. and Glasser, N.F. 1996. *Glacial Geology: ice sheets and landforms*. John Wiley, Chichester.

Benson, L., Burdett, J., Lund, S., Kashgarian, M. and Mensing, S. 1997. Nearly synchronous climate change in the Northern Hemisphere during the last glacial termination. *Nature*, 388, 263–265.

Bentley, C.R. 1987. Antarctic ice streams: a review. *Journal of Geophysical Research*, 92, 8843–8858.

Bentley, C.R. 1997. Rapid sea-level rise soon from West Antarctic Ice Sheet collapse. *Science*, 1077–1078.

Bentley, M.J. 1999. Volume of Antarctic ice at the last glacial maximum, and its impact on global sea level change. *Quaternary Science Reviews*, 18, 1569–1595.

Bentley, M.J. and Anderson, J.B. 1998. Glacial and marine geological evidence for the ice-sheet configuration in the Weddell Sea–Antarctic Peninsula region during the Last Glacial Maximum. *Antarctic Science*, 10, 307–323.

Björnsson, H. 1979. Glaciers in Iceland. *Jokull*, 29, 74–80.

Blunier, T., Chappellaz, J., Schwander, J., Dällenbach, A., Stauffer, B., Stocker, T.F., Raynaud, D., Jouzel, J., Clausen, H.B., Hammer, C.U. and Johnsen, S.J. 1998. Asynchrony of Antarctic and Greenland climate change during the last glacial period. *Nature*, 394, 739–743.

Bolshiyanov, D.Y. and Makeyev, V.M. 1995. The Archipelago Severnaya Zemlya: glaciation, history, environment (in Russian). Gidrometizdat, St. Petersburg, 216 pp.

Bond, G., Broecker, W., Johnsen, S., McManus, J., Labeyrie, L., Jouzel, J. and Bonani, G. 1993. Correlations between climate records from North Atlantic sediments and Greenland ice. *Nature*, 365, 143–147.

Bond, G., Heinrich, H., Broecker, W., Labeyrie, L., McManus, J., Andrews, J., Huon, S., Jantschik, R., Clasen, S., Simet, C., Tedesco, K., Klas, M., Bonani, G. and Ivy, S. 1992. Evidence for massive discharges of icebergs into the North Atlantic Ocean during the last glacial period. *Nature*, 360, 245–249.

Bond, G., Showers, W., Cheseby, M., Lotti, R., Almasi, P., deMenocal, P., Priore, P., Cullen, H., Hajdas, I. and Bonani, G. 1997. A pervasive millennial-scale cycle in North Atlantic Holocene and Glacial climates. *Science*, 278, 1257–1266.

Bond, G.C. and Lotti, R. 1995. Iceberg discharges into the North Atlantic on millennial time scales during the last deglaciation. *Science*, 267, 1005–1010.

Boulton, G.S. 1979. Processes of glacier erosion on different substrata. *Journal of Glaciology*, 23, 15–38.

Boulton, G.S. 1987. A theory of drumlin formation by subglacial sediment deformation. In: Menzies, J. and Rose, J. (eds) *Drumlin Symposium*. A.A. Balkema, Rotterdam. 25–80.

Boulton, G.S. 1993. In: Duff, D. (ed.) *Holmes' Principles of Physical Geology*, Chapman and Hall, London.

Boulton, G.S., Jarvis, J. and Thors, K. 1988. Dispersal of glacially derived sediment over part of the continental shelf off south Iceland and the geometry of the resultant sediment bodies. *Marine Geology*, 83, 193–223.

Boulton, G.S. and Jones, A.S. 1979. Stability of Temperate Ice Caps and Ice Sheets resting on Beds with Deformable Sediment. *Journal of Glaciology*, 24, 29–42.

Boulton, G.S., Baldwin, C.T., Peacock, J.D., McCabe, A.M., Miller, G., Jarvis, J., Horsefield, B., Worsley, P., Eyles, N., Chroston, P.N., Day, T.E., Gibbard, P., Hare, P.E. and von Brunn, V.

1982. A glacio-isostatic facies model and amino acid stratigraphy for the late Quaternary events in Spitsbergen and the Arctic. *Nature*, 298, 437–441.

Boulton, G.S., Jones, A.S., Clayton, K.M. and Kenning, M.J. 1977. A British ice-sheet model and patterns of glacial erosions and deposition in Britain. In: Shotton, R.W. (ed.) *British Quaternary Studies: Recent advances*, Clarendon Press, Oxford, 231–246.

Boulton, G.S., Smith, G.D., Jones, A.S. and Newsome, J. 1985. Glacial geology and glaciology of the last mid-latitude ice sheets. *Journal of the Geological Society of London*, 142, 447–474.

Bowen, D.Q. 1978. *Quaternary Geology*. Pergamon, Oxford, 221pp.

Bowen, D.Q., Rose, J., McCabe, A.M and Sutherland, D.G. 1986. Correlation of Quaternary glaciations in England, Ireland, Scotland and Wales. *Quaternary Science Reviews*, 5, 299–340.

Boyle, E. 1996. Deep water distillation. *Nature*, 379, 679–680.

Boyle, E. and Weaver, A. 1994. Conveying past climates. *Nature*, 372, 41–42.

Broecker, W.S. 1992. The strength of the nordic heat pump. In: Bard, E. and Broecker, W.S. (eds) *The Last Deglaciation: Absolute and Radiocarbon Chronologies*, NATO ASI Series, Series I, *Global Environmental Change*, 2: 344pp. Springer-Verlag, Berlin.

Broecker, W.S. 1994. An unstable superconveyor. *Nature*, 367, 414–415.

Broecker, W.S. and Denton, G.H. 1990. The role of ocean-atmosphere reorganizations in glacial cycles. *Quaternary Science Reviews*, 9, 305–341.

Broecker, W.S., Bond, G. and Klas, M. 1990. A salt oscillator in the glacial Atlantic? 1. The concept. *Paleoceanography*, 5, 469–477.

Brook, E.J., Sowers, T. and Orchardo, J. 1996. Rapid variations in Atmospheric Methane concentration during the past 110 000 years. *Science*, 273, 1087.

Brotchie, J.F. and Silvester, R. 1969. On Crustal Flexure. *Journal of Geophysical Research*, 74, 5240–5252.

Budd, W.F. and Rayner, P. 1990. Modelling global ice and climate changes through the Ice Ages. *Annals of Glaciology*, 23, 23–27.

Budd, W.F. and Smith, I.N. 1979. The growth and retreat of ice sheets in response to orbital radiation changes. *Sea Level, Ice and Climatic Change (proceedings of the Canberra Symposium, December 1979)*, IAHS Publ. no. 131, 369–409.

Budd, W.F., Jenssen, D. and Smith, I.N. 1984. A three dimensional time dependent model of the west Antarctic ice sheet. *Annals of Glaciology*, 5, 29–36.

Bugge, T., Belderson, R.H. and Kenyon, N.H. 1988. The Storegga Slide: *Philosophical Transactions of the Royal Society, London, Series A*, 325, 357–388.

Bush, A.B.G. 2000. A positive climatic feedback mechanism for Himalayan glaciation. *Quaternary International*, 65/66, 3–13.

Calhoun, E.A. 1985. Glaciations of the West Coast Range, Tasmania. *Quaternary Research*, 24, 39–59.

Campy, M. 1986. Glaciations in the Jura Range. *Quaternary Science Reviews*, 5, 403–405.

Chapman, M.R. and Maslin, M.A. 1999. Low-latitude forcing of meridional temperature and salinity gradients in the subpolar North Atlantic and the growth of glacial ice sheets. *Geology*, 27, 875–878.

Chappell, J. and Shackleton, N.J. 1986. Oxygen isotopes and sea level. *Nature*, 324, 137–140.

Chappell, J. 1974. Late Quaternary glacio- and hydro-isostasy, on a layered Earth. *Quaternary Research*, 4, 429–440.

Chorley, R.J., Schumm, S.A. and Sugden, D.E. 1984. *Geomorphology*, Methuen Press, London.

Clapperton, C.M. 1983. The glaciation of the Andes. *Quaternary Science Reviews*, 2, 83–155.

Clapperton, C.M., Hall, M., Mothes, P., Hole, M.J., Still, J.W., Helmens, K.F., Kuhry, P. and Gemmell, A.M.D. 1997. A Younger Dryas ice cap in the Equatorial Andes. *Quaternary Research*, 47, 13–28.

Clark, C.D. 1993. Mega-scale glacial lineations and cross-cutting ice-flow landforms. *Earth Surface Processes and Landforms*, 18, 1–29.

Clark, C.D., Knight, J.K. and Gray, J.T. 2000. Geomorphological reconstruction of the Labrador Sector of the Laurentide Ice Sheet. *Quaternary Science Reviews*, 19, 1343–1366.

Clark, J.A., Farrell, W.E. and Peltier, W.R. 1978. Global changes in post glacial sea-level: a numerical calculation. *Quaternary Research*, 9, 265–287.

Clark, P.U. 1994. Unstable behaviour of the Laurentide Ice Sheet over deforming sediment and its implications for climate change. *Quaternary Research*, 41, 19–25.

Clark, P.U. and Mix, A.C. 2000. Ice sheets by volume. *Nature*, 406, 689–690.

Clark, P.U., Alley, R.B. and Pollard, D. 1999. Northern Hemisphere ice-sheet influences on global climate change. *Science*, 286, 1104–1111.

Clark, P.U., Licciardi, J.M., MacAyeal, D.R. and Jenson, J.W. 1996. Numerical reconstruction of a soft-bedded Laurentide Ice Sheet during the last glacial maximum. *Geology*, 24, 679–682.

Clarke, G.K.C., Collins, S.G. and Thompson, D.E. 1984. Flow, thermal structure, and subglacial conditions of a surge-type glacier. *Canadian Journal of Earth Sciences*, 21, 232–240.

Clemens, S.C. and Tiedemann, R. 1997. Eccentricity forcing of Pliocene-Early Pleistocene climate revealed in a marine oxygen-isotope record. *Nature*, 385, 801–804.

CLIMAP project members. 1976. The surface of the Ice Age Earth. *Science*, 191, 1131–1136.

CLIMAP Project Members. 1981. Seasonal reconstruction of the Earth's surface at the last glacial maximum. *Geological Society of America, Map and Chart Series*, MC-36-1981.

Conchon, O. 1986. Quaternary glaciations in Corsica. *Quaternary Science Reviews*, 5, 429–432.

Conway, H., Hall, B.L., Denton, G.H., Gades, A.M. and Waddington, E.D 1999. Past and future grounding-line retreat of the West Antarctic ice sheet. *Science*, 286, 280–283.

Cuffey, K.M., Conway, H., Gades, A.M., Hallet, B., Lorrain, R., Severinghaus, J.P., Steig, E.J., Vaughn, B. and White, J.W.C. 2000. Entrainment at cold glacier beds. *Geology*, 28, 351–354.

Cuffey, K.N. and Marshall, S.J. 2000. Substantial contribution to sea-level rise during the last interglacial from the Greenland ice sheet. *Nature*, 404, 591–594.

Dahl-Jensen, D., Mosegaard, K., Gundestrup, N., Clow, G.D., Johnson, S.J., Hansen, A.W. and Balling, N. 1998. Past temperatures directly from the Greenland ice sheet. *Science*, 282, 268–271.

Dalgleish, A.N., Boulton, G.S. and Renshaw, E. 2000. The ice age cycle and the deglaciations: an application of nonlinear regression modelling. *Quaternary Science Reviews*, 19, 687–697.

Dansgaard, W., Johnsen, S.J., Møller, J. and Langway Jr., C.C. 1969. One thousand centuries of climatic record from Camp Century on the Greenland Ice Sheet. *Science*, 166, 377–381.

Davis, C.H., Kluever, C.A. and Haines, B.J. 1998. Elevation change of the southern Greenland Ice Sheet. *Science*, 279, 2086–2088.

Dawson, A.G. 1992. *Ice age earth: late Quaternary geology and climate*. Routledge, London, 293pp.

Denham, C.R. and Cox, A. 1971. Evidence that the Lachamp polarity event did not occur 13 300–30 400 years ago. *Earth and Planetary Science Letters*, 13, 181–190.

Denton, G.H. 2000. Does an asymmetric thermohaline-ice-sheet oscillator drive 100 000-yr glacial cycles? *Journal of Quaternary Science*, 15, 301–318.

Denton, G.H. and Hughes, T.J. 1981. The last great ice sheets. John Wiley, New York, 484 pp.

Denton, G.H., Bockheim, J.G., Wilson, S.C. and Stuiver, M. 1989. Late Wisconsin and Early Holocene Glacial History, Inner Ross Sea embayment, Antarctica. *Quaternary Research*, 31, 151–182.

Derbyshire, E., Shi, Y., Li, J., Zheng, B., Li, S. and Wang, J. 1991. Quaternary glaciation of Tibet: the geological evidence. *Quaternary Science Reviews*, 40, 485–510.

Doake, C.S.M. and Vaughan, D.G. 1991. Rapid disintegration of Wordie Ice Shelf in response to atmospheric warming. *Nature*, 350, 780–782.

Dodge, R.E., Fairbanks, R.G., Benninger, L.K. and Maurrasse, F. 1983. Pleistocene Sea Levels from Raised Coral Reefs of Haiti. *Science*, 219, 1423–1425.

Dokken, T. and Hald, M. 1996. Rapid climatic shifts during isotope stages 2–4 in the Polar North Atlantic. *Geology*, 24, 599–602.

Domack, E.W., Jull, A.J.T. and Nakao, S. 1991. Advance of East Antarctic outlet glaciers during the hypsithermal: implications for the volume state of the Antarctic ice sheet under global warming. *Geology*, 19, 1059–1062.

Domack, E.W., Jacobson, E.A., Shipp, S. and Anderson, J.B. 1999. Late Pleistocene–Holocene retreat of the West Antarctic ice-sheet system in the Ross Sea: Part 2 – sedimentologic and stratigraphic signature. *Geological Society of America Bulletin*, 111, 1517–1536.

Domack, E.W., O'Brien, P., Harris, P., Taylor, F., Quilty, P.G., de Santis, L. and Raker, B. 1998. Late Quaternary sediment facies in Prydz Bay, East Antarctica and their relationship to glacial advance onto the continental shelf. *Antarctic Science*, 3, 236–246.

Donglemans, P.W. 1996. Glacial dynamics of the Fennoscandian Ice Sheet: a remote sensing study. Unpublished Ph.D. Thesis. University of Edinburgh.

Dowdeswell, J.A. 1995. Glaciers in the High Arctic and recent environmental change. *Philosophical Transactions of the Royal Society*, 352A, 321–334.

Dowdeswell, J.A. and Scourse, J.D. (eds). 1990. *Glacimarine Environments: Processes and Sediments*. Geological Society Special Publication, 53.

Dowdeswell, J.A. and Williams, M. 1997. Surge-type glaciers in the Russian High Arctic identified from digital satellite imagery. *Journal of Glaciology*, 43, 489–494.

Dowdeswell, J.A. and Collin, R.L. 1990. Fast-flowing outlet glaciers on Svalbard ice caps. *Geology*, 18, 778–781.

Dowdeswell, J.A., Dowdeswell, E.K., Williams, M. and Glazovsky, A.F. 2000. The glaciology of the Russian High Arctic from Landsat imagery. *U.S. Geological Survey Professional Paper*, 1386-F.

Dowdeswell, J.A., Drewry, D.J., Cooper, A.P.R., Gorman, M.R., Liestøl, O. and Orheim, O. 1986. Digital mapping of the Nordaustlandet ice caps

from airborne geophysical investigations. *Annals of Glaciology*, 8, 51–58.
Dowdeswell, J.A., Elverhøi, A., Andrews, J.T. and Hebbeln, D. 1999. Asynchronous deposition of ice-rafted layers in the Nordic seas and North Atlantic Ocean. *Nature*, 400, 348–351.
Dowdeswell, J.A., Kenyon, N.H., Everhøi, A., Laberg, J.S., Hollender, F.-J., Mienert, J. and Siegert, M.J. 1996a. Large-scale sedimentation on the glacier-influenced Polar North Atlantic margins: long-range side-scan sonar evidence: *Geophysical Research Letters*, 23, 3535–3538.
Dowdeswell, J.A., Gorman, M.R., Glazovsky, A.F. and Macheret, Y.Y. 1996b. Airborne radio-echo sounding of the ice caps on Franz Josef Land in 1994: *Materialy Glyatsiologicheskikh Issledovaniy, Khronika*, 80, 248–254.
Dowdeswell, J.A., Maslin, M.A., Andrews, J.T. and McCave, I.N. 1995. Iceberg production, debris rafting and extent and thickness of Heinrich layers (H-1, H-2) in North Atlantic sediments. *Geology*, 23, 301–304.
Dowdeswell, J.A. and Siegert, M.J. 1999. Ice-sheet numerical modeling and marine geophysical measurements of glacier-derived sedimentation on the Eurasian Arctic continental margins. *Geological Society of America Bulletin*, 111, 1080–1097.
Dowdeswell, J.A., Uenzelmann-Neben, G., Whittington, R.J. and Marienfield, P. 1994. The Late Quaternary sedimentary record in Scoresby Sund, East Greenland. *Boreas*, 23, 294–310.
Dowdeswell, J.A., Unwin, B., Nuttall, A.M. and Wingham, D.J. 1999. Velocity structure, flow instability and mass flux on a large Arctic ice cap from satellite radar interferometry. *Earth and Planetary Science Letters*, 167, 131–140.
Drewry, D.J. 1979. Late Wisconsin reconstruction for the Ross Sea region, Antarctica. *Journal of Glaciology*, 24, 231–244.
Drewry, D.J. 1983. *Antarctica: glaciological and geophysical folio*. Scott Polar Research Institute, University of Cambridge.
Drewry, D.J. 1986. *Glacial Geologic Processes*. Arnold, London.
Duplessy, J.C. and Labeyrie, L. 1992. The Norwegian Sea record of the last interglacial to glacial transition. In: Kukla, G.J. and Went, E. (eds) *Start of a Glacial Climate*, (NATO ASI series. Series I: Global Environmental Change; 3), Berlin, Springer-Verlag.
Dyke, A.S. 1999. Last Glacial Maximum and deglaciation of Devon Island, Arctic Canada: support of an Innuitian ice sheet. *Quaternary Science Reviews*, 18, 393–420.
Dyke, A.S. and Prest, V.K. 1987. The late Wisconsin and Holocene history of the Laurentide ice sheet. *Géographie Physique et Quaternaire*, 41, 237–263.
Dyke, A.S., Dredge, L.A. and Vincent, J.-S. 1982. Configuration and dynamics of the Laurentide Ice Sheet during the late Wisconsin maximum. *Géographie Physique et Quaternaire*, 36, 5–14.
Echelmeyer, K., Clark, T.S. and Harrison, W.D. 1991. Surficial glaciology of Jakobshaven Isbræ, West Greenland: Part I. Surface morphology. *Journal of Glaciology*, 37, 368–382.
Echelmeyer, K., Harrison, W.D., Clarke, T.S. and Benson, C. 1992. Surficial glaciology of Jakobshaven Isbræ, West Greenland: Part II. Ablation, accumulation and temperature. *Journal of Glaciology*, 38, 169–181.
Egloff, J. and Johnson, G.L. 1979. Erosional and depositional structures of the southwest Iceland insular margin thirteen geophysical profiles. In: Watkins, J.S., Mondatert, L. and Dickerson, P.W. (eds) *Geological and Geophysical investigations of continental margins*, AAPG, Tulsa, Oklahoma, pp. 43–63.
Ekman, M. 1989. Eustatic changes in sea-level. *Physics and Chemistry of the Earth*, 4, 99–185.
Elderfield, H. and Gannsen, G. 2000. Past temperature and $\delta^{18}O$ of surface waters inferred from forminiferal Mg/Ca ratios. *Nature*, 405, 442–445.
Elias, S.A. 1996. *The Ice-Age history of National Parks in the Rocky Mountains*. Smithsonian Institute Press, Washington. 170 pp.
Elverhøi, A. 1981. Evidence for a Late Wisconsin glaciation of the Weddell Sea. *Nature*, 293, 641–642.
Elverhøi, A. and Solheim, A. 1987. Late Weichselian glaciation of the northern Barents Sea – a discussion. *Polar Research*, 5, 285–287.
Elverhøi, A., Anderson, E.S., Dokken, T., Hebbeln, D., Spielhagen, R., Svendsen, J.I., Sørflaten, M., Rørnes, A., Hald, M. and Forsberg, C.F. 1995. The growth and decay of the Late Weichselian ice sheet in western Svalbard and adjacent areas based on provenance studies of marine sediments: *Quaternary Research*, 44, 303–316.
Elverhøi, A., Fjeldskaar, W., Solheim, A., Nyland-Berg, M. and Rasswurm, L. 1993. The Barents Sea Ice Sheet – a model of its growth and decay during the last ice maximum. *Quaternary Science Reviews*, 12, 863–873.
Elverhøi, A., Norem, H., Andersen, E.S., Dowdeswell, J.A., Fossen, I., Haflidason, H., Kenyon, N.H., Laberg, J.S., King, E.L., Sejrup, H.P., Solheim, A. and Vorren, T. 1997. On the origin and flow behavior of submarine slides on deep-sea fans along the Norwegian–Barents Sea continental margin. *Geo-Marine Letters*, 17, 119–125.
Elverhøi, A., Pfirman, S.L., Solheim, A. and Lassen,

B.B. 1989. Glaciomarine sedimentation in epicontinental seas exemplified by the northern Barents Sea. *Marine Geology*, 85, 225–250.

Emiliani, C. 1993. Milankovitch theory verified. *Nature*, 364, 583–584.

England, J. 1999. Coalescent Greenland and Innuitian ice during the Last Glacial Maximum: revising the Quaternary of the Canadian High Arctic. *Quaternary Science Reviews*, 18, 421–456.

Fairbanks, R.G. 1989. A 17,000-year glacio-eustatic sea level record: influence of glacial melting rates on the Younger Dryas event and deep ocean circulation. *Nature*, 342, 637–643.

Faleide, J.I., Solheim, A., Fiedler, A., Hjelstuen, B.O., Andersen, E.S. and Vanneste, K. 1996. Late Cenozoic evolution of the western Barents Sea–Svalbard continental margin. *Global and Planetary Change*, 12, 53–74.

Farley, K.A. and Paterson, D.B. 1995. A 100-kyr periodicity in the flux of extra-terrestrial ^{3}He to the sea floor. *Nature*, 378, 600–604.

Farrell, W.E. and Clark, J.A. 1976. On post glacial sea-level. *Geophysical Journal of the Royal Astronomical Society*, 46, 647–667.

Fiedler, A. and Faleide, J.I. 1996. Cenozoic sedimentation along the southwestern Barents Sea margin in relation to uplift and erosion of the shelf. *Global and Planetary Change*, 12, 75–93.

Fisher, D.A., Reeh, N. and Langley, K. 1985. Objective reconstructions of the Late Wisconsinan Laurentide ice sheet and the significance of deformable beds. *Géographie Physique et Quaternaire*, 39, 229–238.

Fjeldskaar, W. 1989. Rapid eustatic changes – never globally uniform. In: Collinson, J.D. (ed.) *Correlation in Hydrocarbon Exploration*. Norwegian Petroleum Society, pp. 13–19.

Fjeldskaar, W. 1991. Geoidal-eustatic changes induced by the deglaciation of Fennoscandia. *Quaternary International*, 9, 1–6.

Fjeldskaar, W. and Cathles, L. 1991. Rheology of mantle and lithosphere inferred from postglacial uplift in Fennoscandia. In: Sabadini, R., Lambeck, K and Boschi, E. (eds) *Glacial isostasy, sea level and mantle rheology*, 1–19 (NATO ASI Series C: Mathematics and Physical Sciences 334), Kluwer Academic Publishers, Dordrecht.

Flint, R.F. 1971. *Glacial and Quaternary geology*. Wiley, New York, 892 pp.

Forman, S.L. 1990. Post-glacial relative sea level history of northwestern Spitsbergen, Svalbard. *Geological Society of America Bulletin*, 102, 1580–1590.

Forman, S.L. 1992. Post-glacial relative sea-level history of northwestern Spitsbergen, Svalbard: Alternative interpretation and reply. Reply. *Geological Society of America Bulletin*, 104, 163–165.

Forman, S.L., Lubinski, D., Miller, G.H., Matishov, G., Snyder, J., Myslivets, V. and Korsun, S. 1995. Post-glacial emergence and distribution of Late Weichselian ice sheet loads in the northern Barents and Kara seas, Russia. *Geology*, 23, 574–577.

Forman, S.L., Ingolfsson, O., Gataullin, V., Manley, W.F. and Lokrantz, H. 1999a. Late Quaternary stratigraphy of western Yamal Peninsula, Russia: New constraints on the configuration of the Eurasian ice sheet. *Geology*, 27, 807–810.

Forman, S.L., Lubinski, D.J., Zeeberg, J.J., Polyak, L., Miller, G.H., Matishov, G. and Tarasov, G. 1999b. Postglacial emergence and late Quaternary glaciation on northern Novaya Zemlya, Arctic Russia. *Boreas*, 28, 133–145.

Forman, S.L., Weihe, R., Lubinski, D., Tarasov, G., Korsun, S. and Matishov, G. 1997. Holocene relative sea-level history of Franz Josef Land, Russia. *Geological Society of America Bulletin*, 109, 1116–1133.

Fuhrer, K., Wolff, E.W. and Johnsen, S.J. 1999. Timescales for dust variability in the Greenland Ice Core Project (GRIP) ice core in the last 100,000 years. *Journal of Geophysical Research*, 104, D24, 31,043–31,052.

Fullerton, D.S. 1986. Chronology and correlation of glacial deposits in the Sierra Navada, California. *Quaternary Science Reviews*, 5, 161–169.

Funder, S. and Hansen, H.C. 1996. The Greenland Ice Sheet – a model for its culmination and decay during and after the last glacial maximum. *Bulletin of the Geological Society of Denmark*, 42, 137–152.

Funder, S., Hjort, C., Landvik, J.Y., Nam, S.I., Reeh, N. and Stein, R. 1998. History of a stable ice margin – Greenland during the upper and middle Pleistocene. *Quaternary Science Reviews*, 17, 77–124.

Fyfe, G.J. 1990. The effect of water depth on ice-proximal glaciolacustrine sedimentation: Salpausselkè I, southern Finland. *Boreas*, 19, 147–164.

Gates, W.L. 1976a. Modelling the ice-age climate. *Science*, 191, 1138–1144.

Gates, W.L. 1976b. The numerical simulation of Ice-Age climate with a global general circulation model. *Journal of Atmospheric Sciences*, 33, 1844–1873.

Giovinetto, M.B. and Bentley, C.R. 1985. Surface balance in ice drainage systems of Antarctica. *Antarctic Journal*, 20(4), 6–13.

Glasser, N.F. and Sambrook-Smith, G.H. 1999. Glacial meltwater erosion of the Mid-Cheshire Ridge: implications for ice dynamics during the Late Devensian glaciation of northwest England. *Journal of Quaternary Research*, 14, 703–710.

Glen, J.W. 1955. The creep of polycrystalline ice. *Proceedings of the Royal Society of London*, A228, 519–538.

Goodwin, I.D. and Zweck, C. 2000. Glacio-isostasy and glacial ice load at Law Dome, Wilkes Land, East Antarctica. *Quaternary Research*, 53, 285–293.

GRIP Project members. 1993. Climate instability during the last interglacial period recorded in the GRIP ice core. *Nature*, 364, 203–207.

Grootes, P.M., Stuiver, M., White, J.W.C., Johnsen, S. and Jouzel, J. 1993. Comparison of oxygen isotope records from the GISP2 and GRIP Greenland ice cores. *Nature*, 366, 552–554.

Grosswald, M.G. 1980. Late Weichselian Ice Sheet of Northern Eurasia. *Quaternary Research*, 13, 1–32.

Grosswald, M.G. and Hughes, T.J. 1995. Paleoglaciology's grand unsolved problem. *Journal of Glaciology*, 41, 313–332.

Grousset, F.E., Pujol, C., Labeyrie, L., Auffret, G. and Boelaert, A. 2000. Were the North Atlantic Heinrich events triggered by the behaviour of the European ice sheets? *Geology*, 28, 123–126.

Gualtieri, L., Glushkova, O. and Brigham-Grette, J. 2000. Evidence for restricted ice extent during the last glacial maximum in the Koryak Mountains of Chukotka, far eastern Russia. *Geological Society of America Bulletin*, 112, 1106–1118.

Gupte, S.K., Sharma, P. and Shah, S.K. 1992. Constraints on ice sheet thickness over Tibet during the last 40000 years. *Journal of Quaternary Science*, 7, 283–290.

Hagen, J.O., Liestøl, O., Roland, E. and Jørgensen, T. 1993. Glacier Atlas of Svalbard and Jan Mayen. *Norsk Polarinstitutt Meddelelser*, 129, 141 pp.

Hald, M., Steinsund, P.I., Dokken, T., Korsun, S., Ployak, L. and Aspeli, R. 1994. Recent and Late Quaternary distribution of *Elphidium excavatum clavatum* in Arctic seas. Cushman Foundation Special Publication 32: 141–153.

Hallet, B. 1979. A theoretical model of glacial abrasion. *Journal of Glaciology*, 23, 39–50.

Hallet, B. 1981. Glacial abrasion and sliding: their dependence on the debris concentration in basal ice. *Annals of Glaciology*, 2, 23–28.

Hambrey, M.J. 1994. *Glacial Environments.* UCL Press.

Hamilton, G.S. and Dowdeswell, J.A. 1996. Controls on glacier surging in Svalbard. *Journal of Glaciology*, 42, 157–168.

Hamilton, T.D. 1986. Correlation of Quaternary glacial deposits in Alaska. *Quaternary Science Reviews*, 5, 171–180.

Hanebuth, T., Stattegger, K. and Grootes, P.M. 2000. Rapid flooding of the Sunda Shelf: a late glacial sea-level record. *Science*, 288, 1033–1035.

Hart, J.K. 1996. The relationship between drumlins and other forms of sub-glacial glaciotectonic deformation. *Quaternary Science Reviews*, 16, 93–107.

Hastings, D.W., Russell, A.D. and Emerson, S.R. 1998. Foraminferal magnesium in *Globeriginoides sacculifer* as a paleotemperature proxy. *Paleoceanography*, 13, 161–179.

Hebbeln, D. and Wefer, G. 1991. Effects of ice coverage and ice-rafted material on sedimentation in the Fram Strait. *Nature*, 350, 409–411.

Hebbeln, D., Dokken, T., Andersen, E.S., Hald, M. and Elverhøi, A. 1994. Moisture supply for northern ice-sheet growth during the Last Glacial Maximum. *Nature*, 370, 357–360.

Heinrich, H. 1988. Origin and consequences of cyclic ice rafting in the Northeast Atlantic Ocean during the past 130,000 years. *Quaternary Research*, 29, 143–152.

Henderson, G.M. and Slowey, N.C. 2000. Evidence from U-Th dating against Northern Hemisphere forcing of the penultimate glaciation. *Nature*, 404, 61–66.

Henrich, R., Wagner, T., Goldschmidt, P. and Michels, K. 1995. Depositional regimes in the Norwegian-Greenland Sea: the last two glacial to interglacial transition. *Geologische Rundschau*, 84, 28–48.

Henriksen, S. and Vorren, T.O. 1996. Late Cenozoic sedimentation and uplift history on the mid-Norwegian continental shelf. *Global and Planetary Change*, 12, 171–199.

Hérail, G., Hubschman, J. and Jalut, G. 1986. Quaternary glaciation in the French Pyrenees. *Quaternary Science Reviews*, 5, 397–402.

Hicock, S.R. and Dreimanis, A. 1992. Deforming till in the Great Lakes region: implications for rapid flow along the south-central margin of the Laurentide Ice Sheet. *Canadian Journal of Earth Sciences*, 29, 1565–1579.

Hindmarsh, R.C.A. 1998. Drumlinization and drumlin-forming instabilities: viscous till mechanisms. *Journal of Glaciology*, 44, 293–314.

Hisdal, V. 1985. *Geography of Svalbard.* Norsk Polarinstitutt, Polarhandbok 2, Oslo, 75pp.

Hjelstuen, B.O., et al. 1996, Cenozoic erosion and sediment yield in the drainage area of the Storfjorden Fan: *Global and Planetary Change*, 12, 95–117.

Höfle, C., Edwards, M.E., Hopkins, D.M., Mann, D.H. and Ping, C.-L. 2000. The full-glacial environment of the northern Seward Peninsula, Alaska, reconstructed from the 21,500-year-old Kitluk Paleosol. *Quaternary Research*, 53, 143–153.

Holdsworth, G. 1977. Surges in ice sheets. *Nature*, 269, 588–590.

Holland, M.M., Brasket, A.J and Weaver, A.J. 2000. The impact of rising atmospheric CO_2 on simulated sea ice induced thermohaline circulation variability. *Geophysical Research Letters*, 27, 1519–1522.

Hollin, J.T. and Schilling, D.H. 1981. Late Wisconsin–Weichselian mountain glaciers and ice caps. In: Denton, G.H. and Hughes, T.J. (eds) *The last great ice sheets*, John Wiley, New York. pp. 179–220.

Hooke, R.L. 1998. *Principles of Glacier Mechanics*. Prentice Hall, 248pp.

Hostetler, S.W. 1997. Near to the edge of an ice sheet. *Nature*, 385, 393–394.

Hostetler, S.W., Bartlein, P.J., Clark, P.U., Small, E.E. and Solomon, A.M. 2000. Simulated influences of Lake Agassiz on the climate of central North America 11,000 years ago. *Nature*, 405, 334–337.

Howell, D., Siegert, M.J. and Dowdeswell, J.A. 2000. Numerical modelling of glacial isostasy and ice sheet growth within the Late Weichselian Barents Sea. *Journal of Quaternary Science*, 15, 475–486.

Hubbard, A. 1999. High-resolution modeling of the advance of the Younger Dryas ice sheet and its climate in Scotland. *Quaternary Research*, 52, 27–43.

Hubbard, B. and Hubbard, A. 1998. Bedrock surface roughness and the distribution of subglacially precipitated carbonate deposits: implications for formation at Glacier de Tsanfleuron, Switzerland. *Earth Surface Processes and Landforms*, 23, 261–270.

Hubbard, B., Siegert, M.J. and McCarroll, D. 2000. Spectral evidence of the interaction between glacier bedrock roughness, glacier sliding and erosion. *Journal of Geophysical Research*, 105, 21295–2130.

Hughen, K.A., Overpeck, J.T., Lehman, S.J., Kashgarian, M., Southon, J., Peterson, L.C., Alley, R. and Sigman, D.M. 1998. Deglacial changes in ocean circulation form an extended radiocarbon calibration. *Nature*, 391, 65–68.

Hughes, T.J. 1985. The great Cenozoic ice sheet. *Palaeogeography, Palaeoclimatology, Palaeoecology*, 50, 9–43.

Hughes, T.J. 1987. The marine ice transgression hypothesis. *Geografiska Annaler*, 69, 237–250.

Hulbe, C. 1997. An ice shelf mechanism for Heinrich layer production. *Paleoceanography*, 12, 711–721.

Hulton, N., Sugden, D., Payne, A. and Clapperton, C. 1994. Glacier modeling and the climate of Patagonia during the Last Glacial Maximum. *Quaternary Research*, 42, 10–19.

Hunt, A.G. and Malin, P.E. 1998. Possible triggering of Heinrich events by ice-load-induced earthquakes. *Nature*, 393, 155–158.

Huybrechts, P. 1990a. A 3-D model for the Antarctic Ice Sheet: a sensitivity study on the glacial-interglacial contrast. *Climate Dynamics*, 5, 79–92.

Huybrechts, P. 1990b. The Antarctic ice sheet during the last glacial-interglacial cycle: a three-dimensional experiment. *Annals of Glaciology*, 14, 115–119.

Huybrechts, P. 1992. The Antarctic ice sheet and environmental change: a three dimensional modelling study. *Reports on Polar Research*, (Alfred-Wegener-Institut für Polar und Meeresforschung) 99.

Huybrechts, P. 1993. Glaciological modelling of the Late Cenozoic East Antarctic ice sheet: stability or dynamism? *Geografiska Annaler*, 75, 221–238.

Huybrechts, P., Payne, A. and the EISMINT Intercomparison Group. 1996. The EISMINT benchmarks for testing ice-sheet models. *Annals of Glaciology*, 23, 1–12.

Iken, A., Echelmeyer, K., Harrison, W. and Funk, M. 1993. Mechanisms of fast flow in Jakobshavn Isbræ, West Greenland: Part I. Measurements of temperature and water level in deep boreholes. *Journal of Glaciology*, 39, 15–25.

Iken, A., Röthlisberger, H., Flotron, A. and Haeberli, W. 1983. The uplift of Unteraargletscher at the beginning of the melt season – a consequence of water storage at the bed? *Journal of Glaciology*, 29, 101–119.

Imbrie, J., Boyle, E.A., Clemens, S.C., Duffy, A., Howard, W.R., Kukla, G., Kutzbach, J., Martinson, D.G., McIntyre, A., Mix, A.C., Molfino, B., Morley, J.J., Peterson, L.C., Pisias, N.G., Prell, W.L., Raymo, M.E., Shackleton, N.J. and Toggweiler, J.R. 1992. On the structure and origin of major glaciation cycles. 1. Linear Responses to Milankovitch Forcing. *Paleoceanography*, 7, 701–738.

Jarvis, G.T. and Clarke, G.K.C. 1975. The thermal regime of Trapridge Glacier and its relevance to glacier surging. *Journal of Glaciology*, 14, 235–250.

Johnsen, S.J., Clausen, H.B., Dansgaard, W., Fuhrer, K., Gundestrup, N., Hammer, C.U., Iversen, P., Jouzel, J., Stauffer, B. and Steffensen, J.P. 1992. Irregular glacial interstadials recorded in a new Greenland ice core. *Nature*, 359, 311–313.

Johnston, P.J. and Lambeck, K. 2000. Automatic inference of ice models from postglacial sea level observations: theory and application to the British Isles. *Journal of Geophysical Research*, 105, 13179–13194.

Jones, G.A. 1991a. A stop-start ocean conveyer. *Nature*, 349, 364–365.

Jones, G.A. 1991b. Spatial and temporal distribution of Laurentide and Fennoscandian meltwater during the last deglaciation. *Norsk Geologisk Tidsskrift*, 71, 145–148.

Jones, G.A. and Keigwin, L.D. 1988. Evidence from Fram Strait (78°N) for early deglaciation. *Nature*, 336, 56–59.

Joughin, I., Slawek, T., Fahnestock, M. and Kwok, R. 1996. A mini surge on the Ryder Glacier, Greenland, Observed by Satellite Radar Interferometry. *Science*, 274, 228–230.

Jouzel, J. and Merlivat, L. 1984. Deuterium and oxygen 18 in precipitation: modelling of the isotopic effects during snow formation. *Journal of Geophysical Research*, 89, 11749–11757.

Jouzel, J., Barkov, N.I., Barnola, J.M., Bender, J.M., Chappellaz, J., Genthon, C., Kotlyakov, V.M., Lipenkov, V., Lorius, C., Petit, J.R., Raynaud, D., Raisbeck, G., Ritz, C., Sowers, T., Stievenard, M., Yiou, F. and Yiou, P. 1993. Extending the Vostok ice-core record of palaeoclimate to the penultimate glacial period. *Nature*, 364, 407–412.

Jouzel, J., Raisbeck, G., Benoist, J.P., Yiou, F., Lorius, C., Raynaud, D., Petit, J.R., Barkov, N.I., Korotkevich, Y.S. and Kotlyakov, V.M. 1989. A comparison of deep Antarctic ice cores and their implication for climate between 65,000 and 15,000 years ago. *Quaternary Research*, 31, 135–150.

Kageyama, M. and Valdes, P.J. 2000. Impact of the North American ice-sheet orography on the Last Glacial Maximum eddies and snowfall. *Geophysical Research Letters*, 27, 1515–1518.

Kamb, B. 1970. Sliding motion of glaciers: theory and observation. *Reviews of Geophysics and Space Physics*, 8, 673–728.

Kanfoush, S.L., Hodell, D.A., Charles, C.D., Guilderson, T.P., Mortyn, P.G. and Ninnemann, U.S. 2000. Millennial-scale instability of the Antarctic Ice Sheet during the last glaciation. *Science*, 288, 1815–1818.

Kaufmann, G. and Lambeck, K. 1997. Implications of late Pleistocene Glaciation of the Tibetan Plateau for present-day uplift rates and gravity anomalies. *Quaternary Research*, 48, 267–279.

Kehmkuhl, F. 1998. Extent and spatial distribution of Pleistocene glaciations in Eastern Tibet. *Quaternary International*, 45–46, 123–143.

Kind, N.V. and Leonov, B.N. 1982. The Anthropogen of the Taymyr Peninsula (in Russian). Nauka, Moscow, 183 pp.

King, E.L., Sejrup, H.P., Haflidason, H., Elverhøi, A. and Aarseth, I. 1996. Quaternary seismic stratigraphy of the North Sea Fan: glacially-fed gravity flow aprons, hemipelagic sediments, and large submarine slides: *Marine Geology*, 130, 293–316.

Kleiber, H.P. and Niessen, F. 1999. Late Pleistocene paleo-river channels on the Laptev Sea shelf – implications from sub-bottom profiling. In: Kassens, H., Bauch, H.A., Dmitrenko, I.A., Eicken, H., Hubberten, H.W., Melles, M., Theide, J. and Timokhov, L.A., (eds) *Land-Ocean Systems in the Siberian Arctic: Dynamics and History*, pp. 657–665. Springer, Berlin, Heidelberg, New York.

Kleman, J.C., Hättestrand, C., Borgström, I. and Stroeven, A. 1997. Fennoscandian paleoglaciology reconstructed using a glacial-geological inversion model. *Journal of Glaciology*, 43, 283–299.

Kleman, J.C. and Borgström, I. 1996. Reconstruction of palaeo-ice sheets: the use of geomorphological data. *Earth Surface Processes and Landforms*, 21, 893–909.

Knies, J., Nowaczyk, N., Müller, C., Vogt, C. and Stein, R. 2000. A multiproxy approach to reconstruct the environmental changes along the Eurasian continental margin over the last 150,000 years. *Marine Geology*, 163, 317–344.

Knies, J. and Stein, R. 1998. New aspects of organic carbon deposition and its paleoceanographic implications along the northern Barents Sea margin during the last 30,000 years. *Paleoceanography*, 13, 384–394.

Koç, N., Jansen, E. and Haflidason, H. 1993. Paleoceanographic reconstructions of surface ocean conditions in the Greenland, Iceland and Norwegian Seas through the last 14 ka based on diatoms. *Quaternary Science Reviews*, 12: 115–140.

Krabill, W., Frederick, E., Manizade, S., Martin, C., Sonntag, J., Swift, R., Thomas, R., Wright, W. and Yungel, J. 1999. Rapid thinning of parts of the southern Greenland Ice Sheet. *Science*, 283, 1522–1524.

Krabill, N., Abdalati, W., Frederick, E., Manizade, S., Martin, C., Sonntag, J., Swift, R., Thomas, R., Wright, W. and Yungel, J. 2000. Greenland Ice Sheet: high-elevation balance and peripheral thinning. *Science*, 289, 428–430.

Kuhle, M. 1988. The Pleistocene glaciation of Tibet and the onset of ice ages: an autocycle hypothesis. *GeoJournal*, 17, 457–511.

Kukla, G.J. 2000. The last interglacial. *Science*, 287, 987–988.

Laberg, J.S. and Vorren, T.O. 1996a. The Middle and Late Pleistocene evolution of the Bear Island Trough Mouth Fan. *Global and Planetary Change*, 12, 309–330.

Laberg, J.S. and Vorren, T.O. 1996b. The glacier-fed fan at the mouth of Storfjorden Trough, western Barents Sea: a comparative study. *Geologische Rundschau*, 85, 338–349.

Laberg, J.S. 1996. Late Pleistocene evolution of the sybmarine fans off the western Barents Sea margin. Unpublished PhD Thesis. University of Tromsø, Norway.

Laberg, J.S. and Vorren, T.O. 1995, Late Weichselian debris flow deposits on the Bear Island Trough Mouth Fan. *Marine Geology*, 127, 45–72.

Lambeck, K. 1993a. Glacial rebound of the British Isles – I. Preliminary model results. *Geophysical Journal International*, 115, 941–959.

Lambeck, K. 1993b. Glacial rebound of the British Isles – II. A high-resolution, high precision model. *Geophysical Journal International*, 115, 960–990.

Lambeck, K. 1995. Constraints on the Late Weichselian Ice Sheet over the Barents Sea from Observations of raised shorelines. *Quaternary Science Reviews*, 14, 1–16.

Landvik, J.Y., Bondevik, S., Elverhøi, A., Fjeldskaar, W., Mangerud, J., Siegert, M.J., Salvigsen, O., Svendsen, J.-I. and Vorren, T.O. 1998. Last glacial maximum of Svalbard and the Barents Sea area: ice sheet extent and configuration. *Quaternary Science Reviews*, 17/1-3, 43–75.

Lao, Y., Anderson, R.F., Broecker, W.S., Trumbore, S.E., Hofmann, H.J. and Wolfli, W. 1992. Increased production of cosmogenic ^{10}Be during the Last Glacial Maximum. *Nature*, 357, 576–578.

Lassen, S., Jansen, E., Knudsen, K.L., Kuijpers, A., Kristensen, M. and Chreistensen, K. 1999. Northeast Atlantic sea surface circulation during the past 30-10 C-14 kyr BP. *Paleoceanography*, 14, 616–625.

Le Meur, E. and Huybrechts, P. 1996. A comparison of different ways of dealing with isostasy: examples from modelling the Antarctic ice sheet during the last deglaciation. *Annals of Glaciology*, 23, 309–317.

Lear, C.H., Elderfield, H. and Wilson, P.A. 2000. Cenozoic Deep-sea temperatures and global ice volumes from Mg/Ca in benthic foraminiferal calcite. *Science*, 287, 269–272.

Legarsky, J., Wong, A., Akins, T. and Gogineni, S.P. 1998. Detection of hills from radar data in central-north-Greenland. *Journal of Glaciology*, 44, 182–185.

Legrand, M. and Mayewski, P. 1997. Glaciochemistry of polar ice cores: A review. *Reviews of Geophysics*, 35, 219–243.

Legrand, M., Feniet-Saigne, C., Saltzman, E.S., Germain, C., Barkov, N.I. and Petrov, V.N. 1991. Ice core record of oceanic emission of dimethylsulphide during the last climate cycle. *Nature*, 350, 144–146.

Lehman, S.J. and Keigwin, L.D. 1992. Sudden changes in North Atlantic circulation during the last deglaciation. *Nature*, 356, 757–762.

Letréguilly, A., Reeh, N. and Huybrechts, P. 1991a. The Greenland ice sheet through the last glacial-interglacial cycle. *Global and Planetary Change*, 90, 385–394.

Letréguilly, A., Huybrechts, P. and Reeh, N. 1991b. Steady-state characteristics of the Greenland ice sheet under different climates. *Journal of Glaciology*, 37, 149–157.

Leuenberger, M., Siegenthaler, U. and Langway, C.C. 1992. Carbon isotope composition of atmospheric CO_2 during the last ice age from an Antarctic ice core. *Nature*, 357, 448–490.

Levesque, A.J., Cwynar, L.C. and Walker, I.R. 1997. Exceptionally steep north-south gradients in lake temperatures during the last deglaciation. *Nature*, 385, 423–426.

Levis, S., Foley, J.A. and Pollard, D. 1999. CO_2 climate, and vegetation feedbacks at the Last Glacial Maximum. *Journal of Geophysical Research*, 104, D24, 31,191–31,198.

Lindstrom, D.R. 1989. A study of the Eurasian ice sheet using a combined ice flow and ice shelf numerical model. Unpublished PhD Thesis. University of Chicago, USA.

Lindstrom, D.R. and MacAyeal, D.R. 1986. Paleoclimatic constraints on the maintenance of possible ice-shelf cover in the Norwegian and Greenland Seas. *Palaeoceanography*, 1, 313–337.

Lindstrom, D.R. and MacAyeal, D.R., 1989. Scandinavian, Siberian, and Arctic Ocean Glaciation: Effect of Holocene Atmospheric CO_2 variations. *Science*, 243, 628–631.

Long, A.J., Roberts, D.H. and Wright, M.R. 1999. Isolation basin stratigraphy and Holocene relative sea-level change on Arveprinsen Ejland, Disko Bugt, West Greenland. *Journal of Quaternary Science*, 14, 323–345.

Lorius, C., Jouzel, J., Ritz, C., Merlivat, L., Barkov, N.I., Korotkevitch, T.S. and Kotlyakov, V.M. 1985. A 150,000 year climate record from Antarctic ice. *Nature*, 316, 591–596.

Loutre, M.F. and Berger, A. 2000. No glacial-interglacial cycle in the ice volume simulated under a constant astronomical forcing and a variable CO_2. *Journal of Geophysical Research*, 27, 783–786.

Lowe, J.J. and Walker, M.J.C. 1997. *Reconstructing quaternary environments* (2nd edition). Longman, Harlow. 446pp.

MacAyeal, D.R. 1993. Binge-purge oscillations of the Laurentide Ice Sheet as a cause of the North Atlantc's Heinrich events. *Palaeoceanography*, 8, 775–784.

Macdonald, R. 1996. Awakenings in the Arctic. *Nature*, 380, 286–287.

Mahaffy, M.W. 1976. A Three Dimensional Numerical Model of Ice Sheets: Tests on the Barnes Ice Cap, Northwest Territories. *Journal of Geophysical Research*, 81, 1059–1066.

Makeyev, V.M., Arslanov, K.A. and Garutt, V.E. 1979. The ages of mammoths from the Severnaya Zemlya Archipelago and some problems of the Late Pleistocene Paleogeography (in Russian). *Doklady Academy Nauk SSSR* 245(2), 421–424.

Mangerud, J. 1991. The last ice age in Scandinavia. *Striae*, 34, 15–30.

Mangerud, J. and Svendsen, J.I. 1992, The last interglacial/glacial period on Spitsbergen, Svalbard. *Quaternary Science Reviews*, 11, 633–664.

Mangerud, J., Svendsen, J.I. and Astakhov, V.I. 1999. Age and extent of the Barents and Kara ice sheets in Northern Russia. *Boreas*, 18, 46–80.

Mangerud, J., Dokken, T., Hebbeln, D., Heggen, B., Ingolfsson, O., Landvik, J.Y., Mejhal, V., Svendsen, J.I. and Vorren, T.O. 1998. Fluctuations of the Svalbard-Barents Sea Ice Sheet during the last 150,000 years. *Quaternary Science Reviews*, 17, 11–42.

Mangerud, J., Jansen, E. and Landvik, J. 1996. Late Cenozoic history of the Scandinavian and Barents ice sheets. *Global and Planetary Change*, 12, 11–26.

Markgraf, V., Baumgaartner, T.R., Bradbury, J.P., Diaz, H.F., Dunbar, R.B., Luckman, B.H., Selter, G.O., Swetnam, T.W. and Villalba, R. 2000. Paleoclimate reconstruction along the Pole-Equator-Pole transect of the Americas (PEP1). *Quaternary Science Reviews*, 19, 125–240.

Marsella, K.A., Bierman, P.R., Davis, P.T. and Caffee, M.W. 2000. Cosmogenic ^{10}Be and ^{26}Al ages for the Last Glacial Maximum, eastern Baffin Island, Arctic Canada. *Geological Society of America Bulletin*, 112, 1296–1312.

Marshall, S.J. and Clarke, G.K.C. 1997a. A continuum mixture model of ice stream thermodynamics in the Laurentide Ice Sheet 1. Theory. *Journal of Geophysical Research*, 102, 20599–20613.

Marshall, S.J. and Clarke, G.K.C. 1997b. A continuum mixture model of ice stream thermodynamics in the Laurentide Ice Sheet 2. Application to the Hudson Strait Ice Stream. *Journal of Geophysical Research*, 102, 20615–20637.

Marshall, S.J. and Clarke, G.K.C. 1999. Modeling North American freshwater runoff through the last glacial cycle. *Quaternary Research*, 52, 300–315.

Marshall, S.J., Clarke, G.K.C., Dyke, A.S. and Fisher, D.A. 1996. Geologic and topographic controls on fast flow in the Laurentide and Cordilleran Ice Sheets. *Journal of Geophysical Research*, 101, 17827–17839.

Marshall, S.J., Tarasov, L., Clarke, G.K.C. and Peltier, W.R. 2000. Glaciological reconstruction of the Laurentide Ice Sheet: physical processes and modelling challenges. *Canadian Journal of Earth Sciences*, 37, 769–793.

Marsiat, I. and Valdes, P. in press. Sensitivity of the Northern Hemisphere climate of the Last Glacial Maximum to Sea Surface Temperatures. *Climate Dynamics*.

Mazaud, A., Vimeux, F. and Jouzel, J. 2000. Short fluctuations in Antarctic isotope records: a link with cold events in the North Atlantic? *Earth and Planetary Science Letters*, 177, 219–225.

McConnell, J.R., Arthern, R.J., Mosley-Thompson, E., Davis, C.H., Bales, R.C., Thomas, R. and Kyne, J.D. 2000. Changes in Greenland ice sheet elevation attributed primarily to snow accumulation variability. *Nature*, 406, 877–879.

Meier, M.F. and Post, A.S. 1969. What are glacier surges? *Canadian Journal of Earth Sciences*, 6, 807–817.

Melles, M., Siegert, C., Hahne, J. and Hubberten, H.W. 1996. Klima- und Umweltgeschischte des nördlichen Mittelsibriens im Spartquatär – erste Ergebnisse. *Geowissenschaften*, 14, 376–380.

Millar, D.H.M. 1981. Radio-echo layering in polar ice sheets and past volcanic activity. *Nature*, 292, 441–443.

Miller, M.F. and Mabin, M.C.G. 1998. Antarctic Neogene landscapes – in the refrigerator or in the Deep Freeze? *GSA Today*, 8, 1–8.

Mix, A.C. and Ruddiman, W.F. 1984. Oxygen-isotope analyses and Pleistocene ice volumes. *Quaternary Research*, 21, 1–20.

Møller, J.J. 1992. Late Weichselian glacial maximum on Andøya, North Norway. *Boreas*, 21, 1–13.

Möller, P., Bolshiyanov, D.Y. and Bergsten, H. (1999). Weichselian geology and palaeoenvironmental history of the Taymyr Peninsula, Siberia, indicating no glaciation during the last global glacial maximum. *Boreas*, 28, 92–114.

Montoya, M., Crowley, T.J. and von Storch, H. 1998. Temperatures at the last interglacial simulated by a coupled ocean-atmosphere climate model. *Palaeoceanography*, 13, 170–177.

Morse, D.L., Waddington, E.D. and Steig, E.J. 1998. Ice age storm trajectories inferred from radar stratigraphy at Taylor Dome, Antarctica. *Geophysical Research Letters*, 25, 3383–3386.

Muller, R.A. and MacDonald, G.J. 1997. Glacial cycles and astronomical forcing. *Science*, 215–218.

Näslund, J.O., Fastook, J.L. and Holmlund, P. 2000. Numerical modelling of the ice sheet in western Dronning Maud Land, East Antarctica: impacts of present, past and future climates. *Journal of Glaciology*, 46, 54–66.

Newnham, R.M. and Lowe, D.J. 2000. Fine-resolution pollen record of late-glacial climate reversal from New Zealand. *Geology*, 28, 759–762.

Nørgaard Pedersen, N., Spielhagen, R.F., Thiede, J. and Kassens, H. 1998. Central Arctic surface ocean environment during the past 80,000 years. *Paleoceanography,* 13, 193–204.

Nye, J.F. 1957. The distribution of stress and velocity in glaciers and ice sheets. *Proceedings of the Royal Society, Series A*, 239, 113–133.

Nye, J.F. 1970. Glacier sliding without cavitation in a linear viscous approximation. *Proceedings of the Royal Society of London*, 315A, 381–403.

O'Cofaigh, C., Lemmen, D.S., Evans, D.J.A. and Bednarski, J. 1999. Glacial landform-sediment assemblages in the Canadian High Arctic and their implications for late Quaternary glaciation. *Annals of Glaciology*, 28, 195–201.

O'Cofaigh, C., England, J. and Zreda, M. 2000. Late Wisconsinan glaciation of southern Eureka Sound: evidence for extensive Innuitian ice in the Canadian High Arctic during the Last Glacial Maximum. *Quaternary Science Reviews*, 19, 1319–1341.

Oerlemans, J. 1993. Evaluating the role of climate cooling in iceberg production and the Heinrich events. *Nature*, 364, 783–786.

Orvis, K.H. and Horn, S.P. 2000. Quaternary glaciers and climate on Cerro Chirripo, Costa Rica. *Quaternary Research*, 54, 24–37.

Østrem, G. and Haakensen, N. 1993. Glaciers of Norway. *U.S. Geological Survey Professional Paper*, 1386-E, 63–109.

Paterson, W.S.B. 1972. Laurentide Ice Sheet: estimated volumes during Late Wisconsin. *Reviews of Geophysics and Space Physics*, 10, 885–917.

Paterson, W.S.B. 1994. *The Physics of Glaciers* (third edition). Pergamon Press, Oxford.

Pavlidis, Y.A. 1992. The scale of the last glaciation in the arctic basin. *Oceanology*, 32, 352–365.

Pavlidis, Y.A., Dunayev, N.N. and Shcherbakov, F.A. 1997. The Late Pleistocene paleogeography of arctic Eurasian shelves. *Quaternary International*, 41/42, 3–9.

Payne, A.J. 1995. Limit cycles and the basal thermal regime of ice sheets. *Journal of Geophysical Research*, 100, 4249–4263.

Payne, A.J. and Baldwin, D.T. 1999. Thermomechanical modelling of the Scandinavian Ice Sheet: implications for ice stream formation. *Annals of Glaciology*, 28, 83–89.

Payne, A.J. and Dongelmans, P.W. 1997. Self organisation in the thermomechanical flow of ice sheets. *Journal of Geophysical Research*, 102, 12,219–12,234.

Payne, A.J., Sugden, D.E. and Clapperton, C.M. 1989. Modeling the growth and decay of the Antarctic Peninsula Ice Sheet. *Quaternary Research*, 31, 119–134.

Peltier, W.R. 1988b. Global Sea Level and Earth Rotation. *Science*, 240, 895–901.

Petit, J.R., Basile, I., Leruyuet, A., Raynaud, D., Lorius, C., Jouzel, J., Stievenard, M., Lipenkov, V.Y., Barkov, N.I., Kudryashov, B.B., Davis, M., Saltzman, E. and Kotlyakov, V. 1997. Four climate cycles in Vostok ice core. *Nature*, 387, 359–360.

Petit, J.R., Jouzel, J., Raynaud, D., Barkov, N.I., Barnola, J.M., Basile, I., Benders, M., Chappellaz, J., Davis, M., Delaygue, G., Delmotte, M., Kotlyakov, V.M., Legrand, M., Lipenkov, V.Y., Lorius, C., Pepin, L., Ritz, C., Saltzman, E. and Stievenard, M. 1999. Climate and atmospheric history of the past 420,000 years from the Vostok ice core, Antarctica. *Nature*, 399, 429–436.

Pflaumann, U., Duprat, J., Pujol, C. and Labeyrie, L.D. 1996. SIMMAX: A modern analog technique to deduce Atlantic sea surface temperatures from planktonic foraminifera in deep-sea sediments. *Paleoceanography*, 11(1), 15–35.

Pinot, S., Ramstein, G., Marsiat, I., de Vernal, A., Peyton, O., Duplessy, J.-C. and Weinelt, M. 1999. Sensitivity of the European LGM climate to North Atlantic sea-surface temperature. *Geophysical Research Letters*, 26, 1893–1896.

Polyak, L., Forman, S.L., Herlihy, F.A., Ivanov, G. and Krinitsky, P. 1997. Late Weichselian deglacial history of the Svyataya (Saint) Anna Trough, northern Kara Sea, Arctic Russia. *Marine Geology*, 143, 169–188.

Polyak, L., Gataullin, V., Okuneva, O. and Stelle, V. 2000. New constraints on the limits of the Barents-Kara ice sheet during the last glacial maximum based on borehole stratigraphy from the Pechora Sea. *Geology*, 28, 611–614.

Pope, P.G. and Anderson, J.B. 1992. Late Quaternary glacial history of the northern Antarctic Peninsula's western continental shelf: Evidence from the marine record. *Antarctic Research Series*, 57, 63–91.

Porter, S.C. 1975. Equilibrium-line altitudes of Late Quaternary glaciers in the Southern Alps, New Zealand. *Quaternary Research*, 5, 27–47.

Porter, S.C. 1986. Glaciation of Mauna Kea, Hawaii. *Quaternary Science Reviews*, 5, 181–182.

Post, A. 1969. Distribution of surging glaciers in western North America. *Journal of Glaciology*, 8, 229–240.

Press, W.H., Flannery, B.P., Teukolsky, S.A. and Vetterling, W.T. 1989. Numerical Recipes (The Art of Scientific Computing). Cambridge University Press, 702pp.

Pudsey, C.J., Barker, P.F. and Larter, R.D. 1994. Ice

sheet retreat from the Antarctic Peninsula. *Continental Shelf Research*, 14, 1647–1675.

Punkari, M. 1993. Modelling of the dynamics of the Scandinavian ice sheet using remote sensing and GIS methods. In: Aber, J.S. (ed.) *Glaciotectonics and mapping glacial deposits*, Proceedings of the INQUA commission of formation and properties of glacial deposits. Canadian Plains Research Center, University of Regina.

Punkari, M. 1995. Glacial flow systems in the zones of confluence between the Scandinavian and Novaya Zemlya ice sheets. *Quaternary Science Reviews*, 14, 589–603.

Punkari, M. 1997. Glacial and glaciofluvial deposits in the interlobate areas of the Scandinavian ice sheet. *Quaternary Science Reviews*, 16, 741–753.

Ram, M. and Koenig, G. 1997. Continuous dust concentration profile of pre-Holocene ice from the Greenland Ice Sheet Project 2 ice cores: Dust stadials, interstadials and the Eemian. *Journal of Geophysical Research*, 102, 26641–26648.

Rasmussen, T.L., van Weering, T.C.E. and Labeyrie, L. 1996. High resolution stratigraphy of the Faeroe–Shetland Channel and its relation to North Atlantic paleoceanography: the last 87 kyr. *Marine Geology*, 131, 75–88.

Raymond, C.F. 1987. How do glaciers surge? *Journal of Geophysical Research*, 92, 9121–9134.

Rebesco, M., Camerlenghi, A., de Santis, L., Domack, E. and Kirby, M. 1998. Seismic stratigraphy of Palmer Deep: a fault bounded late Quaternary sediment trap on the inner continental shelf, Antarctic Peninsula Pacific margin. *Marine Geology*, 151, 89–110.

Richards, D.A., Smart, P.L. and Edwards, R.L. 1994. Sea levels for the last glacial period constrained using 230Th ages of submerged speleothems. *Nature* 367, 357–360.

Richards, D.A., Beck, W., Donahue, D.J., Edwards, R.L., Silverman, B.W. and Smart, P.L. (submitted). Millennial-scale fluctuations of atmospheric radiocarbon from 20 to 11 ka based on a speleotherm from the Bahamas. *Science*.

Ridgewell, A.J., Watson, A.J. and Raymo, M.E. 1999. Is the spectral signature of the 10 kyr glacial cycle consistent with the Milankovitch origin? *Paleoceanography*, 14, 437–440.

Roberts, B.L. 1991. Modeling the Cordilleran Ice Sheet. *Géographie physique et Quaternaire*, 45, 287–299.

Roe, G.H. and Allen, M.R. 1999. A comparison of competing explanations for the 100,000-yr ice age cycle. *Geophysical Research Letters*, 26, 2259–2262.

Romanovski, N.N. 1993. Basic understanding of cryogenesis of the lithosphere (in Russian). MSU Publication, Moscow, 336 pp.

Rose, K.E. 1979. Characteristics of ice flow in Marie Byrd Land, Antarctica. *Journal of Glaciology*, 24, 63–75.

Ruddiman, W.F. and McIntyre, A. 1977. Late Quaternary Surface Ocean Kinematics and Climate Change in the High-Latitude North Atlantic. *Journal of Geophysical Research*, 82, 3877–3887.

Rutberg, R.L., Hemming, S.R. and Goldstein, S.L. 2000. Reduced North Atlantic Deepwater flux to the glacial southern ocean inferred from neodymium isotope ratios. *Nature*, 405, 935–938.

Sættem, J., Poole, D.A.R., Ellingsen, K.L. and Sejrup, H.P. 1992. Glacial geology of outer Bjørnøyrenna, southwestern Barents Sea. *Marine Geology*, 103, 15–51.

Salvigsen, O. and Slettemark, Ø. 1995. Past glaciation and sea levels on Bjørnøya, Svalbard. *Polar Research*, 14, 245–252.

Scherer, R.S., Aldahan, A., Tulaczyk, S., Possnert, G., Engelhard, H. and Kamb, B. 1998. Pleistocene collapse of the West Antarctic Ice Sheet. *Science*, 281, 82–85.

Schluchter, C. 1986. The Quaternary glaciations of Switzerland, with special reference to the northern Alpine foreland. *Quaternary Science Reviews*, 5, 413–419.

Schmitz, W.J. Jr. 1995. On the interbasin-scale thermohaline circulation. *Reviews of Geophysics*, 33, 151–173.

Schytt, V. 1993. Glaciers of Sweden. *U.S. Geological Survey Professional Paper*, 1386-E, 111–125.

Seidov, D. and Maslin, M.A. 1999. North Atlantic deep water circulation collapse during Heinrich events. *Geology*, 27, 23–26.

Sejrup, H.P., Larsen, E., King, E.L., Haflidason, H. and Nesje, A. 2000. Quaternary glaciations in southern Fennoscandia: evidence from southwestern Norway and the northern North Sea region. *Quaternary Science Reviews*, 19, 667–685.

Sejrup, H.P., Haflidason, H., Aarseth, I., Forsberg, C.F., King, E., Long, D. and Rokoengen, K. 1994. Late Weichselian glaciation history of the northern North Sea. *Boreas*, 23, 1–13.

Sexton, .D.J., Dowdeswell, J.A., Solheim, A. and Elverhøi, A. 1992. Seismic architecture and sedimentation in northwest Spitsbergen fjords. *Marine Geology*, 103, 53–68.

Shackleton, N.J. 1987. Oxygen isotopes, ice volume and sea level. *Quaternary Science Reviews*, 6, 183–190.

Sharp, M., Dowdeswell, J.A. and Gemmell, J.C. 1989. Reconstructing past glacier dynamics and erosion from glacial geomorphic evidence: Snowdon, North Wales. *Journal of Quaternary Science*, 4, 115–130.

Sharp, M.J. 1988. Surging glaciers: behaviour and mechanisms. *Progress in Physical Geography*, 12, 349–370.

Shaw, J., Faragini, D.M., Kvill, D.R. and Rians, R.B. 2000. The Athabasca fluting field, Alberta Canada: implications for the formation of large-scale fluting (erosional lineations). *Quaternary Science Reviews*, 2000, 959–980.

Shaw, J. and Gilbert, R. 1992. Evidence for large-scale subglacial meltwater flood events in southern Ontario and northern New York State. *Geology*, 20, 90–92.

Shaw, J., Kvill, D. and Rains, B. 1989. Drumlins and catastrophic subglacial floods. *Sedimentary Geology*, 62, 177–202.

Shennan, I., Lambeck, K., Horton, B., Innes, J., Lloyd, J., McArthur, J., Purcell, T. and Rutherford, M. 2000. Late Devensian and Holocene records of relative sea-level changes in northwest Scotland and their modelling implications for glaciohydro-isostatic modelling. *Quaternary Science Reviews*, 19, 1103–1135.

Shipp, S., Anderson, J. and Domack, E. 1999. Late Pleistocene-Holocene retreat of the West Antarctic ice-sheet system in the Ross Sea: Part 1 – Geophysical results. *Geological Society of America Bulletin*, 111, 1486–1516.

Shoemaker, E.M. 1992. Water sheet outburst floods from the Laurentide Ice Sheet. *Canadian Journal of Earth Sciences*, 29, 1250–1264.

Shoemaker, E.M. 1994, Correspondence. Reply to comments on 'Subglacial floods and the origin of low-relief ice-sheet lobes'. *Journal of Glaciology*, 40, 201–202.

Shoemaker, E.M. 1995. On the meltwater genesis of Drumlins. *Boreas*, 24, 3–10.

Shoemaker, E.M. 1999. Subglacial water-sheet floods, drumlins and ice-sheet lobes. *Journal of Glaciology*, 45, 201–213.

Siegert, M.J. and Dowdeswell, J.A. (submitted). Late Weichselian iceberg, meltwater and sediment production from the Eurasian High Arctic ice sheet: results from numerical ice-sheet modelling. *Marine Geology*.

Siegert, M.J. 1997. Quantitative reconstructions of the last glaciation of the Barents Sea: a review of ice-sheet modelling problems. *Progress in Physical Geography*, 21, 200–229.

Siegert, M.J. 2000. Radar evidence of water-saturated sediments beneath the central Antarctic ice sheet. In: Maltman, A., Hambrey, M. and Hubbard, B. (eds) *The deformation of glacial materials*, Special Publication of the Geological Society.

Siegert, M.J. and Dowdeswell, J.A. 1995. Numerical modeling of the Late Weichselian Svalbard–Barents Sea Ice Sheet. *Quaternary Research*, 43, 1–13.

Siegert, M.J. and Fjeldskaar, W. 1996. Isostatic uplift in the Late Weichselian Barents Sea: implications for ice sheet growth. *Annals of Glaciology*, 23, 352–358.

Siegert, M.J. and Bamber, J.L. (in press). Subglacial water at the heads of Antarctic ice-stream tributaries. *Journal of Glaciology*.

Siegert, C., Derevyagin, A.Y., Shilova, G.N., Hermichen, W.-D. and Hiller, A. 1999. Paleoclimatic indicators from permafrost sequences in the eastern Taymyr Lowland. In: Kassens, H., Bauch, H.A., Dmitrenko, I.A., Eicken, H., Hubberten, H.W., Melles, M., Theide, J. and Timokhov, L.A. (eds) *Land-Ocean Systems in the Siberian Arctic: Dynamics and History*, pp. 477–499. Springer, Berlin, Heidelberg, New York.

Siegert, M.J., Dowdeswell, J.A. and Melles, M. 1999. Late Weichselian glaciation of the Eurasian High Arctic. *Quaternary Research*, 52, 273–285.

Siegert, M.J., Dowdeswell, J.A., Hald, M. and Svensen, J.I. (in press). Modelling the Eurasian ice sheet through a full (Weichselian) glacial cycle. *Global and Planetary Change*.

Siegert, M.J., Hodgkins, R. and Dowdeswell, J.A. 1998. A chronology for the Dome C deep ice-core site through radio-echo layer correlation with the Vostok ice core, Antarctica. *Geophysical Research Letters*, 25, 1019–1022.

Sjöberg, L.E. 1991. Fennoscandian land uplift – an introduction. *Terra Nova*, 3, 356–357.

Solheim, A., Russwurm, L., Elverhøi, A. and Berg, M.N. 1990. Glacial geomorphic features in the northern Barents Sea: direct evidence for grounded ice and implications for the pattern of deglaciation and late glacial sedimentation. In: Dowdeswell, J.A. and Scource, J.D. (eds) *Glacimarine Environments: Processes and Sediments*, pp. 253–268. Geological Society Special Publication 53.

Steig, E.J., Brook, E.J., White, J.W.C., Sucher, C.M., Bender, M.L., Lehman, S.L., Morse, D.L., Waddington, E.D. and Clow, G.D. 1998. Synchronous climate changes in Antarctica and the North Atlantic. *Science*, 282, 92–95.

Stein, R., Nam, S., Grobe, H. and Hubberten, H. 1996. In: Andrews, J.T., et al. (eds) Late Quaternary Paleoceanography of the North Atlantic Margins. *Geological Society of London Special Publication*. 111, 275–287.

Stephens, B.B. and Keeling, R.F. 2000. The influence of Antarctic sea ice on glacial-interglacial CO_2 variations. *Nature*, 404, 171–174.

Stuiver, M. and Reimer, P.J. 1993. Extended ^{14}C data base and revised CALIB 3.0 ^{14}C age calibration program. *Radiocarbon*, 35, 215–230.

Stuiver, M., Kromer, B., Becker, B. and Ferguson, C.W. 1986. Radiocarbon age calibration back to 13,300 years BP and the ^{14}C age matching of the German oak and US Bristlecone Pine chronologies. *Radiocarbon*, 28:2B, 969–979.

Stuiver, M. and Grootes, P.M. 2000. GISP2 Oxygen isotope ratios. *Quaternary Research*, 53, 277–284.

Sugden, D. and John, B.S. 1976. *Glaciers and Landscape*. Arnold, London.

Sugden, D.E. 1968. The selectivity of glacial erosion in the Cairngorm Mountains, Scotland. *Transactions of the Institute of British Geographers*, 45, 79–92.

Sugden, D.E. 1977. Reconstruction of the morphology, dynamics and thermal characteristics of the Laurentide Ice Sheet at its maximum. *Arctic and Alpine Research*, 9, 21–47.

Sugden, D.E. 1978. Glacial Erosion by the Laurentide Ice Sheet. *Journal of Glaciology*, 20, 367–391.

Sugden, D.E. 1982. *Arctic and Antarctic*. Blackwell, Oxford. 472pp.

Svendsen, J.I., Astakhov, V.I., Bolshiyanov, D.Y., Dowdeswell, J.A., Gataullin, V., Hjort, C., Hubberten, H., Larsen, E., Mangerud, J., Möller, P., Saarnisto, M. and Siegert, M.J. 1999. Maximum extent of the Eurasian ice sheets in the Barents and Kara Sea region during the Late Weichselian. *Boreas*, 28, 234–242.

Svensson, A., Biscaye, P.E. and Grousset, F.E. 2000. Characterisation of late glacial continental dust in the Greenland Ice Core Project ice core. *Journal of Geophysical Research*, 105, 4637–4656.

Svensson, N.O. 1991. Post glacial land uplift patterns of South Sweden and the Baltic Sea region. *Terra Nova*, 3, 369–378.

Taylor, J. 1999. Large-scale sedimentation and ice sheet dynamics in the Polar North Atlantic. Unpublished PhD Thesis. University of Bristol, England.

Tushingham, A.M. and Peltier, W.R. 1991. A new Global Model of late Pleistocene deglaciation based upon geophysical predictions of post glacial relative sea level change. *Journal of Geophysical Research*, 96, 4497–4523.

Tveranger, J., Astakhov, V. and Mangerud, J. 1995. The margin of the last Barents–Kara Ice Sheet at Markhida, northern Russia. *Quaternary Research*, 44, 328–340.

Tveranger, J., Astakhov, V., Mangerud, J., and Svendsen, J.I. 1998. Signature of the last shelf-centered glaciation at a key section in the Pechora basin, Arctic Russia. *Journal of Quaternary Science*, 13, 189–203.

Tziperman, E. 1997. Inherently unstable climate behaviour due to weak thermohaline ocean circulation. *Nature*, 592–595.

van der Veen, C.J. 1996. Tidewater calving. *Journal of Glaciology*, 42, 375–385.

van der Veen, C.J. 1999. *Fundamentals of Glacier Dynamics*. Balkema, Rotterdam, 462 pp.

Vasil'chuk, Y., Punning, J.-M. and Vasil'chuk, A. 1997. Radiocarbon ages of mammoths in northern Eurasia: implications for population development and Late Quaternary environment. *Radiocarbon*, 39, 1–18.

Vaughan, D.G. and Doake, C.S.M. 1996. Climate driven retreat of ice shelves on the Antarctic Peninsula. *Nature*, 379, 328–331.

Velichko, A.A., Kononov, Y.M. and Faustova, M.A. 1997. The last glaciation of Earth: size and volume of ice-sheets. *Quaternary International*, 41/42, 43–51.

Veyret, Y. 1986. Quaternary glaciations in the French Massif Central. *Quaternary Science Reviews*, 5, 395–396.

Vogt, P.R., Crane, K. and Sundvor, E. 1993. Glacigenic mudflows on the Bear Island submarine fan. *EOS, Transactions of the American Geophysical Union*, 74, 449, 452, 453.

Vorren, T.O., Laberg, J.S., Blaume, F., Dowdeswell, J.A., Kenyon, N.H., Mienert, J., Rumohr, J. and Werner, F. 1998. The Norwegian–Greenland Sea continental margins: morphology and Late Quaternary sedimentary processes and environment. *Quaternary Science Reviews*, 17, 273–302.

Vorren, T.O., Lebesbye, E., Andreassen, K. and Larsen, K.-B. 1989. Glacigenic sediments on a passive continental margin as exemplified by the Barents Sea. *Marine Geology*, 85, 251–272.

Vorren, T.O., Hald, M. and Lebesbye, E. 1988. Late Cenozoic Environments in the Barents Sea. *Paleoceanography*, 3, 601–612.

Wagner, G., Beer, J., Laj, C., Kissel, C., Masarik, J., Muscheler, R. and Synal, H.A. 2000. Chlorine-36 evidence for the Mono Lake event in the Summit GRIP ice core. *Earth and Planetary Science Letters*, 181, 1–6.

Walder, J.S. 1982. Stability of sheet flow of water beneath temperate glaciers and implications for glacier surging. *Journal of Glaciology*, 28, 273–293.

Walder, J.S. 1994. Correspondence. Comments on 'Subglacial floods and the origin of low-relief ice-sheet lobes'. *Journal of Glaciology*, 40, 199–200.

Walsh, J.E. and Crane, R.G. 1992. A comparison of GCM simulations of Arctic Climate. *Geophysical Research Letters*, 19, 29–32.

Webb, R.S., Rind, D.H., Lehman, S.J., Healy, R.J. and Sigman, D. 1997. Influence of ocean heat transport on the climate of the Last Glacial Maximum. *Nature*, 385, 695.

Weertman, J. 1957. Deformation of floating ice shelves. *Journal of Glaciology*, 3, 38–42.

Weertman, J. 1973. Creep of ice. In: Whalley, E., Jones, S.J. and Gold, L.W. (eds) *Physics and Chemistry of Ice*, Royal Society of Canada, Ottawa. pp. 320–337.

Weidick, A. 1985. The ice cover of Greenland. *Gletscher-Hydrologiske Meddelelser*, 85/4.

Weidick, A. 1995. Satellite Image Atlas of the World. Greenland. *U.S. Geological Survey Professional Paper*, 1386-C, 141 pp.

Whillans, I.M. 1976. Radio-echo layers and the recent stability of the West Antarctic Ice Sheet. *Nature*, 264, 152–155.

White, S.E. 1986. Quaternary glacial stratigraphy and chronology of Mexico. *Quaternary Science Reviews*, 5, 201–205.

Willis, I.C. 1995. Intra-annual variations in glacier motion: a review. *Progress in Physical Geography*, 19, 61–106.

Winograd, I.J., Coplen, T.B., Landwehr, J.M., Riggs, A.C., Ludwig, K.R., Szabo, B.J., Kolesar, P.T. and Revesz, K.M. 1992. Continuous 500,000-year climate record from vein calcite in Devil's Hole, Nevada. *Science*, 258, 255–260.

Winograd, I.J., Landwehr, J.M., Ludwig, K.R., Coplen, T.B. and Riggs, A.C. 1997. Duration and structure of the past four interglaciations. *Quaternary Research*, 48, 141–154.

Wolfe, A.P. and King, R.H. 1999. A palaeolimnological constraint to the extent of the last Glaciation on northern Devon Island, Canadian High Arctic. *Quaternary Science Reviews*, 18, 1563–1568.

Wu, P. and Johnston, P. 2000. Can deglaciation trigger earthquakes in N. America? *Journal of Geophysical Research*, 27, 1323–1326.

Yokoyama, Y., Lambeck, K., De Deckker, P., Johnston, P. and Fifield, L.K. 2000. Timing of the Last Glacial Maximum from observed sea-level minima. *Nature*, 406, 713–716.

Yu, E.-F., Francois, R. and Bacon, M.P. 1996. Similar rates of modern and last-glacial ocean thermohaline circulation inferred from radiochemical data. *Nature*, 379, 689–694.

Zielinski, G.A., Mayewski, P.A., Meeker, D.L., Whitlow, S. and Twickler, M.S. 1996. A 110,000-yr record of explosive volcanism from the GISP2 (Greenland) ice core. *Quaternary Research*, 44, 109–118.

Zreda, M., England, J., Phillips, F., Elmore, D. and Sharma, P. 1999. Unblocking of the Nares Strait by Greenland and Ellesmere ice-sheet retreat 10,000 years ago. *Nature*, 398, 139–142.

Index